Flow and Sediment Transport in Compound Channels

The Experiences of Japanese and UK Research

Flow and Sediment Transport in Compound Channels

The Experiences of Japanese and

UK Research

Ian K. McEwan and Syunsuke Ikeda

2007

CRC Press
Taylor & Francis Group
Boca Raton London New York

CRC Press is an imprint of the
Taylor & Francis Group, an **informa** business

CRC Press
Taylor & Francis Group
6000 Broken Sound Parkway NW, Suite 300
Boca Raton, FL 33487-2742

First issued in hardback 2017

© IAHR 2008
CRC Press is an imprint of Taylor & Francis Group, an Informa business

No claim to original U.S. Government works

ISBN-13: 978-9-8105-9363-6 (pbk)
ISBN-13: 978-1-1384-7524-3 (hbk)

**Visit the Taylor & Francis Web site at
http://www.taylorandfrancis.com**

**and the CRC Press Web site
http://www.crcpress.com**

Other IAHR Monographs in the series:

(published by IAHR- for more information and to order visit www.iahr.org or contact the Secretariat at iahr@iahr.org)

Fluvial Processes monograph	M. Selim Yalin A. M. Ferreira da Silva, 2001, 197 pages.
Fluvial Processes Solutions Manual	M. Selim Yalin A. M. Ferreira da Silva, 2001, 83 pages.
Hydraulicians in Europe 1800–2000	W. H. Hager, 2003, 774 pages.
Water Engineering in Ancient Civilizations - 5,000 years of history	P-L. Viollet, F. M. Holly, 2007, 322 pages.

(published by Taylor and Francis — Balkema, www.taylorandfrancis.com/)
IAHR Members receive a 10% discount on all IAHR publications

Physical Processes and Chemical Reactions in Liquid Flows	F. S. Rhys/A. Gyr, 1998, 200 pages.
Mudflow Rheology and Dynamics	Ph. Coussot, 1997, 270 pages.
Hyperconcentrated Flow	Zhaohui Wan/ZhaoyinWang, 1994, 320 pages.
Turbulence in open-channel Flows	I. Nezu/H. Nakagawa, 1993, 286 pages.
Turbulence Models and Their Application in Hydraulics	W. Rodi, 1993 3rd edition, 116 pages.
Mobile Barrages and Intakes on Sediment Transporting Rivers	M. Bouvard, 1992, 322 pages.
Debris Flow	T. Takahashi, 1991, 156 pages.

The International Association of Hydraulic Engineering and Research (IAHR), founded in 1935, is a worldwide independent organisation of engineers and water specialists working in fields related to hydraulics and its practical application. Activities range from river and maritime hydraulics to water resources development and eco-hydraulics, through to ice engineering, hydroinformatics and continuing education and training. IAHR stimulates and promotes both research and it's application, and by so doing strives to contribute to sustainable development, the optimisation of world water resources management and industrial flow processes. IAHR accomplishes its goals by a wide variety of member activities including: working groups, research agenda, congresses, specialty conferences, workshops and short courses; Journals, Monographs and Proceedings; by involvement in international programmes such as UNESCO, WMO, IDNDR, GWP, ICSU, The World Water Forum; and by co-operation with other water-related (inter)national organisations.

Acknowledgements

The UK-Japan joint seminar was conducted under the financial support by the UK Engineering and Physical Sciences Research Council (EPSRC, Grant number GR/L78475) and the Hokkaido River Disaster Prevention Research Center (HRD-PRC). Their support and encouragement are gratefully acknowledged. The British Council also funded the initial visit to Japan during which the suggestion to organise the seminar was first made. The first meeting of the seminar was held at Tokyo Institute of Technology in April 1998.

Fig. 1 The first meeting of the joint seminar at Tokyo Institute of Technology on April 16, 1998. Top Row from left to right E. M. Valentine, K. Shiono, Y. Kawahara, Y. Shimizu, Unknown, S. Aya, T. Tsujimoto, B. B. Willetts and P. R. Wormleaton. Bottom row from left to right, G. Pender, I. Nezu, S. Ikeda, I. K. McEwan, S. Fukuoka, D. W. Knight and Y. Toda.

The second meeting was held at Ross Priory, near Loch Lomond, in October 1998, where we benefited from the beautiful setting and accommodation provided by the University of Strathclyde.

The third meeting of the joint seminar returned to Japan and was hosted in Hokkaido by Dr. K. Hoshi at HRDPRC and Dr. Y. Watanabe at the Hokkaido Development Bureau. The hospitality that was shown and the very keen interest in the work were notable. The visit included a helicopter trip to observe Ishikari River. Dr. Hoshi best understood the importance of this international activity. It is our greatest regret that he passed away in December 11, 2006, before this book was

Fig. 2 The second seminar at Ross Priory on October 15, 1998 Front row from left to right Y. Shimizu, S. Ikeda, G. Pender, I. K. McEwan, D. W. Knight and, P. R. Wormleaton. Middle Row from left to right Y. Kawahara, S. Aya, B. B. Willetts, K. Shiono and E. M. Valentine. Back row from left to right T. Tsujimoto and A. Watanabe.

published. It is our hope that this volume represents a progress on flow and sediment transport in compound channels and that it will attract a new generation to study and further enhance our knowledge of this fascinating and important phenomenon which the text describes.

S. Ikeda and I. K. McEwan, February 2008

Contents

Chapter 1

Introduction

I. K. McEwan and S. Ikeda

1.1.   Introduction

Predicting the behaviour of rivers in flood conditions presents a major challenge
to hydraulic engineers. Floods can be catastrophic, damaging property and taking
human life, particularly, but not exclusively, in developing countries. The challenge to hydraulic engineers is to engage with these issues and promote a form of
river engineering, which is sympathetic to the environment while taking all reasonable steps to ensure the success of flood defence measures. The reward to the
profession in accepting this challenge is a position at the heart of global development and the satisfaction of contributing to human well being in both the developing and the developed world.

This collective monograph is a first step towards meeting this challenge. It
comes at a time when concerns about the impact of global warming are multiplying and with this, a growing awareness that changes in climate may change
the pattern and severity of flood events. The extent and effects of climate change
on flood events is beyond the scope of book because the assumption, implicit
throughout the text, is that the volumetric discharge is known and that the task at
hand is to predict the behaviour under this discharge. However future uncertain-
ties do provide an important context for this volume as it emphasizes the need for
work on compound channel hydraulics. The methodologies and physical insights
that are described are therefore essential to the current and future management of
river channels.

The text is organised as follows. Chapter 2 presents the state-of-the-art of
research into "Flow structure in compound channels". Researchers in both Japan
and the UK have been particularly active in this area over the last two decades
and contributors to the chapter have drawn on their knowledge of this work to
provide a statement of the progress that has been made. Chapter 3 is a parallel
account of work on sediment processes in compound channels. By necessity, this
work is less well developed that the account of flow structure because, in a large
measure, a deeper understanding of the flow structure is a pre-requisite to treating sediment problems in compound channels. Chapter 3, therefore, documents
current progress and identifies issues that must be resolved in future research.
Chapter 4 describes current capabilities in numerical modelling. Increasingly large

and complex numerical models can be applied to treat engineering problems, not least, those concerned with flow in compound channels. This chapter details and summarizes the variety of approaches that may be taken and provides a series of examples of state-of-the-art simulations. Chapter 5 moves firmly towards issues that affect the design and operation of compound channels. Any visitor to Japan will quickly recognize that the engineering of river channels for flood prevention is advanced and that the Japanese experience can benefit others. Chapter 5 provides access to some of the methods and concepts which have underpinned this activity.

1.2. Background to the Text

The idea for the joint seminar was developed by Professor Syunsuke Ikeda and Dr Ian McEwan in 1996 while the latter was visiting Japan with support from the British Council. In 1997, the UK Engineering and Physical Sciences Research Council (EPSRC) awarded a grant to enable seven UK academics to participate in the joint seminar. Parallel support for the seven Japanese participants was obtained from the Hokkaido River Disaster Prevention Research Centre.

The Participants in the Joint Seminar were

S Ikeda, Tokyo Institute of Technology	I K McEwan, University of Aberdeen
S Fukuoka, Hiroshima University	D W Knight, University of Birmingham
I Nezu, Kyoto University	G Pender, Heriot-Watt University
T Tsujimoto, Nagoya University	K Shiono, Loughborough University
S Aya, Osaka Institute of Technology	E M Valentine[1], Charles Darwin University (Australia)
Y Kawahara, Kagawa University	B B Willetts, University of Aberdeen
Y Shimizu, Hokkaido University	P R Wormleaton, QMW, University of London

The first meeting of the group was held in Tokyo and Kyoto in April 1998 with three further meetings taking place at Ross Priory, near Loch Lomond in October 1999, Sapporo, Hokkaido in July 2000 and Obihiro, Hokkaido in September 2001. During the Kyoto meeting the group agreed that the primary aim of the joint seminar was to publish a milestone document recording the outcome of the seminar. The expressed focus of the Seminar was to draw together the comprehensive studies conducted in UK and Japan during the 1980's and 1990's into a single source. In very broad terms these two centres have each made major contributions to the development of our knowledge of compound channel flow.

The aim of the monograph is to provide a comprehensive state-of-the-art description of the work carried out in the UK and Japan on "Flow and sediment transport in compound channels." It therefore describes research which has been

---

[1]Formerly of the University of Newcastle, UK.

conducted, primarily over the last two decades, and which has yielded a fairly detailed picture of the important behaviours of compound channels and produced a number of engineering prediction methods which ought to be widely adopted in practice. This text will inevitably highlight areas where our knowledge is sparse and it is hoped that it will spur others on in the task of filling such gaps. Clearly, however, much remains to be done and these efforts will require a concerted global effort in the present and the future.

The concept of bi-national groups of researchers meeting together intermittently over a period of some years is, of course, not new. Our Group has drawn both inspiration and experience from the successful US/Japanese Groups. The format is one that we would commend to all, not least the funding agencies. The interaction that is a fundamental part of a joint seminar can produce tangible outcomes in the form of collaborative publications. However, as important, is the exposure to new ideas and approaches to problems, which, writing from the UK and Japanese perspectives, can be expected to inform and influence research for years to come. In doing so, this benchmarking process promotes quality and relevance in research.

Chapter 2

Flow Structure

D. W. Knight, S. Aya, S. Ikeda, I. Nezu and K Shiono

2.1. Introduction

The understanding of flow structure in straight and meandering compound channels is a prerequisite for the development of appropriate mathematical models and design methods for practical engineering application. This chapter deals with rigid boundary fixed plan form channels in an attempt to simplify the analysis prior to dealing with loose boundary channels and sediment processes in Chapter 3. In recent years, the knowledge gained from empirical studies using rigid boundary compound channels has been significant, and is now summarized under various headings. These relate to: the different types of channel that have been studied (straight, curvilinear and meandering), different flow structures (planform and longitudinal vortices, turbulence and secondary flows), flow resistance (velocity and boundary shear stress distributions) and stage-discharge relationships. Various measurement techniques and their link to the development of a variety of 1-D and 2-D models are also described.

When the water level in a river is below the level of the adjacent floodplain, the river is said to be flowing inbank, whereas when the water level is above the bankfull level, the river is said to be flowing out-of-bank or overbank. As a river changes from inbank to overbank flow, there is a significant increase in the complexity of the flow behaviour, even for relatively straight reaches. The difference in velocity between the main channel and the floodplain flows may produce strong lateral shear layers, which typically lead to the generation of organized planform vortices induced by inflectional point instability, as illustrated in Figs. 2.1–2.3. Figure 2.1 shows large planform vortices observed in the Tone River during the 1981 flood, and Fig. 2.2 illustrates schematically this type of large-scale vortex structure associated with shear flows in overbank flow.

Another aspect of turbulence, typically observed in such channels, is the intermittent upward flow motion at the interface between the main channel and the floodplains, induced by the imbalance of the normal Reynolds stresses originating from the complex nature of the channel cross section. This aspect of turbulence is closely related with secondary flow of Prandtl's 2nd kind, which is often observed in the form of a series of helicoidal vortices rotating in the streamwise direction.

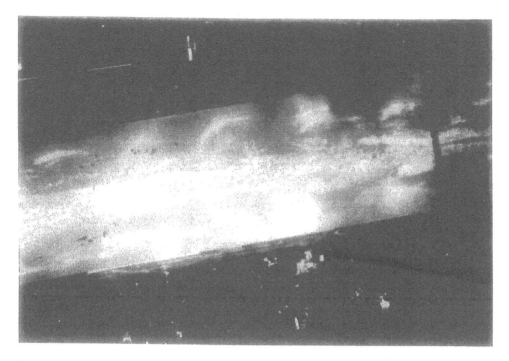

Fig. 2.1 Horizontal vortices observed in Tone River during the 1981 flood. The flow is from right to left (Courtesy, Ministry of Construction of Japan).

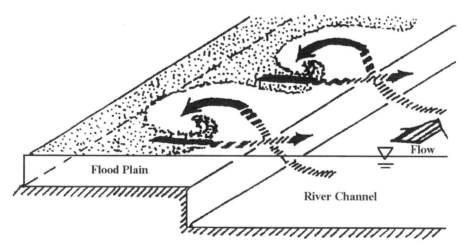

Fig. 2.2 Large-scale vortex structure associated with overbank flow.

These streamwise vortices are generally present in all turbulent flows in straight conduits with a non-circular cross section, due to the anisotropic turbulence. The interaction of planform and streamwise vortices, and their effect on key hydraulic parameters, such as velocity and boundary shear stress, are shown schematically

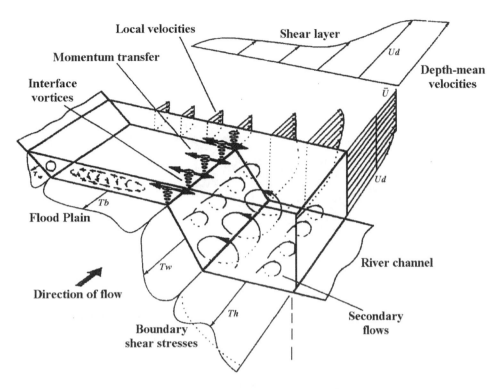

Fig. 2.3 Hydraulic parameters associated with overbank flow (after Knight & Shiono, 1996).

for overbank flow in Fig. 2.3. The flow structure in straight compound channels is thus more complex than that in simple channels, as significant planform and streamwise vortices are present, with organized coherent structure, rotating about either vertical or horizontal axes.

For overbank flow in a meandering channel, the flow structure is even more complex, as the flow over the floodplain may find a more direct route downstream, short circuiting the main river channel, and creating different flow structures, as shown schematically in Figs. 2.4 and 2.5. Whereas inbank flow may be treated as predominately one-dimensional (1-D) in the streamwise direction, despite some three-dimensional flow structures being present, overbank flow must therefore be treated differently, because other 3-D flow structures are clearly important.

It is these 3-D flow mechanisms that make the analysis of overbank flows inherently more difficult than the analysis of inbank flows. Anderson *et al.* (1996) provide a general review of floodplain processes, Ashworth *et al.* (1996) a general review of coherent structures in open channel flow, and Knight & Shiono (1996) a specific review of river channel and floodplain hydraulics.

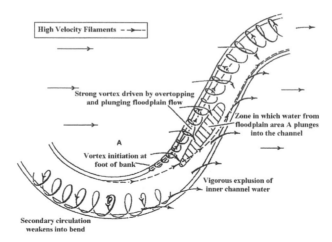

Fig. 2.4 Flow features within a flooded meandering channel (after Sellin *et al.*, 1993).

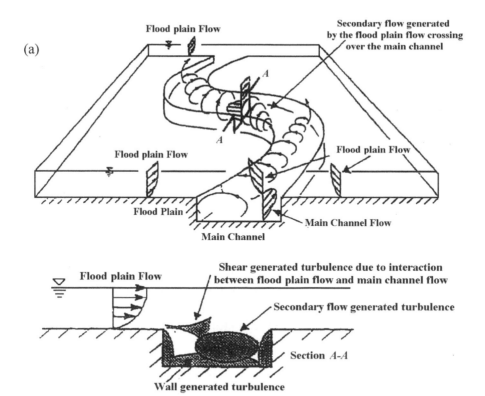

Fig. 2.5 Flow structures associated with overbank flow in a meandering channel. (a) Secondary flow generation mechanism. (b) Main contributions to turbulence energy production in cross-over section (after Shiono & Muto, 1998).

2.2. Straight Compound Channels

2.2.1. Three-dimensional approach and governing equations

In this section, the characteristics of turbulence in straight compound channels are described and the governing equations given. Organized planform vortices were first described by Sellin (1964), based on laboratory studies, and later by Kinoshita (1984), based on field observations from the Tone River, Japan, as shown in Fig. 2.1. This aerial view shows an array of horizontal vortices observed during the 1981 flood event in the Tone River, in which two rows of planform vortices are seen along both sides of the floodplain. Since then, many researchers have studied the instability induced nature of organized planform vortices, e.g. Tamai *et al.* (1986a, 1986b), Fukuoka & Fujita (1989), Chu *et al.* (1991), Ikeda (1995), Ikeda & Kuga (1997), Nezu & Nakayama (1997) and Nezu, Onitsuka & Iketani (1999), among others.

The governing equations of primary mean flow are now described briefly. The time averaged primary flow, U, and the time averaged secondary flows, V and W, are described by the following time averaged 3-D RANS (Reynolds Averaged Navier-Stokes) equations for fully-developed uniform turbulent flows, i.e., $\partial/\partial x = 0$, in straight open channels (Bradshaw, 1987; Nezu & Nakagawa 1993):

$$V\frac{\partial U}{\partial y} + W\frac{\partial U}{\partial z} = g\sin\theta - \frac{1}{\rho}\frac{\partial p}{\partial x} + \frac{\partial(-\overline{uv})}{\partial y} + \frac{\partial(-\overline{uw})}{\partial z} + \nu\nabla^2 U \tag{2.1}$$

$$V\frac{\partial V}{\partial y} + W\frac{\partial V}{\partial z} = -\frac{1}{\rho}\frac{\partial p}{\partial y} + \frac{\partial(-\overline{v^2})}{\partial y} + \frac{\partial(-\overline{vw})}{\partial z} + \nu\nabla^2 V \tag{2.2}$$

$$V\frac{\partial W}{\partial y} + W\frac{\partial W}{\partial z} = -g\cos\theta - \frac{1}{\rho}\frac{\partial p}{\partial z} + \frac{\partial(-\overline{vw})}{\partial y} + \frac{\partial(-\overline{w^2})}{\partial z} + \nu\nabla^2 W \tag{2.3}$$

The continuity equation is

$$\frac{\partial V}{\partial y} + \frac{\partial W}{\partial z} = 0 \tag{2.4}$$

in which, u, v, and w denote the velocity fluctuations in the x-direction (stream-wise), y-direction (spanwise with the origin at the side wall), and z-direction (vertical with the origin at the bed), p is the mean pressure, θ is the angle of channel slope to the horizontal axis, ν is the kinematic viscosity and ∇^2 is the Laplacian operator. Equation (4.1) is for the streamwise direction, and contains the primary flow terms, whereas (4.2) and (4.3) are in the other co-ordinate directions and contain the secondary flow terms. From (4.3), we can obtain under certain circumstances

$$p \cong \rho g(H - z)\cos\theta \tag{2.5}$$

where H is the flow depth. The mean pressure, p, is therefore assumed to be approximately hydrostatic. From (4.1), (4.4) and (4.5), the following primary-order equation is obtained:

$$\frac{\partial UV}{\partial y} + \frac{\partial UW}{\partial z} = gS_f + \frac{\partial(\tau_{yx})}{\partial y} + \frac{\partial}{\partial z}\left(\frac{\tau_{zx}}{\rho}\right) \tag{2.6}$$

$$\frac{\tau}{\rho} = -\overline{uw} + \nu\frac{\partial U}{\partial z} \quad \text{and} \quad \frac{\tau_{yx}}{\rho} = -\overline{uv} + \nu\frac{\partial U}{\partial y} \tag{2.7}$$

where $S_f = \sin\theta - \cos\theta(dH/dx)$ is the energy gradient, and τ is the total shear stress (= Reynolds shear stress + viscous stress). If the flow depth H is constant, S_f is equal to the channel bed slope $S_o = \sin\theta$. Integrating (4.6a) from $z = 0$ to $z = h'$ (local water depth) gives:

$$\frac{\tau_b}{\rho} = gS_f h' + h'\frac{d}{dy}(T - J) \tag{2.8}$$

where

$$J = \frac{1}{h'}\int_0^{h'}(UV)dz, \qquad T = \frac{1}{h'}\int_0^{h'}(\overline{\tau_{yx}}/\rho)dz \tag{2.9}$$

where τ_b is the boundary shear stress (wall or bed). In the case of a compound channel, h' is equal to H in the main channel, and is equal to $(H - h)$ in the flood-plains, where $h =$ bankfull height. The depth ratio, Dr, is given by $= (H - h)/H$ and is important in describing the character of a compound channel flow. If the spanwise mean momentum, UV, and the turbulent momentum, i.e., the spanwise Reynolds stress, $-\overline{uv}$, can be measured with LDA, then the boundary or wall shear stress $\tau_b = \rho U_*^2$ and the local friction velocity U_* may be evaluated from (4.7).

The governing equations of secondary flow (i.e. the vorticity equation) may be obtained by cross differentiating the secondary equations (4.2) and (4.3), to eliminate the pressure term. This yields the equation of streamwise vorticity in the following form (e.g. see Perkins, 1970; Bradshaw, 1987):

$$\underset{A}{V\frac{\partial\Omega}{\partial y}} + \underset{B}{W\frac{\partial\Omega}{\partial z}} = \underset{C}{\frac{\partial^2}{\partial y\partial z}\left(\overline{w^2} - \overline{v^2}\right)} + \underset{D}{\left(\frac{\partial^2}{\partial y^2} - \frac{\partial^2}{\partial z^2}\right)\overline{vw}} + \nu\nabla^2\Omega \tag{2.10}$$

where

$$\Omega = \frac{\partial V}{\partial z} - \frac{\partial W}{\partial y} \tag{2.11}$$

2.2.1.1. Applying the governing equations

One example of the effect of anisotropic turbulence and large-scale planform vortices on Reynolds stresses in compound channels with overbank flow is shown in Figs. 2.6–2.8 (Knight and Shiono, 1990). These are taken from the straight channel results of the FCF Phase A data, which are available from HR Wallingford

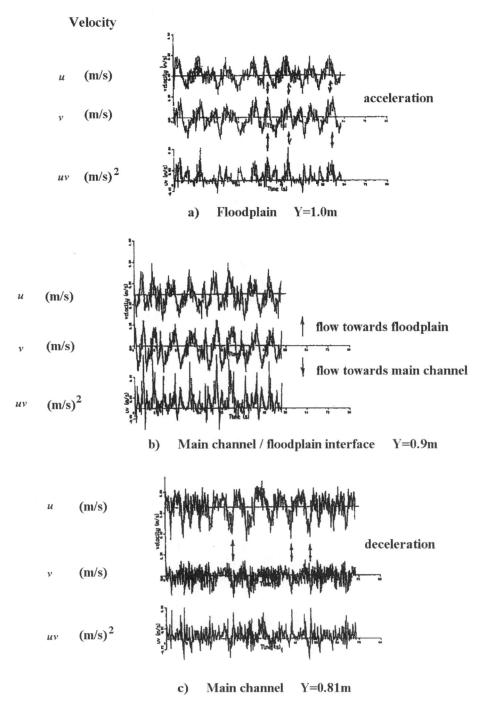

Fig. 2.6 Temporal variation of turbulence data for overbank flow from the Flood Channel Facility with $Dr = 0.152$.

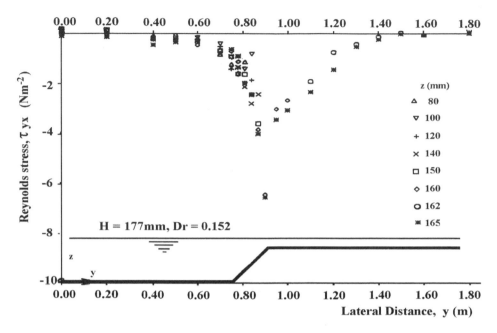

Fig. 2.7 Lateral variation of Reynolds stresses, τ_{yx} recorded in the FCF.

(Knight, 1992), but not temporal variations. Figure 2.6 shows the temporal variations in mean streamwise and transverse velocities, U and V, together with the corresponding temporal variations in local Reynolds stress, $\tau_{yx}/\rho = -\overline{uv}$, on a vertical plane, xz, for 3 lateral positions $y = 0.8, 0.9 \ \& \ 1.0\,$m, defined in Figure 2.6. At this location, where the interaction process between the river and its floodplain is at its most intense, the periodicity of the large planform vortices is self-evident. The corresponding local Reynolds stresses, $\tau_{yx} = -\rho\overline{uv}$ and $\tau_{zx} = -\rho\overline{uw}$ (on vertical and horizontal planes) are shown in Figs. 2.7 and 2.8 respectively for one half of this symmetric channel.

The equations for the vertical distributions of local Reynolds stress, τ_{zx}, and lateral distribution of depth-averaged Reynolds stress, $\overline{\tau}_{yx}$, were derived from (4.1) by Knight, Samuels & Shiono (1990) in a somewhat different form to (4.7) as

$$\tau_{zx} = \rho g\,(H-z)\sin\theta + \int_z^H \frac{\partial \tau_{yx}}{\partial y}dz - \int_z^H \frac{\partial\,(\rho UV)}{\partial y}dz + \rho UW \qquad (2.12)$$

$$\overline{\tau}_{yx} = (\rho UV)_d - \frac{1}{H}\int_0^y \left[\rho g H S_o - \tau_b\sqrt{\left(1 + \frac{1}{s^2}\right)}\right]dy \qquad (2.13)$$

in which the viscous terms are negligibly small. The overbar indicates a depth-averaged value and $s = $ side slope of channel (1:s, vertical: horizontal).

Figure 2.7 shows the lateral distributions of local Reynolds stress, τ_{yx}, at different depths, z, for various y values. At the channel centreline, $y = 0$, $\tau_{yx} = 0$, as would be expected for a symmetric channel. For $0.5 < y < 1.5\,$m, there is evidence

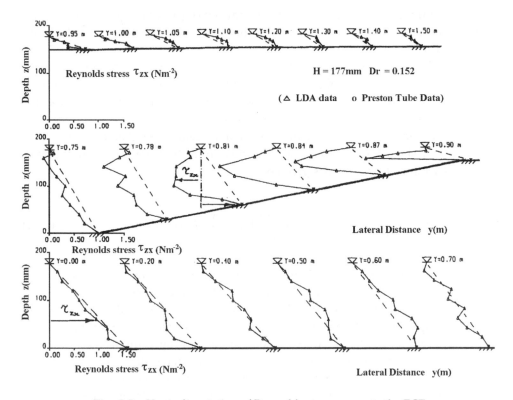

Fig. 2.8 Vertical variation of Reynolds stresses, τ_{zx} in the FCF.

of a strong shear layer region, in which these Reynolds stresses are significant, reaching a maximum at $y = 0.90$ m, i.e. at the top of the main channel bank and floodplain/main channel interface. Figure 2.8 shows the corresponding τ_{zx} values. It can be seen that the vertical distribution of τ_{zx} is approximately linear at the centreline of the channel, $y = 0$, as would be expected from the first term in (4.7) or (4.11), and when remaining terms with derivatives with respect to y are zero.

In the vicinity of the main river channel bank ($y = 0.81$ m), the distributions are highly non-linear, and are in agreement with the influence of the various terms in (4.11). In particular it should be noted the logarithmic velocity law is no longer valid in most of this region. This has significant implications for determining the local boundary shear stress, which is also shown in Fig. 2.8 for all lateral positions, measured by both LDA and Preston tube techniques. At large values of y ($y > 1.50$ m), i.e. outside the region in which τ_{yx} is non-zero (see Fig. 2.7), but still within the region where vorticity and secondary flows are influential, the vertical distributions return to a linear form. The experimental closed duct data of Meyer & Rehme (1994) confirm the nature of these distributions and the 3-D spatial plots of these stresses shown in Knight, Samuels & Shiono (1990) and Shiono & Knight (1991).

D. W. Knight *et al.*

Another important feature of Fig. 2.7 is that the depth-averaged values of τ_{yx}, defined in (4.9) as $\bar{\tau}_{yx}$, are related to the local resistance coefficients, which are discussed further in Section 2.5.

Re-arranging (4.12) gives the apparent shear stress (ASS) acting on any vertical interface at a given lateral position within the cross section, where the ASS is defined as the depth-averaged apparent shear stress acting on a vertical plane, i.e. the apparent shear force, ASF, divided by the length of the vertical interface. The ASF may be calculated from measured boundary shear stress data by subtracting the boundary shear force from the weight force up to that particular lateral position. Thus from (4.12) the ASS is given by

$$\text{ASS} = \frac{1}{H} \int_0^y \left[\rho g H S_o - \tau_b \sqrt{\left(1 + \frac{1}{s^2} \right)} \right] dy = (\rho U V)_d - \bar{\tau}_{yx} \qquad (2.14)$$

Figures 2.9–2.12 show some boundary shear stress, depth-averaged apparent shear stress, depth-averaged secondary flow and depth-averaged Reynolds stress data from the FCF data of Shiono & Knight (1991). The relative contributions of the depth-averaged Reynolds stress term, $\bar{\tau}_{yx}$, and the secondary flow term, $(\rho U V)_d$, to the apparent shear stress (ASS) are seen to be roughly comparable in magnitude, highlighting the importance of vortex structures in resistance studies. Recent numerical work by Thomas & Williams (1995), using data from Test Series 22 from the FCF and a large eddy simulation (LES), confirms this finding. There is also some evidence for these vertical distributions of Reynolds stress in the field data of Babaeyan-Koopaei (2002).

It is apparent from Figs. 2.11 and 2.12 that the lateral spread or influence of the Reynolds stresses is less than the lateral influence of the secondary flow/vortical

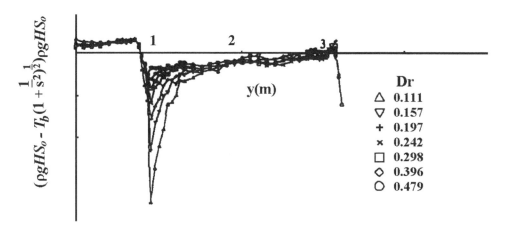

Fig. 2.9 Lateral variation of boundary shear stress, τ_b for different Dr with $B/b = 4.2$ with $b/h = 5.0$ and $s = 1.0$ (see Fig. 2.89).

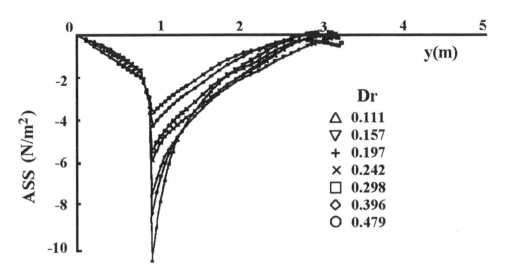

Fig. 2.10 Lateral variation of depth-mean apparent shear stress (ASS) for different *Dr*.

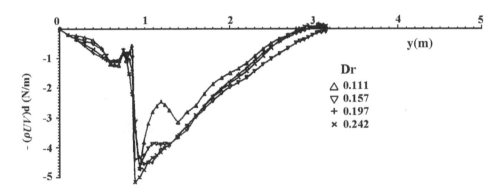

Fig. 2.11 Lateral variation of apparent stress $(\rho UV)_d$ and force per unit length $H(\rho UV)_d$ due to secondary flows for different *Dr*.

structures. The lateral extent of shear layers, based on either depth-averaged velocity or boundary shear stress, have been quantified by Rhodes & Knight (1995b) for a range of depths using compound duct data. However, it should be remembered that temporal averaging of the governing equations might in fact obscure certain phenomena, as highlighted by Meyer & Rehme (1994), Shiono & Knight (1990) and Tamai, Asaeda & Ikeda (1986a,1986b). The stress terms in (4.1) include both mean and fluctuating components, but that once decomposed into Reynolds stress and secondary flow terms, the term $(\rho UV)_d$ includes at least two effects. These have been represented in Fig. 2.3 as both vorticity in the vertical direction at the floodplain/main channel interface and longitudinal vorticity on the

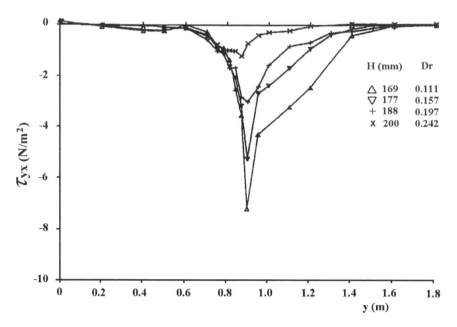

Fig. 2.12 Lateral variation of depth-averaged Reynolds stress, $\overline{\tau}_{yx}$, for four relative depths in the range $0.111 \leq Dr \leq 0.242$.

floodplain. In the former there are some known frequency effects that need to be dealt with as described in Section 2.2.3.1.

However, the term $(\rho UV)_d$ includes both vorticity effects and is therefore a convenient general 'sink' term representing both influences. This appears to be a fortuitous way of representing the vorticity in the river/floodplain process in depth-averaged models.

2.2.2. Two-dimensional approach and depth-averaged models

In this section, the characteristics of turbulence in straight compound channels are described and the governing equations are also given. In natural geophysical flows, since the lateral scale, characterized by the channel width, is usually much larger than the vertical scale (typically the depth), then the flow field can be treated as two-dimensional as a first approximation. It therefore follows that organized planform vortices, and their associated turbulence structure, predominate in the flow field for compound channels with large width to depth ratios.

2.2.2.1. Turbulence model

As described previously, when the width to depth ratio is large the turbulence is anisotropic and the flow field may be split into two parts, i.e., the large scale planform vortices and the small scale isotropic turbulence, scaled by the depth.

This idea was first presented by Nadaoka & Yagi (1993), who proposed a new turbulence model, called the SDS-2DH model. In this model, the planform scale organized vortices are solved directly by integrating the depth-averaged continuity and momentum equations. The SDS (sub-depth scale) turbulence is solved with a transport equation for the turbulence kinetic energy, in which turbulent kinetic energy production by both bed friction and lateral shear are associated with the movement of the planform vortices. The equations are as follows:

Depth-averaged continuity equation

$$\frac{\partial \eta}{\partial t} + \frac{\partial (U_d H)}{\partial x} + \frac{\partial (V_d H)}{\partial y} = 0 \tag{2.15}$$

Depth-averaged momentum equation in the x-direction

$$\frac{DU_d}{Dt} = -g\frac{\partial \eta}{\partial x} + \frac{\partial}{\partial x}\left(2\nu_t \frac{\partial U_d}{\partial x} - \frac{2}{3}k\right) + \frac{\partial}{\partial y}\left(\nu_t\left(\frac{\partial U_d}{\partial y} + \frac{\partial V_d}{\partial x}\right)\right)$$
$$- \frac{C_f}{H}U_d\sqrt{U_d^2 + V_d^2} \tag{2.16}$$

Depth-averaged momentum equation in the y-direction

$$\frac{DV_d}{Dt} = -g\frac{\partial \eta}{\partial y} + \frac{\partial}{\partial x}\left(\nu_t\left(\frac{\partial V_d}{\partial x} + \frac{\partial U_d}{\partial y}\right)\right) + \frac{\partial}{\partial y}\left(2\nu_t \frac{\partial V_d}{\partial y} - \frac{2}{3}k\right)$$
$$- \frac{C_f}{H}V_d\sqrt{U_d^2 + V_d^2} \tag{2.17}$$

where D is the total derivative, x is the longitudinal coordinate, y is the transverse coordinate, t is time, H is the depth of flow, U_d is the depth-averaged velocity component in the x-direction, V_d is the depth-averaged velocity component in the y-direction, η is the elevation of the free surface from a reference level, k is the depth-averaged turbulent kinetic energy for the SDS turbulence, C_f is a bed friction coefficient, g is gravitational acceleration, and ν_t is the isotropic eddy viscosity associated with SDS turbulence.

The transport equation for SDS turbulence kinetic energy is

$$\frac{Dk}{Dt} = \frac{\partial}{\partial x}\left(\frac{\nu_t}{\sigma_k}\frac{\partial k}{\partial x}\right) + \frac{\partial}{\partial y}\left(\frac{\nu_t}{\sigma_k}\frac{\partial k}{\partial y}\right) + P_{kh} + P_{kv} - \varepsilon \tag{2.18}$$

$$P_{kh} = \nu_t\left[2\left(\frac{\partial U_d}{\partial x}\right)^2 + 2\left(\frac{\partial V_d}{\partial y}\right)^2 + \left(\frac{\partial U_d}{\partial y} + \frac{\partial V_d}{\partial x}\right)^2\right] \tag{2.19}$$

$$P_{kv} = \left[C_f(U_d^2 + V_d^2)\right]^{1.5}\Big/l \tag{2.20}$$

$$\varepsilon = C_\mu^{3/4}k^{3/2}\Big/l \tag{2.21}$$

$$\nu_t = C_\mu k^2/\varepsilon \tag{2.22}$$

$$l = \alpha H \tag{2.23}$$

where σ_k is an empirically determined numerical constant ($= 1.0$), ε is turbulent kinetic energy dissipation, P_{kh}, P_{kv} are turbulence kinetic energy production due to horizontal shear and vertical shear, respectively, l is a characteristic turbulence length, C_μ is an empirical constant ($= 0.09$), and α is a numerical constant ($= 0.1$). The Chezy coefficient, C_f, is calculated by the following expression:

$$C_f = gn^2/H^{1/3} \tag{2.24}$$

where n is Manning's roughness coefficient.

2.2.2.2. Analytical model

As a result of the turbulence measurements described briefly in Section 2.2.1, some analytical solutions to the 2-D depth-integrated momentum equation were obtained by Shiono & Knight (1988 & 1991) for certain compound channel flows. The development of 2-D models was undertaken in order to predict the transverse variation of depth-averaged velocity and boundary shear stress within river channels of any cross section shape. Once these variations are known, it is possible to obtain better estimates of the stage-discharge relationship (conveyance capacity) and sediment transport processes (transport rate, erosion and deposition). In order to obtain an analytical solution, the depth-averaged momentum equation has to be solved for steady uniform turbulent flow in the streamwise direction. The development of these analytical solutions is now briefly described, using the notation originally given in Shiono & Knight (1988, 1991) where an overbar indicates either a time-averaged or a depth averaged parameter.

The equation for the longitudinal streamwise component of momentum of a fluid element may be combined with the continuity equation to give

$$\rho \left[\frac{\partial UV}{\partial y} + \frac{\partial UW}{\partial z} \right] = \rho g S_0 + \frac{\partial}{\partial y}(-\rho \overline{uv}) + \frac{\partial}{\partial z}(-\rho \overline{uw}) \tag{2.25}$$

$$\text{(I)} \qquad\qquad \text{(II)} \qquad \text{(III)} \qquad\qquad \text{(IV)}$$

The physical meaning of the terms in equation (4.27) are: (I) = secondary flow term, (II) = weight component term, (III) = vertical plane Reynolds stress term, and (IV) = horizontal plane Reynolds stress term. If the viscous stress is negligibly small, and omitted in (2.7), then for uniform flow, i.e. $dH/dx = 0$, equation (4.27) becomes equivalent to (4.6a).

Integrating (4.27) over the depth of water, the depth-averaged momentum equation becomes

$$\frac{\partial H(\rho UV)_d}{\partial y} = \rho g H S_0 + \frac{\partial H \overline{\tau}_{yx}}{\partial y} - \tau_b \left(1 + \frac{1}{s^2}\right)^{1/2} \tag{2.26}$$

where τ_b, is the bed shear stress, and s is the side slope (1:s, vertical: horizontal). The depth-averaged terms are defined by

$$(\rho UV)_d = \frac{1}{H}\int_0^H (\rho UV)dz \quad \text{and} \quad \overline{\tau}_{yx} = \frac{1}{H}\int_0^H (-\rho \overline{uv})\,dz \tag{2.27}$$

Differentiating (4.12) with respect to y also gives (4.28).

Using the Darcy-Weisbach friction factor and adopting an eddy viscosity approach, as defined by the following equations

$$f = \frac{8\tau_b}{\rho U_d^2}, \quad U_* = \left(\frac{1}{8}f\right)^{1/2} U_d \quad \text{and} \quad \overline{\tau}_{yx} = \rho \overline{\varepsilon}_{yx} \frac{\partial U_d}{\partial y},$$

$$\overline{\varepsilon}_{yx} = \lambda U_* H, \quad \overline{\varepsilon}_{yx} = \lambda H \left(\frac{1}{8}f\right)^{1/2} U_d \qquad (2.28)$$

Equation (4.28) becomes

$$\rho g H S_o - \frac{1}{8}\rho f U_d^2 \left(1 + \frac{1}{s^2}\right)^{1/2} + \frac{\partial}{\partial y}\left\{\rho \lambda H^2 \left(\frac{f}{8}\right)^{1/2} U_d \frac{\partial U_d}{\partial y}\right\}$$

$$= \frac{\partial}{\partial y}\left[H \left(\rho U V\right)_d\right] \qquad (2.29)$$

It may be noticed from Equation (4.28) that the local shear velocity, $U_* = \sqrt{\tau_b/\rho}$ is affected by the free shear layer turbulence and the secondary flow. In regions of high lateral shear it might be argued that the U_* in (2.28) should be replaced by the primary or shear velocity difference between the two regions (e.g. see DELV model given by Wormleaton, 1988). However, in the interests of simplicity and because of its common usage by hydraulic modellers the form of (2.28) is retained with λ being regarded as a "catch all" parameter to describe various three dimensional effects.

In an earlier paper, Shiono & Knight (1988) assumed that the right hand term of Equation (4.31) was zero to make a linear differential equation and obtained analytical solutions to it for channels of various shape. The predicted results of the transverse distribution of depth mean velocity was in good agreement with the experimental data, but the predicted boundary shear stress was not as good as the predicted velocity. The experimental results indicated that the effect of secondary flow on boundary shear stress, even in a straight channel, was significant as also shown by Knight, Yuen & Alhamid (1994). As shown in Fig. 2.11, Shiono & Knight (1991) further demonstrated that, for the particular cases considered, the shear stress term due to secondary flow in Equation (4.28) decreases approximately linearly either side of a maximum value that occurs at the edge of the floodplain and the main channel. As a result, with a first-order approximation to the data, the lateral gradient of secondary flow force per/unit length of the channel may be written as

$$\frac{\partial H(\rho U V)_d}{\partial y} = \Gamma_{mc} \quad \text{or} \quad \Gamma_{fp}(= \text{non zero}) \qquad (2.30)$$

where the subscripts mc and fp refer to the main channel and floodplain respectively. This allows Equation (4.31) to become a second order linear differential equation and to be solved analytically. The analytical solution to (4.31) may then be expressed for a constant depth, H, domain as

$$U_d = \left\{ A_1 \exp(\gamma y) + A_2 \exp(-\gamma y) + \frac{8gS_0 H}{f}(1 - \beta) \right\}^{0.5} \tag{2.31}$$

where

$$\gamma = \left(\frac{2}{\lambda}\right)^{0.5} \left(\frac{f}{8}\right)^{0.25} \frac{1}{H}, \quad \beta = \frac{\Gamma}{\rho g S_0 H}$$

and for a linear-side-slope domain as

$$U_d = \left\{ A_3 \zeta^{\alpha_1} + A_4 \zeta^{-\alpha_1 - 1} + \omega \zeta + \eta \right\}^{0.5} \tag{2.32}$$

where

$$\alpha_1 = -\frac{1}{2} + \frac{1}{2}\left\{ 1 + \frac{s(1 + s^2)^{0.5}}{\lambda} (8f)^{0.5} \right\}^{0.5},$$

$$\omega = \frac{gS_0}{\frac{(1+s^2)^{0.5}}{s}\frac{f}{8} - \frac{\lambda}{s^2}\left(\frac{f}{8}\right)^{0.5}}, \quad \eta = \frac{\Gamma}{\rho\frac{(1+s^2)^{0.5}}{s}\frac{f}{8}}$$

and ζ is the depth function on the side-slope domain (e.g. $\zeta = H - (y - b)/s$ for the main-channel side slope, where b = semi width of main channel bed) and A_1 to A_4 are constants.

The solutions give the lateral variation of depth-mean velocity and boundary shear stress in a channel of any shape provided that its geometry can be described by a number of linear boundary elements provided the friction factor, f, the dimensionless eddy viscosity parameter, λ, and the secondary flow parameter, Γ, are prescribed for each sub-area.

2.2.3. Experimental studies on flow structure

2.2.3.1. Planform vortices

Many laboratory studies have been made of flow in straight compound channels (e.g. Imamoto *et al.*, 1982; Knight & Demetriou, 1983; Myers, 1978; Sellin, 1964; Shiono & Knight, 1991; Townsend, 1968; Wormleaton *et al.*, 1982). In most of these the flow structure and momentum transfer have been examined by varying the channel geometry and in particular the ratio of the flood plain depth to the main channel depth (e.g. Nezu, Onitsuka & Iketani, 1999; Ikeda *et al.*, 1994, 2000). Herein the laboratory experiments by Ikeda *et al.* (2000) are shown, in which a tilting flume with 1.2 m width and 12 m length was used. The longitudinal bed slope of the flume was 0.00046. A floodplain with a height of 5 cm was placed along one side of the flume, and a continuous side slope with a maximum angle of 48 degree (side slope 1:0.9) was installed between the main channel and the floodplain. The total width of the side slope was 15 cm. A summary of the key hydraulic parameters used in the experiments is given in Table 2.1.

Table 2.1 Major hydraulic variables of laboratory tests in which the ratio of the flood plain depth and the main channel depth was changed (Ikeda et al., 2000).

| Case | Without vegetation | | | | | | With bank vegetation | | | Without vegetation and suspended sediment |
	A	B	C	D	E	F	G	H	I	J
Main channel depth h_m (cm)	5.75	6.00	6.25	6.50	7.00	7.50	6.00	6.50	7.50	6.00
Flood plain depth h_f (cm)	0.83	1.08	1.33	1.58	2.08	2.58	1.08	1.58	2.58	1.00
Depth ratio h_f/h_m	0.144	0.180	0.213	0.243	0.297	0.344	0.180	0.243	0.344	0.180
Maximum velocity in main channel u_m (cm/s)	31.4	31.3	31.4	33.1	34.8	38.7	30.7	32.1	32.9	22.6
Minimum velocity in flood plain u_f (cm/s)	5.4	8.6	12.8	14.0	16.3	21.5	11.0	13.5	19.9	7.6
Minimum velocity in vegetation zone u_v (cm/s)							4.9	5.89	7.5	

Planform vortices were observed in a laboratory flume for a depth ratio, Dr, of 0.18, as shown in Fig. 2.13(a), and these are seen to be very similar to the vortices shown in Fig. 2.1. The wavelength of the vortices can be estimated reasonably well with a linear stability analysis, which predicts the wavelength of the maximum growth rate (Ikeda & Kuga, 1997), as well as by numerical calculation using the SDS-2DH turbulence model. The results suggest that the planform vortices are induced by the shear instability. For a relatively large depth ratio of 0.344, the periodic planform vortices disappear, and active intermittent boils were observed, as seen in Fig. 2.13(b). Similar results were reported by Nezu *et al.* (1999), in which it has been observed that the intermittent boils become stronger as the depth ratio increases.

A conditional sampling technique was used to measure the structure of planform vortices (Fig. 2.14), for which a signal from a capacity-type wave gauge was used as a trigger to know their location. The profile of the planform vortices was skewed, with strong movement towards the main channel in the frontal region of the vortex, and mild movement in the rear region. PIV and PTV techniques have been employed to obtain the instantaneous flow field of the planform vortices (e.g., Nezu, Nakagawa & Saeki, 1994; Nezu, Saeki & Nakagawa, 1994; Ikeda *et al.*, 1995b), but the results are not yet sufficiently detailed to determine the structure.

Figure 2.15(a) shows the calculated instantaneous 2-D velocity field obtained using a frame moving with the temporally-averaged velocity at the interface between the main channel and the floodplain, and Fig. 2.15(b) shows the vorticity field associated with the planform vortices. The plan view of the vortices

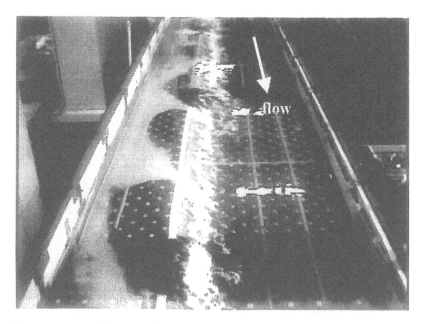

Fig. 2.13   (a) Horizontal vortices for Case B for which the depth ratio is 0.18 (see Table 2.1).

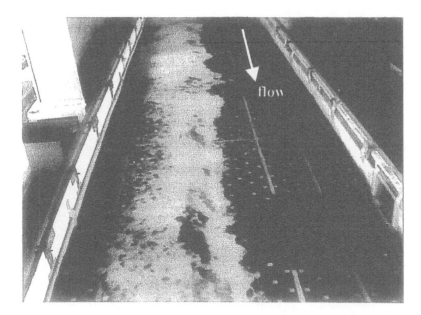

Fig. 2.13 (b) Horizontal vortices for Case F for which the depth ratio is 0.344 (see Table 2.1).

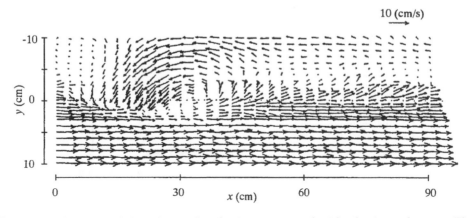

Fig. 2.14 Plan view of the velocity distributions associated with a horizontal vortex. The depth ratio, Dr, is 0.167.

shows unsymmetrical behaviour in the streamwise direction, i.e., the maximum fluid velocity locates upstream of the geometrical centre of the vortices. Furthermore, the vortices are elongated in the streamwise direction, which yields a correlation between u and v, thereby being a major influence in maintaining the lateral Reynolds stress, $-\overline{uv}$.

The instantaneous free surface elevation observed at the boundary between the main channel and the floodplain is depicted in Fig. 2.16(a). The elevation of the

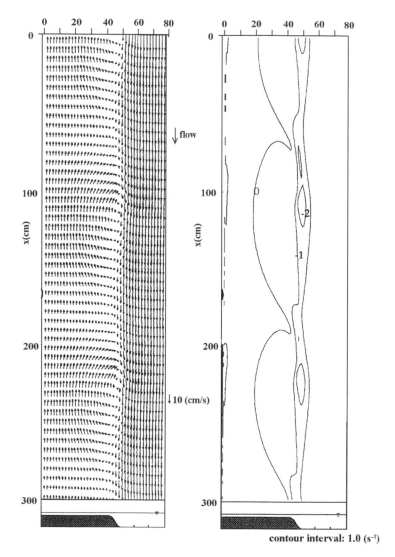

Fig. 2.15 Plan view of the 2-D velocity fields calculated using a SDS-2DH turbulence model.

free surface is low when the centre of a vortex passes through the gauging point. Figure 2.16(b) shows the results of numerical calculation corresponding to Fig. 2.16(a), and agrees reasonably well with the measured variation.

The temporally-averaged flow field is important from an engineering point of view, and Fig. 2.17 shows the results of measurements in the laboratory flume using 2-D laser-Doppler velocimetry, together with the prediction by numerical calculation. The measured fluid velocities are smaller than the calculated values in the main channel close to the floodplain. This discrepancy is closely related to the existence of intermittent boils, which transport near-bed fluid with small

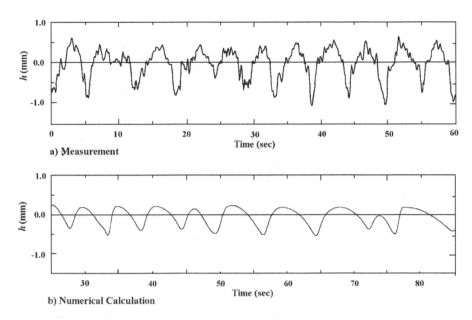

Fig. 2.16 Temporal variations of free surface elevation at the boundary of main channel and flood plain; a) measurement, b) numerical calculation by SDS-2DH model (Case B, Table 2.1).

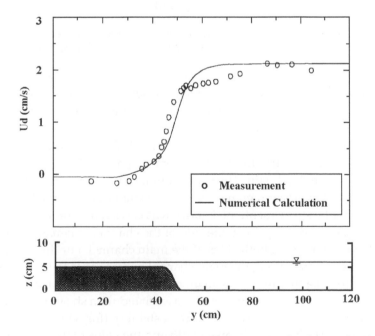

Fig. 2.17 Temporally-averaged flow field, in which open circles denote depth-averaged streamwise velocity measured and solid line indicated calculated velocity (Case B, Table 2.1).

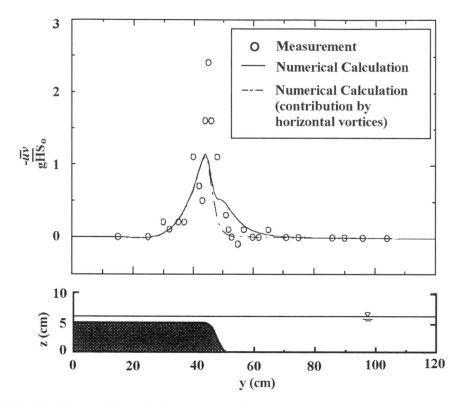

Fig. 2.18 The lateral Reynolds stress, in which open circles denote measurement by laser-doppler velocimetry and solid lines are numerical calculation (Case B, Table 2.1).

momentum towards the free surface, yielding lower fluid velocities there. The effect of such boils was not included in the present numerical calculation.

The corresponding depth-averaged Reynolds stresses are depicted in Fig. 2.18, which are similar to those shown in Figs. 2.7 and 2.12. The numerical calculation reveals that the major contribution to the lateral Reynolds stress, $-\overline{uv}$, is the planform vortices on the floodplain. The contribution of the SDS turbulence is almost negligible in the shear layer region of the main channel close to the floodplain.

The experimental study mentioned above was carried out for a two-stage channel with a single floodplain along one side of the channel, whereas rivers usually have two floodplains along both sides of the main channel. For this case, two vortex streets are generated along the main channel, which leads to the generation of another stability, like that in the Karman vortex street (Ikeda & Kuga, 1997). It is expected that if the wavelength of vortices associated with shear instability agrees with the wavelength of two-row vortex street stability (for which the wavelength is $b/0.281$, where b is the main channel width and the value of 0.28 was given theoretically by von Karman), then Fig. 2.19 shows the spectrum density distributions of the temporal free surface fluctuations associated with horizontal vortices. The measurements were conducted at the edge of the main channel and the floodplain,

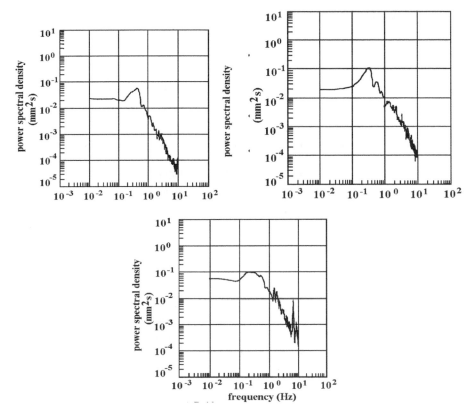

Fig. 2.19 Spectrum density distribution of free surface variation for three cases (a) B-60, (b) B-30 and (c) B-10. See Table 2.2 for further details.

7 m downstream from the entrance of the compound channel, where the wavelength of the vortices reached a statistical equilibrium value. It is evident that the spectrum has a sharp peak for Case B-30 (see Table 2.2), for which the main channel width is 30 cm. The result indicates that the vortex streets are very stable. The observation of vortices by eye has revealed that they are arrayed in a staggered manner in planform. In Case B-10, where the main channel width is 10 cm, the spectrum is broad banded and the peak is not so sharp.

The experimental results for the depth-averaged Reynolds stress at the main channel/floodplain interface are shown in Fig. 2.20. This shows that the Reynolds stresses increase linearly with b/H, for each depth tested, and reach a maximum value at around $b/H = 5$, which corresponds to Case B-30, where H is the depth of flow in the main channel. The Reynolds stresses then decrease and reach an equilibrium value at higher b/H values. This indicates that the vortex streets become independent of each other when b/H is large (i.e. larger than about 8 in this experiment). The results indicate that the momentum exchange rate at the boundary of the main channel and the floodplains becomes largest when the vortex streets are stable and arrayed in a staggered manner. The depth-averaged Reynolds stresses

D. W. Knight *et al.*

Table 2.2 Major hydraulic variables of laboratory tests in which the ratio between the main channel width and the depth was changed (Ikeda & Kuga, 1997).

Case	Main channel depth H (cm)	Main channel width B (cm)	Flood plain width Bs (cm)	Reynolds number in main channel Re	Froude number in main channel Fr
A-60	5.70	60.0	30.0	24500	0.575
A-50	5.70	50.0	35.0	24700	0.579
A-40	5.70	40.0	40.0	24600	0.577
A-30	5.70	30.0	45.0	24300	0.571
A-20	5.70	20.0	50.0	23300	0.546
A-10	5.70	10.0	55.0	18000	0.423
B-60	6.00	60.0	30.0	27200	0.591
B-55	6.00	55.0	32.5	27400	0.595
B-50	6.00	50.0	35.0	27500	0.598
B-45	6.00	45.0	37.5	27600	0.600
B-40	6.00	40.0	40.0	27300	0.594
B-35	6.00	35.0	42.5	26900	0.585
B-30	6.00	30.0	45.0	26400	0.573
B-25	6.00	25.0	47.5	26200	0.570
B-20	6.00	20.0	50.0	26100	0.567
B-15	6.00	15.0	52.5	23500	0.511
B-10	6.00	10.0	55.0	21800	0.474
B-5	6.00	5.0	57.5	18000	0.391
C-60	6.30	60.0	30.0	29500	0.596
C-50	6.30	50.0	35.0	29700	0.601
C-40	6.30	40.0	40.0	29900	0.605
C-30	6.30	30.0	45.0	29300	0.592
C-20	6.30	20.0	50.0	28500	0.575
C-10	6.30	10.0	55.0	24000	0.485

shown in Fig. 2.20 for an asymmetric channel should be compared with those in Fig. 2.12 for one of the symmetric FCF channels.

More recent experimental studies by Bousmar (2002), van Prooijen (2004) and Bousmar & Zech (1999, 2002, 2004) have highlighted their characteristic nature. The difficulties of including them numerical models based on the RANS equations have been highlighted by Tominaga & Knight (2004).

2.2.3.2. Longitudinal vortices

The nature of time-averaged longitudinal vortices in straight compound channels was investigated by Nezu (1996), Nezu, Onitsuka & Iketani (1999) and Nezu, Onitsuka & Sagara (1999) through a series of laboratory tests. Three kinds of experiments were conducted in a 10 m long, 40 cm wide and 30 cm deep tilting flume. Smooth acrylic boxes were placed on the right side of the channel as the floodplain. The hydraulic conditions are summarized in Tables 2.3 to 2.5, in which, S_o

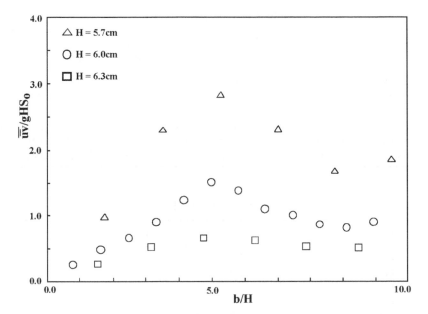

Fig. 2.20 The depth-averaged Reynolds stress at the boundary of the main channel and the flood plain as a function of b/H.

is the bed slope, H is the water depth in the main channel, h' is the water depth on the floodplain, B_f is the floodplain width, B is the channel width, D is the height of the floodplain from the main-channel bed, Q is the flow discharge and U_m is the bulk mean velocity. Furthermore, $\mathrm{Re} \equiv 4RU_m/v$ is the Reynolds number, and $Fr \equiv U_m/\sqrt{gH}$ is the Froude number in which R is the hydraulic radius. An innovative 2-W argon-ion four beam LDA system (Dantec) was used for the $u - v$ measurements by setting the probe above the free surface, and for the $u - w$ measurements by setting the probe normal to the sidewall. Low Froude number flows were needed in order to accurately measure the $u - v$ components with no surface waves present.

Group S of the experiments was intended to reveal the effects of floodplain width, B_f, on secondary flow and turbulence, with the flow depth, H, kept constant at $H/D = 2.0$ (Table 2.3). Group R was undertaken to investigate the effects

Table 2.3 Group S (smooth).

Type	S	H (cm)	B_f/B	h/H	$Q(Vs)$	U_m (cm/s)	U_{max} (cm/s)	Re	Fr
S-A	1/5000	10.0	0.5	0.5	2.6	8.37	11.16	4255	0.085
S-B	1/5000	10.0	0.375	0.5	3.1	9.03	11.91	4999	0.091
S-C	1/5000	10.0	0.25	0.5	3.3	9.34	12.41	5671	0.094
S-D	1/5000	10.0	0.25 + 0.375	0.5	2.7	9.99	12.33	5866	0.094

Table 2.4 Group R (rough).

Type	S	H (cm)	B_f/B	h/H	ks (cm)	Q(Vs)	U_m (cm/s)	U_{max} (cm/s)	Re	Fr
R1-A	1/4000	10.0	0.5	0.5	0.1	2.6	8.47	11.61	4196	0.085
R1-B	1/4000	10.0	0.675	0.5	0.1	3.1	9.20	11.91	5024	0.093
R1-C	1/4000	10.0	0.25	0.5	0.1	3.3	9.42	12.41	5679	0.095
R2-A	1/3000	10.0	0.5	0.5	1.25	2.6	8.50	13.02	4205	0.086
R2-B	1/3000	10.0	0.375	0.5	1.25	3.1	9.34	13.52	5097	0.093
R2-C	1/3000	10.0	0.25	0.5	1.25	3.3	9.99	12.33	5866	0.095

Table 2.5 Group H.

Case	S	H(cm)	B_f/B	H/D	Q_t/s	U_{max} (cm/s)	Fr	Re $\times 10^6$
H6	1/3000	6.0		1.2	1.44	13.4		1.3
H7	1/4000	7.0		1.4	2.06	14.9		1.8
H8	1/5000	8.0	0.5	1.6	2.73	15.6	0.20	2.3
H10	1/6000	10.0		2.0	4.20	16.4		3.1
H15	1/7500	15.0		3.0	8.37	19.0		6.2

of roughness elements (glass beads) on secondary flow and turbulence (Table 2.4). Both the main-channel and floodplain beds were initially smooth. The diameter of glass beads used, d, were 0.1 cm and 1.25 cm. These glass beads were attached most densely on only the floodplain (i.e. Nikuradse's equivalent sand roughness $k_s = d$), but the bed of the main channel was kept smooth. The flow depth was also kept constant at $H/D = 2.0$ again. Groups S and R were conducted by Nezu (1996). Group H was undertaken to investigate the effects of the floodplain depth on secondary flow and turbulence (Table 2.5). The flow depth ratio, was changed systematically, i.e., $H/D = 1.2, 1.4, 1.6, 2.0$ and 3.0, while retaining a smooth condition for the floodplain bed (Nezu, Onitsuka & Sagara, 1999b).

Figure 2.21 shows the vector description of time-averaged secondary flow that was obtained from the 2-D LDA measurements for all cases of Group S, i.e. with smooth floodplains (Table 2.3). Figure 2.22 shows the corresponding numerical results that were computed by a 3-D algebraic stress model (ASM), which was developed for compound open channel flows by Naot, Nezu & Nakagawa (1993a, 1993b) from that for 3-D rectangular open channel flows by Naot & Rodi (1982). Of particular significance is a strong up-flow of secondary flow from the junction between the main channel and the floodplain, towards the free surface. As the floodplain width, B_f, gets smaller, the up-flow tends to incline towards the main channel, which may suggest an effect from the opposite side wall of the main channel, i.e. an effect of the aspect ratio, B_m/H. The numerical calculations are in good agreement with the LDA data, although the latter are somewhat larger near the junction. Cellular secondary flows are seen on the floodplain, and down-flows occur in the centre of the main channel. Figure 2.23 shows comparisons of LDA

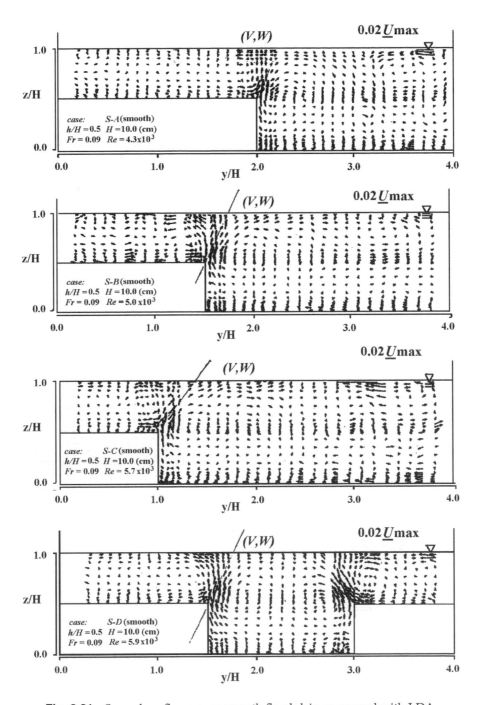

Fig. 2.21 Secondary flows over smooth floodplain, measured with LDA.

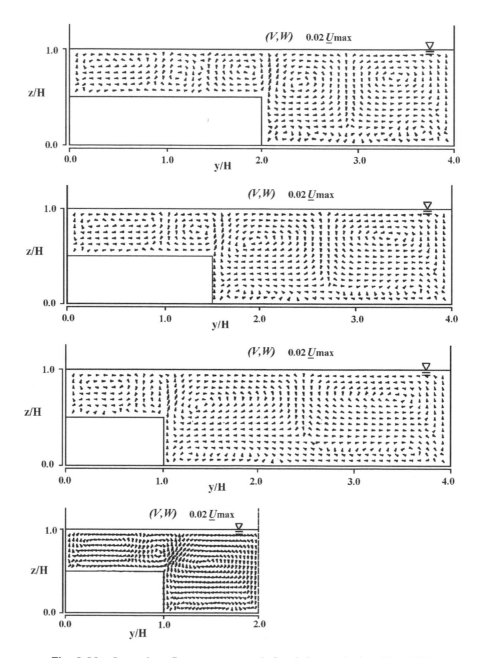

Fig. 2.22 Secondary flows over smooth floodplain, calculated by ASM.

isovels with the calculations of 3-D ASM. In general the two agree well, particularly with regard to the following features.

One is the bulge from the junction towards the free surface, and the other is the velocity dip phenomenon, in which the maximum velocity appears below the free

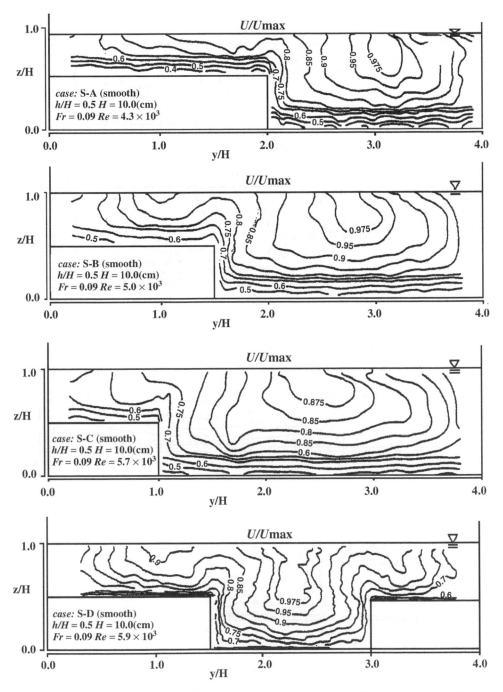

Fig. 2.23 (a) Isovel lines of $U(y, z)$ for smooth flood plain derived from experimental data. These may be compared to the equivalent calculated data in Fig. 2.23(b).

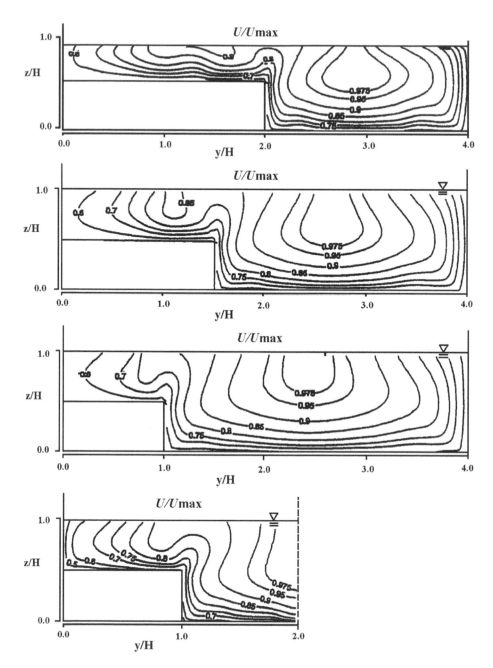

Fig. 2.23 (b) Calculated isovel lines of $U(y, z)$ for smooth flood plains.

surface due to the small aspect ratio, i.e. $B_m/H < 5$. These bulges in the isovels are produced by the secondary flows. It should be noted that the bulge in the measured isovels are the largest for $B_f/B = 0.375$, i.e. the case S-B. The wavy isovel lines on the floodplain are due to the effects of cellular secondary flows, discussed from a theoretical standpoint by Nezu & Nakagawa (1993). These distributions of primary flow, U, and secondary flows V and W, are in good agreement with the LDA database obtained in earlier papers by Tominaga, Nezu & Kobatake (1989) and Tominaga & Nezu (1991) who used a 2-D LDA system (TSI/Kanomax). The corresponding results for rough floodplains (Table 4) are presented in full in Nezu (1996). The effect of floodplain roughness on secondary flow is relatively small. Similar to the smooth floodplain case, the bulges in the isovel lines from the junction are the largest in the case of R1-B ($B_f/B = 0.375$ again). In contrast, the bulges from the junction are relatively small for large roughness of $k_s = 1.25$ cm.

Figure 2.24 shows the distribution of primary mean velocity, $U(y,z)$, against the spanwise direction, y/B, at the elevation of $z/h' = 0.75$, i.e. at 3/4 of the flood-plain depth. The value of U/U_{max} attains a peak on the floodplain and becomes minimum near the junction in case S (smooth) and case R1 (small roughness), both of which are "deep" flow category, i.e., $H/D = 2.0$, as mentioned later. These val-ley features of $U(y)$ are due to the low momentum of secondary flow arising from the junction. However, the value of $U(y)$ appears to increase monotonically with an increase of y/B for the case R2 (large roughness). This is because the effect of floodplain roughness appears directly over the floodplain.

Figures 2.25 and 2.26 show some examples of the vertical component of mean velocity, $W(x,y)$, at the elevation of $z/h' = 0.25$ for small and large floodplain

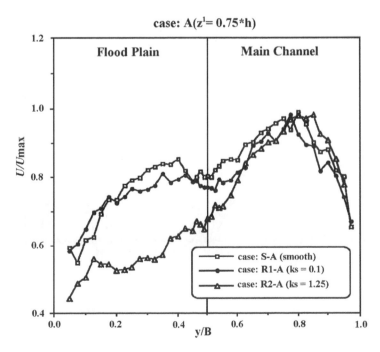

Fig. 2.24 Primary velocity $U(z)$ against the spanwise direction at the elevation $y/h = 0.75$.

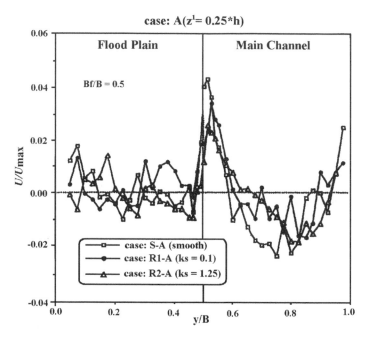

Fig. 2.25 Vertical component of velocity, Case A, large floodplain width, $B_f/B = 0.50$.

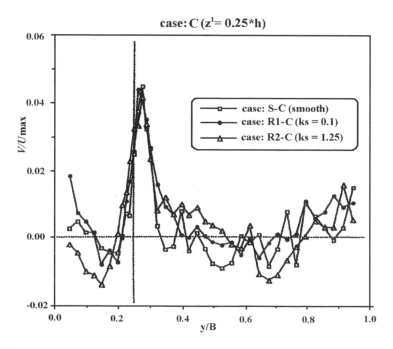

Fig. 2.26 Vertical component of velocity, Case C, large floodplain width, $B_f/B = 0.25$.

widths, i.e., $B_f/B = 0.25$ and 0.5, respectively. In the geometry of case C, i.e. $B_f/B = 0.25$, it should be noted that the value of W, normalized by the maximum velocity U_{max}, attains a distinct maximum, i.e. $W/U_{max} = 0.044$ near the junction just inside the main channel, irrespective of floodplain roughness. In contrast, an effect of floodplain roughness appears in the geometry of case A, i.e. $B_f/B = 0.5$, where the peak value of W/U_{max} tends to decrease with an increase in the roughness size k_s. The peak value for smooth floodplains is $W/U_{max} = 0.044$, whereas for the largest roughened floodplains it is $W/U_{max} = 0.026$. Of particular significance is the wavy distribution of W/U_{max} over the floodplain. This again suggests that cellular secondary flows appear over the floodplain, as mentioned previously.

The effect of flow depth on secondary flow and isovels were also investigated (Table 2.5) by Nezu, Onitsuka & Sagara (1999). Figure 2.27 shows the velocity vectors of the secondary flow normalized by the maximum mean velocity, U_{max}, for all cases of Group H. In the case of H6 ($H/D = 1.2$) and H7 ($H/D = 1.4$), a set of secondary-flow cells may be observed near the junction and in the corner of the main channel. These secondary flow patterns agree well with those in rectangular open channel flows that were measured by Nezu & Rodi (1985). The upper cell is called the "*free-surface vortex*" and the lower cell is called the "*bottom vortex*" in rectangular open channel flows. Nezu & Nakagawa (1984) indicated that the strength of the free-surface vortex in open channel flows is larger than that in closed channel flows, because of the "free surface" condition in the case of open channel flows. The size of the free-surface vortex increases with an increase in flow depth, as seen from cases H6 and H7, and a clear circulation of the main secondary-flow cells can be observed in cases H8 ($H/D = 1.6$), H10 ($H/D = 2.0$) and H15 ($H/D = 3.0$). The main secondary flow runs from the junction of floodplain to the free surface in the main channel, and the strength of the secondary currents decreases in proportion to the distance from the junction edge. This fact suggests that such secondary flows are generated not by anisotropy of turbulence due to the free surface, but by the junction itself in compound open channels.

Figure 2.28 shows the corresponding isovel lines of the mean velocity, $U(x, y)$, normalized by the maximum velocity U_{max}. In the case of H6 ($H/D = 1.2$) and H7 ($H/D = 1.4$), the isovel lines near the side wall of the junction are almost parallel to the side wall and the isovel lines near the floodplain bed are also parallel to the bed. In contrast, in the case of H8 ($H/D = 1.6$), the isovel lines bulge from the junction edge toward the free surface of the main channel. This feature is more marked in the case of H10 ($H/D = 2.0$) and H15 ($H/D = 3.0$). These phenomena are generated by the secondary flow, as pointed out by Tominaga, Nezu & Kobatake (1989), Tominaga & Nezu (1991), Shiono & Knight (1991), Nezu (1996) and others, all using LDA measurements.

The maximum velocity point is located near the surface in the case of H6 ($H/D = 1.2$), as this approximately corresponds to a rectangular open channel with $B_m/D = 5$. However, the position of the maximum velocity point moves below the free surface with an increase in the flow depth, because the influence of the secondary flows increase with increasing flow depth, H. In the case of H15

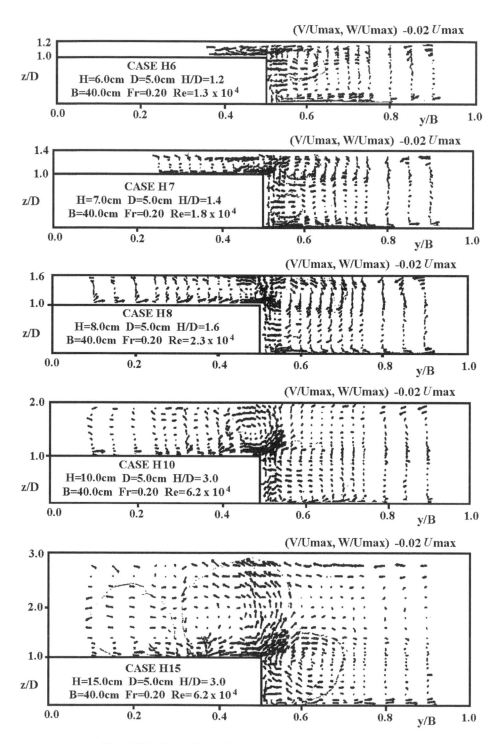

Fig. 2.27 Secondary Flows for Group H (see Table 2.5).

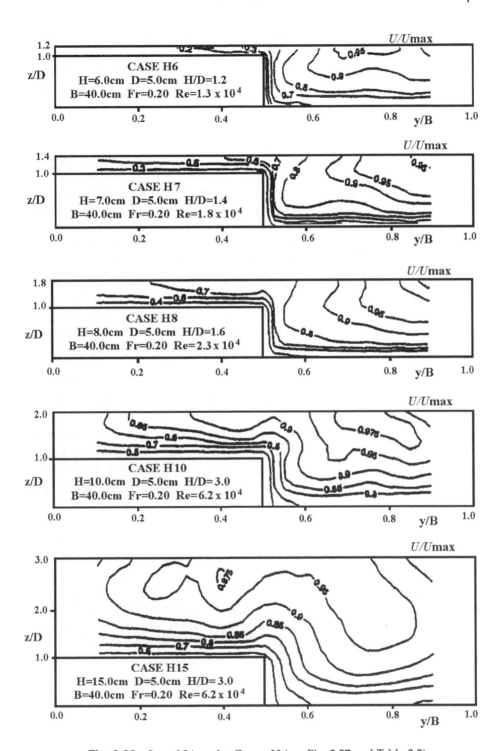

Fig. 2.28 Isovel Lines for Group H (see Fig. 2.27 and Table 2.5).

($H/D = 3.0$), the maximum velocity points are located well below the free surface in both the main-channel and floodplain. This phenomenon may also be caused by the secondary flows, because low momentum flow is transported from the main channel sidewall and the floodplain bed to the free water surface by the secondary currents.

Similar features may be found in many sets of data from compound air ducts, such as those found in Knight & Lai (1985), Lai & Knight (1988) and Rhodes & Knight (1995), and available at www.flowdata.bham.ac.uk. Figures 2.29 and 2.30 show the effect of duct depth and streamwise vorticity on the isovels of the stream-wise component of velocity, U, the depth-averaged velocity, U_d, (symbol Δ) and the boundary shear stress, τ_b (symbol x). In these plots the isovels are non- dimensionalised by the section-mean velocity, U_A, the boundary shear stresses by the section mean, τ_o ($= -R\partial p/\partial x$), the momentum flux (+) by the mean value, M_o ($= \rho U_A^2 A$) and the kinetic energy ($\blacktriangledown$) by the mean value E_o ($= M_o U_A/2$).

For large depths, Dr ($= (H - h)/H$) > 0.3, and large floodplain widths, $B/b > 3$, as shown in Fig. 2.29, there are local velocity maxima on each floodplain, comparable in magnitude to that in the main channel. The maximum velocity filament in the main channel is seen to be depressed below the nominal 'free surface' by the strong secondary flow at the re-entrant corners, which tends to bring low momentum fluid up to the surface, thereby reducing the discharge considerably in the upper layers of the main channel. The values of U_d and τ_b on the floodplain and main channel are roughly similar, indicating little net lateral momentum transfer between these regions.

Figure 2.30 shows just one data set of a whole series, at a low relative depth of $Dr = 0.05$. In this case there is not only a strong lateral shear layer present, but now the longitudinal vortices are manifest in the distortion of the isovels in the lower main channel, due to the low aspect ratio. For these particular experiments the aspect ratio of the main channel was deliberately fixed at 2.0, in order to provide the maximum secondary flow effect for one series of tests with different depths, floodplain widths and roughnesses. It is interesting to note that U_d (Δ) is sensibly constant in the central region of the main channel, whereas $\tau_b(x)$ is much affected by longitudinal vortices. Figure 2.31 shows some lateral distributions of local streamwise velocity, U, at various z elevations. Near the bed, the influence of secondary currents is apparent with a double peak in local velocity, but at the surface the velocity has a single maximum at the centreline, as shown in Fig. 2.30. These data, along with that shown in Lai (1986), Rhodes (1991) and Meyer & Rehme (1994), repay detailed study for the insights they give into the structure of flow in compound channels.

2.2.3.3. Turbulence intensities and anisotropy

Anisotropic turbulence near the junction generates secondary flow, which is time-dependent and occurs intermittently, and such time-dependent features cannot be resolved by RANS equations using simple turbulence models, such as the one

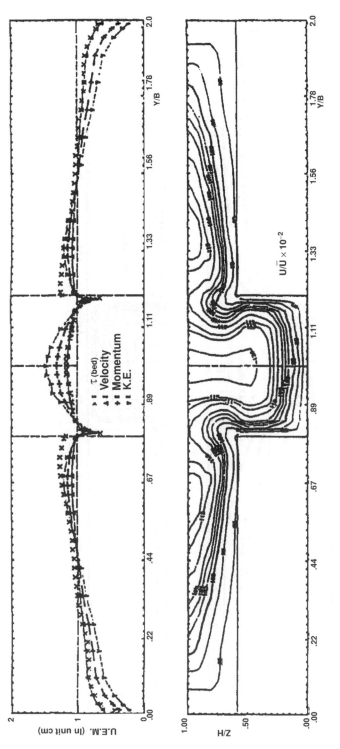

Fig. 2.29 Isovels and boundary shear stresses ($B/b = 4.94$, large depth).

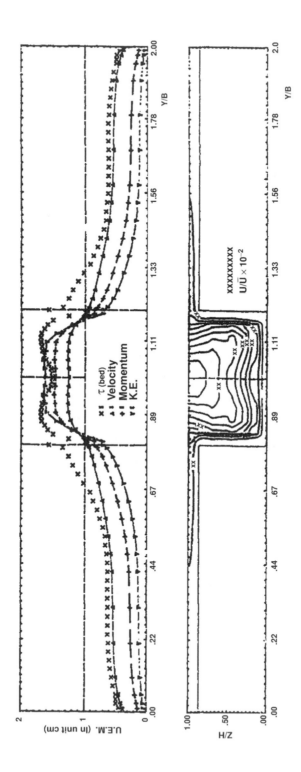

Fig. 2.30 Isovels and boundary shear stresses ($B/b = 4.94$, low depth).

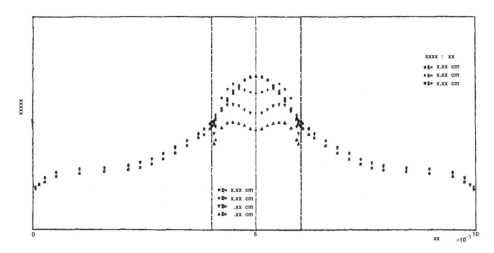

Fig. 2.31 Streamwise velocity distributions ($B/b = 4.94$, low depth).

described in Section 2.2.2. The time-averaged structure of secondary flow is explained by anisotropic turbulent intensities, whereby the spanwise component v of turbulent intensity is quite different from its vertical component w, as seen in Fig. 2.32. These characteristics produce a complicated 3-D flow structure in compound open channels. Although a time-dependent structure of secondary flow in open channel flow is not yet fully understood, Tominaga, Nezu & Kobatake (1989) and Tominaga & Nezu (1991) successfully conducted accurate turbulence measurements in compound open channel flows with a two-component fibre-optic laser Doppler anemometer (Kanomax-TSI) revealing the structure of time-averaged secondary flows. Shiono & Knight (1991) conducted similar turbulence measurements with a LDA in a large-scale Flood Channel Facility, the results of which have been briefly described in Section 2.2.1.

Numerical simulation has been developed in order to predict the field of complex turbulence such as occurs in compound open channel flows. Naot, Nezu & Nakagawa (1993a, 1993b) have extended the Naot and Rodi (1982) model for flow in a single rectangular open channel to flow in a compound open channel, and have developed a 3-D algebraic stress model (ASM) suitable for compound open channel flows. The 3-D ASM calculation produced results that agreed well with the LDA data of Tominaga & Nezu (1991). Cokljat & Younis (1994 & 1995) overcame a weak point of ASM, that the magnitude of anisotropic turbulence was somewhat underpredicted (compare Fig. 2.21 with Fig. 2.22), and developed a Reynolds stress model (RSM), which was also compared with the LDA data of Tominaga & Nezu (1991).

Nezu & Naot (1995) conducted turbulence measurements in compound open channel flows with variable-depth floodplain junction, using an innovative two-component LDA (Dantec) system, and compared them with the results of a 3-D ASM. Furthermore, Nezu (1996) compared new LDA database with calculations

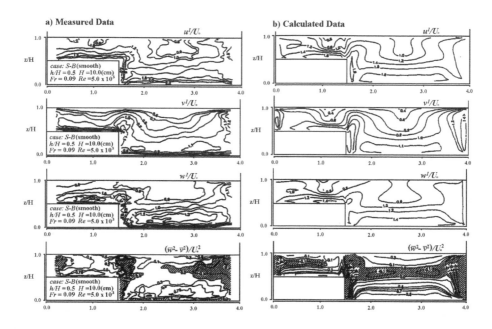

Fig. 2.32 Contours of Turbulence Intensities u', v', w', and $(w^2 - v^2)$ in the Case of S-B.

using 3-D ASM in various compound open channel flow geometries with smooth and rough floodplains. Nezu *et al.* (1999) have more recently revealed some important effects of floodplain depth on secondary flow and planform vortices using both LDA and particle-image velocimetry (PIV).

Figure 2.32 shows an example of contours of all three components of turbulence intensities, u', v' and w', and the anisotropy of turbulence, $(\overline{v^2} - \overline{w^2})$, for the case S-B, i.e. $B_f/B = 0.375$. Similar results were obtained in the other cases. The turbulence quantities are normalized by the averaged friction velocity, $\overline{U}_*$, along the bed. The features related to u' are similar to that of U; the minimum intensity appears below the free surface. The behaviour of v is also similar to that of u. In contrast, w decreases rapidly near the free surface at the junction, indicating that the intensities of the turbulence are especially strong in this region. The gradient of this turbulence anisotropy, i.e. the term B of (2.10) generates secondary flow, as shown in Figs. 2.21 and 2.32. The calculations using 3-D ASM of Naot *et al.* (1993) appear to predict quantitatively the overall characteristics of the turbulence quite well.

Figure 2.33 shows the contours of primary Reynolds stress, $-\overline{uv}$ and $-\overline{uw}$. It should be noticed that these Reynolds stresses change sign near the junction, e.g. the sign of $-\overline{uv}$ is negative on the floodplain side, but positive on the main channel side. The spanwise Reynolds stress $-\overline{uv}$ implies a strong interaction between the main channel and the floodplain, as indicated by (2.9) and explained in Nezu, Nakagawa & Abe (1995).

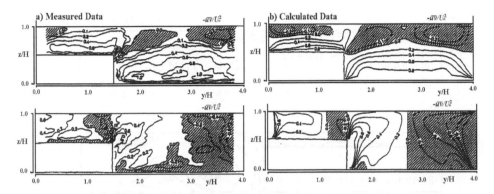

Fig. 2.33 Contours of Reynolds Stresses, $-\overline{uv}/\bar{U}_*^2$ in the Case of S-B.

In the same manner, Fig. 2.34 shows the contours of spanwise Reynolds stress $-\overline{uv}$ in Group H. The value of $-\overline{uv}$ has a maximum near the free surface in the case of H6 ($H/D = 1.2$). This is caused by the apparent shear, i.e. planform vortex, between the main channel and the floodplain. The maximum value near the free surface decreases with an increase in the flow depth. Such a maximum is not observed in the case of H10 ($H/D = 2.0$) and H15 ($H/D = 3.0$). These features coincide well with the results of Fig. 2.33 ($H/D = 2.0$), which are generated by secondary flow. The generation rate G of the turbulent kinetic energy by the Reynolds stresses is defined as follows:

$$G = (-\overline{uv})\frac{\partial U}{\partial y} + (-\overline{uw})\frac{\partial U}{\partial z} + (-\overline{vw})\left(\frac{\partial V}{\partial z} + \frac{\partial W}{\partial y}\right). \qquad (2.34)$$

The first term on the right hand side is caused by the shear in the vertical direction, the second term is caused by shear in the horizontal direction and the third term is caused by secondary flow in the longitudinal direction. Figure 2.35 shows the contours of the generation rate, $GH/\overline{U}_*^3$, normalized by the flow depth H in the main channel and the friction velocity, $\overline{U}_*$. The maximum value appears near the free surface at the junction in the case of H6 ($H/D = 1.2$).

This is caused by the planform vortices, which are generated by the shear between the floodplain and the main channel. In the case of H7 ($H/D = 1.4$) and H8 ($H/D = 1.6$), the maximum value can be seen both near the floodplain side wall and the main channel bed. This feature is one of the characteristics of wall turbulence. In contrast, in the case of H10 ($H/D = 2.0$) and H15 ($H/D = 3.0$) the maximum value appears not near the wall but far away from the junction. This phenomenon is connected with secondary flow from the junction.

The dissipation rate, ε, of turbulent kinetic energy was evaluated from the Kolmogoroff's $-5/3$ power law because the spectral distribution of $u(t)$ obeyed the $-5/3$ power law at all measuring positions for all cases. Figure 2.36 shows contours of the dissipation rate, $\varepsilon H/\overline{U}_*^3$. The patterns of the dissipation rate are

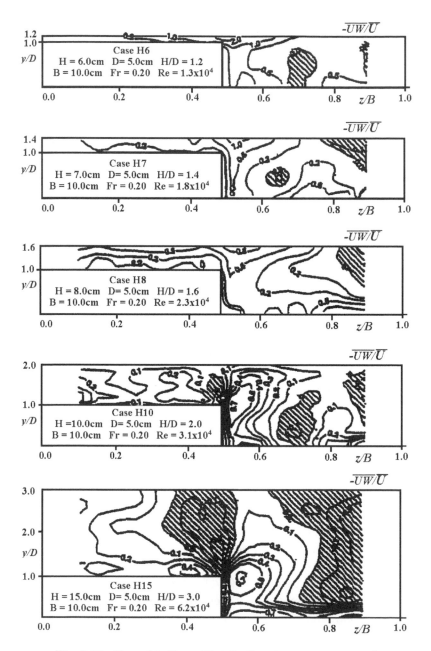

Fig. 2.34 Reynolds Stress Distributions in a cross-stream plane.

similar to those of the generation rate. However, the dissipation rate may be larger than the generation rate in strongly 3-D regions, as seen in the case of H-15 in Fig. 2.35. In these cases the k–ε model is not appropriate because of the anisotropy of the flow.

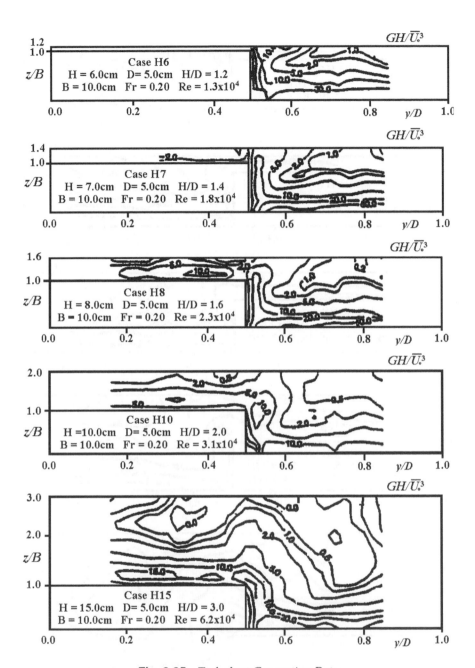

Fig. 2.35 Turbulent Generation Rate.

2.2.3.4. Interaction between main channel and floodplain

Figure 2.37 shows an example of the values of J, T and $(T - J)$ which are defined by Equation (2.9). The second term on the right hand side of (4.7) implies an interaction between the main channel and the floodplain, which is often referred to as an

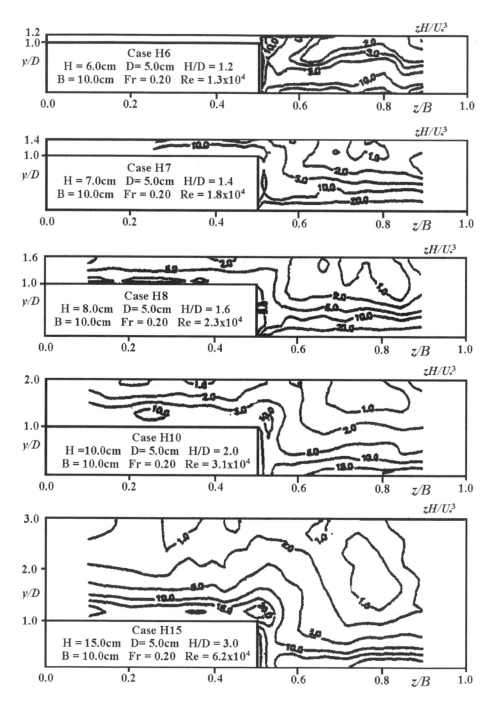

Fig. 2.36 Turbulent Dissipation Rate in a cross-stream plane.

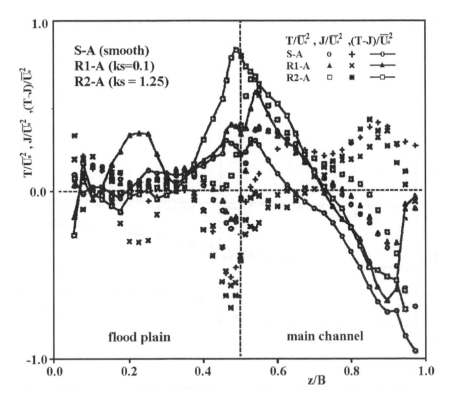

Fig. 2.37 Contributions of Secondary-current Term J and Reynolds–stress Term T to Wall Shear Stress; Interaction between main channel and flood plain (see Equation 2.9).

"apparent shear stress" in compound channel flow work. Figure 2.37 shows that this apparent shear stress increases near the junction, and furthermore that it also increases with increasing floodplain roughness. These features can be explained by (4.7) and are also in agreement with the results of Thomas & Williams (1995a,b). Of particular significance is the 'valley feature' in the spanwise direction, y, of the primary mean velocity $U(y,z)$ near the junction in the case of deep floodplain depth, as shown in Fig. 2.28. The depth-averaged primary mean velocity $U_d(y)$ can be calculated easily from $U(y,z)$. The spanwise distribution of $U_d(y)$ for experiments H6, H8 and H18 are shown in Fig. 2.38, in which the vertical axis is shifted by 10 cm/s for each case. Ikeda *et al.* (1992, 1995) have proposed a theoretical curve for $U_d(y)$, obtained by considering the inflectional (shear) instability of velocity between the main channel and the floodplain. These theoretical curves are also shown in Fig. 2.38, for comparison with the experimental data. Good agreement between the two is evident in the case of H6. However, the inflectional instability theory cannot explain the valley feature of $U_d(y)$, which is due to secondary flow from the junction, as indicated in Fig. 2.27. These results therefore suggest that stronger planform vortices are generated due to inflectional instability of the

D. W. Knight *et al.*

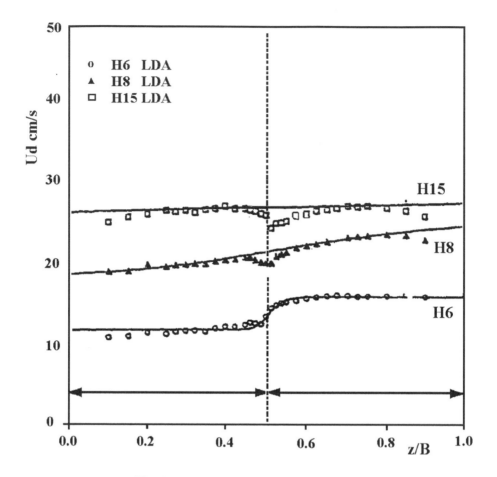

Fig. 2.38 Depth-averaged primary velocity.

shear layer in the case of shallow floodplain flows, whereas stronger streamwise secondary flows are generated due to anisotropy of turbulence near the junction in the case of deep floodplain flows. These are discussed further in the next section.

2.2.3.5. Link between planform and longitudinal vortices

Using a flow visualization technique with hydrogen bubbles, intermittent boils in compound open channels were first recognized by Imamoto *et al.* (1984, 1990, 1992). They found that the boils occur intermittently at the edge of the floodplain, that they move towards the free surface (see Fig. 2.39), and that the boils and the organized planform vortices are the major agencies to exchange fluid momentum between the main channel and the floodplain flows.

Ikeda *et al.* (1995a) studied the relationship between the boils and the planform vortices using 2-D argon-laser anemometry and hot-film anemometry in tandem. A conditional sampling technique was employed to study 3-D profiles of the flow.

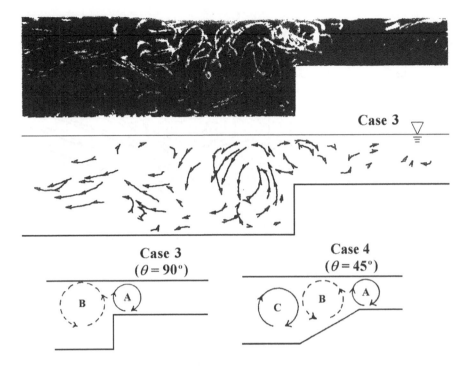

Fig. 2.39 Photograph of instantaneous motion of secondary flow visualized by the neutral buoyancy tracer method in a compound open channel, and path lines of tracers.

It was found that there was some organized structure to the 3-D time-averaged flow, and that the flow is predominantly two-dimensional over the floodplain, with some 3-D structure in the region close to the interface with the main channel. Figure 2.40 shows that in this region there is an upward component of flow, near the centre of the planform vortices, and that the flow is moving towards the flood-plain in the frontal area of the vortex and in the reverse direction in the rear of the vortex. The secondary flow of the 2nd kind is thus closely related with this struc-ture, and the temporal average of the velocity components for one period of vortex at a cross section reveals the existence of the secondary flow near the boundary, as has been shown previously in Figs. 2.21, 2.23 and 2.27. It is therefore suggested that the boils and the secondary flow of the 2nd kind are closely related with the three-dimensionality of the organized planform vortices.

Instantaneous velocity vectors and details of plan vorticity have already been given in Section 2.2.3.1. Nezu *et al.* (1999) obtained the instantaneous velocity vec-tor $\{u(x,y), w(x,y)\}$ at the elevation $z = H - h/2$ for Group H by using PIV. They then analysed the vorticity, ω, defined as

$$\omega = \frac{\partial u}{\partial y} - \frac{\partial v}{\partial x} \quad \text{i.e. } \omega > 0 \tag{2.35}$$

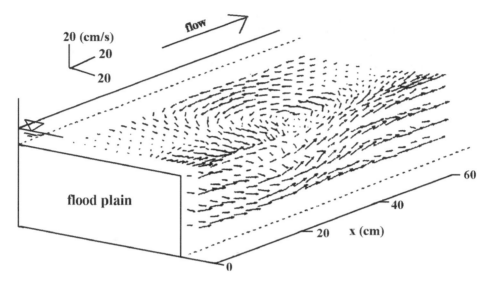

Fig. 2.40 Bird's eye view of 3D velocity vectors associated with horizontal organized vortex.

Figures 2.41 and 2.42 show some examples of the instantaneous velocity vector (u, v) and the vorticity ω for H6 and H15 respectively. The other cases of H7, H8 and H10 are available in Nezu, Onitsuka & Iketami (1999). It should be noted that the flow patterns are quite different between the two cases. In the case of shallow floodplain depth, i.e. H6, a clockwise vortex appears due to the horizontal shear between the main channel and floodplain and it evolves downstream. This verifies that the vorticity is positive in most of the area of the compound channel.

In contrast, in the case of deep floodplain depth, i.e. H15, the value of vorticity is positive in part of the main channel, but negative in part of the floodplain. It therefore follows that one clockwise vortex appears in the main channel and one counter-clockwise vortex appears on the floodplain, as seen clearly in Fig. 2.42.

However, these vortices in deep floodplain depth are weaker than those in shallow floodplain depth because the former vorticity is much smaller than the latter one, as judged from Figs. 2.41 and 2.42. In the light of this, Nezu *et al.* (1999) have proposed a new schematized 3-D flow structure for compound channel flows, as shown in Fig. 2.43. In the case of shallow floodplain flow, a vortex street is generated due to an inflectional instability between the main channel and floodplain, as indicated in Fig. 2.17. See also Ikeda *et al.* (1995). The secondary flow from the junction is not so important as compared with planform vortices, because the horizontal distribution of primary mean velocity $U(y)$ has not a valley feature near the junction. The wall (bed) shear stress on the floodplain increases towards the junction, due to the horizontal interaction between the main channel and the floodplain, as seen in Fig. 2.30.

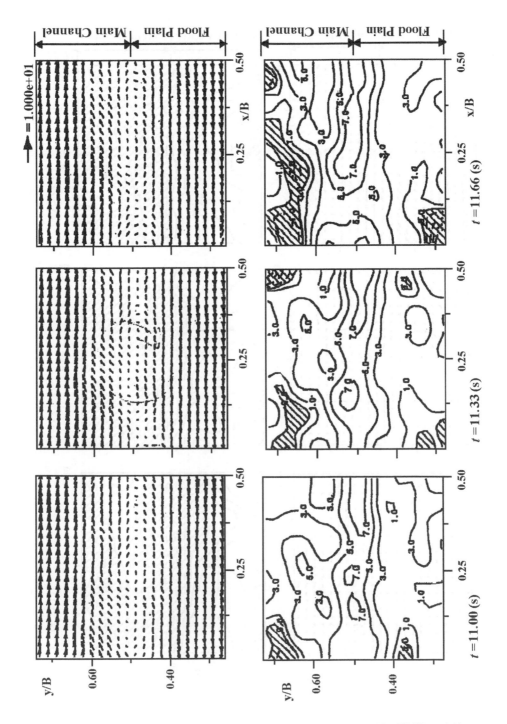

Fig. 2.41 Instantaneous Velocity Vector and Vortices in case H6 ($H/D = 1.2$).

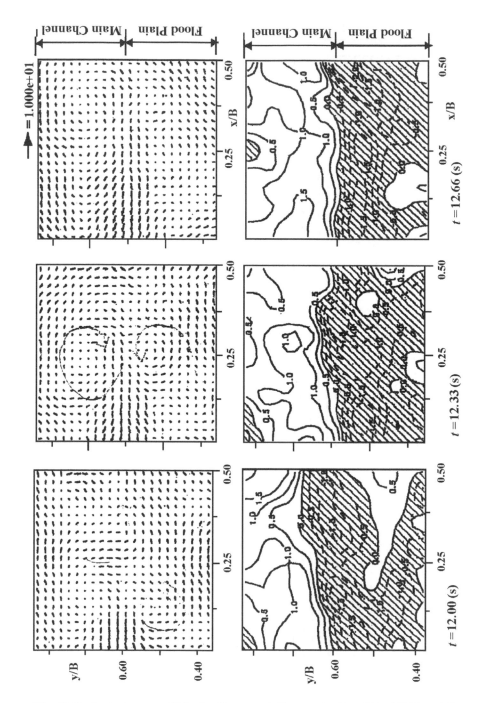

Fig. 2.42 Instantaneous Velocity Vector and Vortices in Case H15 ($H/D = 3.0$).

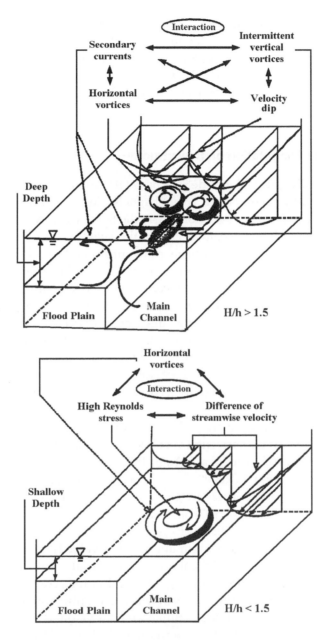

Fig. 2.43 (a) Flow fields in Shallow Depth Flows and (b) Flow fields in Deep Depth Flows.

In contrast, in the case of deep floodplain flow depth, secondary flow from the junction governs the 3-D flow structure in compound open channel flow. This secondary flow is generated by the anisotropy of turbulence in the shear layer between the main channel and the floodplain, as seen in Fig. 2.32. Low momentum and energy are transported from the junction by these secondary currents. As a result the horizontal distribution of primary mean velocity $U(y)$ has a valley feature near the junction as also has the turbulence. See Figs. 2.24, 2.29 and 2.33. Two weak vortex lines appear on both the floodplain and the main channel, as shown in Fig. 2.40. According to Nezu *et al.* (2000), the shallow floodplain depth category of Fig. 2.41(a) is $H/D < 1.5$, whereas the deep floodplain depth category is $H/D > 1.5$.

2.2.3.6. Effect of vegetation on flow structure

Vegetation usually grows alongside riverbanks between the main channel and the floodplains in natural compound river channels. The vegetation produces additional resistance to flow in this region where the flow structure is important. Since the smallest fluid velocity usually occurs in the vegetation layer, the direction of rotation of planform vortices is possibly different along the both sides of the layer. This provides an interesting phenomenon from an engineering point of view as well as science. The shear layers produced along each side of the band of vegetation will yield two rows of planform vortex streets along both sides. The vortex streets may interact with each other as described previously, or may merge with each other to grow into larger vortices. The behaviour of these vortices is important in that they can affect the exchange of fluid momentum between the main channel flow and the flow over the floodplains. They may also be influential in the lateral transport of suspended sediments from the main channel to the floodplains (Ikeda, 1999; James, 1987).

The effects of bank vegetation on flow in compound channels have been studied extensively regarding horizontal vortices (e.g. Fukuoka *et al.*, 1993, 1994, 1995; Ikeda *et al.*, 1992, 2000; Nadaoka & Yagi, 1993; Nezu & Onitsuka, 1998a; Tsujimoto & Kitamura, 1994). Ikeda *et al.* (1992) performed a linear stability analysis, and found that planform vortices are generated by an inflectional instability in the shear flows. Nadaoka & Yagi (1993) have applied their SDS-2DH turbulence model to calculate the horizontal vortices. The major advantage of this turbulence model is that it can predict the wavelength and the 2-D velocity field without assuming the spatial modes of velocity component. It can also calculate the temporal development of vortices through merging of individual vortices. Fukuoka *et al.* (1994) have performed laboratory tests in which a longitudinal band of permeable meshes was placed in the middle of a flume, and it was found that two vortex rows are generated along the both sides of permeable meshes and they can interact with each other. They performed a stability analysis by examining the major two modes of velocity fluctuation associated with the horizontal vortices. They also applied the results to the Tone River (see Fig. 2.1 and Section 5.2.4), and concluded that the flow pattern observed in Tone River is caused by the organized structure of planform vortices. However, these studies have treated flows with flat beds, and are not necessarily applicable to compound channels with bankside vegetation.

In contrast, Nezu & Onitsuka (1998b, 2001) have recently carried out LDA & PIV measurements in partly vegetated open-channel flows. They found that near the free surface a large-scale secondary flow extends from the main channel to the vegetated zone, indicating strongly that suspended sediment transport takes place from the main channel to the vegetated zone.

Ikeda *et al.* (2000) have also employed the SDS-2DH turbulence model to predict the behaviour of planform vortices for flows in compound channels with bankside vegetation. They also conducted a series of experiments to measure the depth-averaged flow velocity and the wavelength of the horizontal vortices in order to validate the results of numerical calculations (see Table 2.1).

The flume used by Fukuoka *et al.* (1994) and Ikeda *et al.* (2000) is the one that has been previously described. Some results obtained by Ikeda *et al.* (2000) are now used to explain the effect of vegetation on the structure of flow.

The flow patterns for Runs H and I, visualized by aluminum powder, are shown in Fig. 2.44. Organized planform vortices are seen to exist even for Run I, for which the depth ratio, $(H - h)/H$ is 0.344. For this large depth ratio, without bank vegetation (Run F, refer to Fig. 2.13(b)), the organized vortices normally disappear and

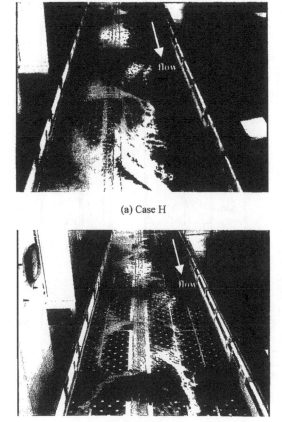

(a) Case H

(b) Case I

Fig. 2.44 Flow visualizations for Runs H and I, in which vegetation band is installed along the side bank (see Table 2.1).

intermittent boils predominate in the flow field, as described before. The difference is apparently caused by the existence of the vegetation band along the bank. The velocity distribution becomes more uniform over the vertical column, due to the drag induced by the vegetation, and the situation increases two-dimensionality of flow. The existence of vegetation thus suppresses the occurrence of boils.

The numerical computation has suggested that two rows of vortex streets, that have different directions of rotation, are generated along both sides of the vegetation band at the initial stage of growth, and then the scale of the vortices increases by merging with each other vortex street (Fig. 2.45). At the same time, two rows of vortex streets begin to interact with each other to form large vortices as depicted in

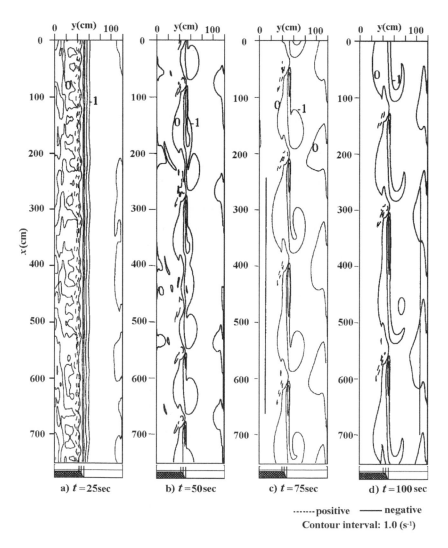

a) $t=25$ sec   b) $t=50$ sec   c) $t=75$ sec   d) $t=100$ sec

------ positive   —— negative
Contour interval: 1.0 (s⁻¹)

Fig. 2.45 Temporal development of horizontal vortices along both sides of the vegetation band (numerical calculation).

Fig. 2.46(a). It is seen that organized planform vortices, which are widened across the vegetation band, are thus formed. This result agrees with the flow visualization as seen in Fig. 2.44. The rotation of these vortices is predominantly negative, which suggests that the vortex street formed in the main channel is dominant (Fig. 2.46(b)). However, the positive rotation still remains on the flood plain, though it is weak.

Figure 2.47 shows the lateral distributions of depth-averaged flow velocity and Reynolds stress. The predicted values are depicted by solid lines, and the open

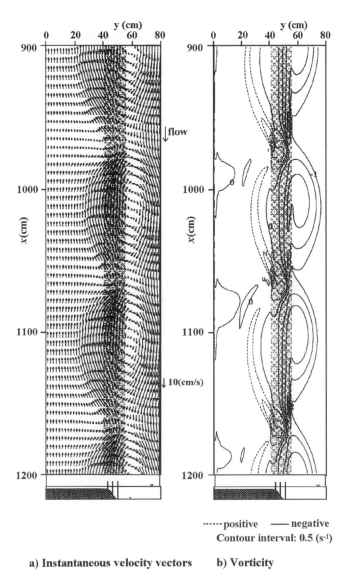

······positive   —— negative

Contour interval: 0.5 (s⁻¹)

a) Instantaneous velocity vectors   **b) Vorticity**

Fig. 2.46 (a) Instantaneous velocity vectors and (b) vorticity seen from a frame moving with the fluid velocity at the boundary of the main channel and the flood plain (Case H).

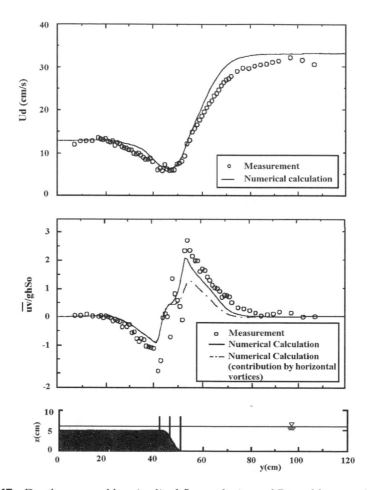

Fig. 2.47   Depth-averaged longitudinal flow velocity and Reynolds stress (Case H).

circles are the measurements obtained by LDA. Almost 70% of the Reynolds stresses are attributed to the organized planform vortices, indicating that these vortices play a dominant role in lateral transport of fluid momentum. These plan-form vortices are also a major element in the erosion of the main channel bed, as well as being important for the transport and deposition of suspended materials in flood plains (Ikeda, 1999). The transport of sediment is treated in Chapter 3.

2.3.   Curvilinear Compound Channels

2.3.1.   Experimental studies

In curved channels, the secondary flow of Prandtl's 1st kind adds more complexity to the flow field. To study this complexity a series of laboratory tests (Ikeda *et al.*, 2001) were conducted in a uniformly curved channel, which was 1.0 m in width and had an inner bend radius of 4.0 m, as shown in Fig. 2.48. The deflection angle

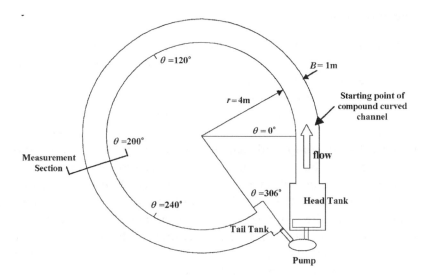

Fig. 2.48 Plan view of the curved channel used.

of the curve was 306 degrees, and the bed was fixed with a streamwise slope of 0.002. The flat bed was covered with uniform coarse sand of 1.0 mm diameter, and subsequently attached to the bed by glue. Two floodplains were installed along the both sides of the channel, the height and the width of which were 5 cm and 15 cm respectively.

The velocity components were measured using a classical Pitot-tube to which a thin silk thread was attached to indicate the direction of flow, the thread being deflected by the secondary flow of the 1st kind. A two-component electro-magnetic flowmeter was also used to measure the flow velocities. The major hydraulic variables are summarized in Table 2.6. Run A1, for which the depth ratio is 0.231, is employed as a typical 'shallow' case, because the intermittent boils are known not to be influential on the planform vortices in straight compound channels when the depth ratio is smaller than 0.25. Run A3, for which the depth ratio was 0.50, is also described below as a corresponding typically 'deep' case.

Table 2.6 Major hydraulic variables of laboratory tests in a compound curved channel.

Case			A1	A2	A3
Discharge	Q	(l/s)	17.9	22.1	37.7
Main channel depth	h_m	(cm)	6.5	7.5	10.0
Flood plain depth	h_f	(cm)	1.5	2.5	5.0
Depth ratio	h_f/h_m		0.231	0.333	0.500
Mean velocity	U	(cm/s)	25.3	33.9	42.4
Main channel width	b	(cm)	70.0	70.0	70.0
Flood plain width	B_s	(cm)	15.0	15.0	15.0

2.3.2. Flow structure

Figure 2.49 shows the free surface for Run A1, visualized by fine polystyrene beads sprayed on to the surface upstream. A pattern of planform vortices is clearly visible along the edge of the main channel and the floodplain near the outer bank. On the other hand, planform vortices are not so apparent near the inner bank. Figure 2.50 shows the corresponding visualization for Run A3, in which planform vortices are still evident along the outer bank at this larger depth. As described previously, for a straight compound channel with a similar depth ratio, the planform vortices disappear, but this is clearly not the case for Run A3 where vortices are seen to exist along both inner and outer banks. Observation by eye has revealed that the rotation of these vortices is not unidirectional, but sometimes counter-rotating vortices have appeared. These phenomena found in this curved compound channel for a large depth ratio is entirely different than what were observed in straight compound channels. The reason is explained later in relation to the existence of secondary flows of the 1st kind.

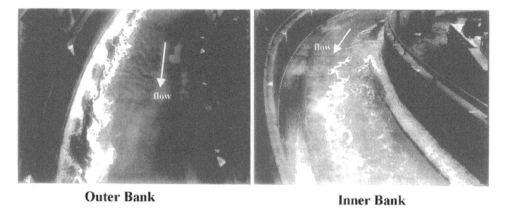

Outer Bank   **Inner Bank**

Fig. 2.49   Visualization of the free surface for Run A1 for which the depth ratio is 0.231.

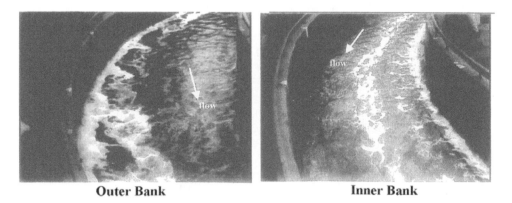

Outer Bank   **Inner Bank**

Fig. 2.50   Visualization of the free surface for Run A3 for which the depth ratio is 0.5.

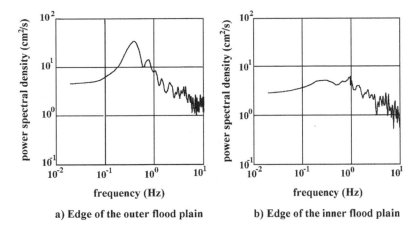

a) Edge of the outer flood plain   b) Edge of the inner flood plain

Fig. 2.51 Power spectrum of longitudinal fluid velocity for Run A1 measured at a) the edge of the outer flood plain and b) the edge of the inner flood plain.

The power spectra of turbulence, observed by electro-magnetic flowmeter, support the above observations. Figures 2.51(a) and 51(b) show the power spectra for Run A1, measured at the edges of the outer floodplain and the inner floodplain respectively. A sharp peak is seen in the outer floodplain data. The spectrum has a wider distribution for the inner floodplain, suggesting that the degree of organization is significantly less. In Fig. 2.52, for which the power spectra for Run A3 are depicted, a clear peak exists for the inner floodplain, entirely different from that for Run A1. The depth-averaged distributions of the streamwise velocity component are shown in Fig. 2.53 for both runs. The maximum fluid velocity appears

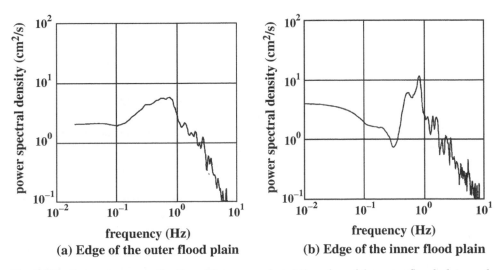

(a) Edge of the outer flood plain   (b) Edge of the inner flood plain

Fig. 2.52 Power spectrum for Run A3 measured at a) the edge of the outer flood plain and b) the edge of the inner flood plain.

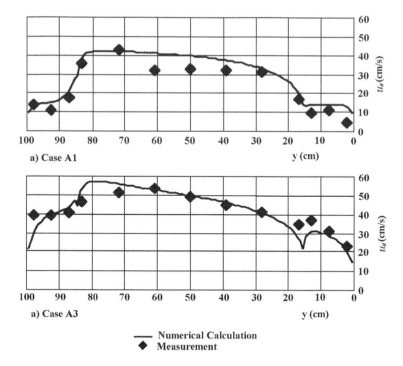

Fig. 2.53 The depth-averaged distributions of the longitudinal velocity component, in which solid lines indicate numerical calculation.

in the main channel close to the outer floodplain in both cases, caused by the net lateral transport of fluid momentum due to the existence of secondary flow of the 1st kind. The solid lines in the Fig. 2.53 indicate the results of a 3-D model with a one-equation turbulence model in which the effect of secondary flow is included. A remarkable feature is the depression of fluid velocity near the edge of the inner flood plain, the reason for which is the transport of small fluid momentum near the bottom and the sidewall towards the free surface by the strong secondary flow. This secondary flow becomes stronger as the depth of flow increases, and the clear kink in the fluid velocity for the case of Run A3 suggests that clockwise and counter-clockwise vortices can exist together near the edge of the inner floodplain. This can explain the co-existence of planform vortices with different senses of rotation near the inner edge, as described previously.

The spatial distributions of the longitudinal velocity and the lateral velocity components of the secondary flow are depicted in Fig. 2.54 for Run A3. The vertical distributions of longitudinal fluid velocity become more distorted in the central region of the main channel, i.e., the maximum fluid velocity occurs below the free surface. This is typically seen in curved channels, which is caused by the intense momentum exchange between the upper layer and the lower layer of flow due to the existence of secondary flow. The secondary flow component is found to be stronger in the outer floodplain than in the inner floodplain.

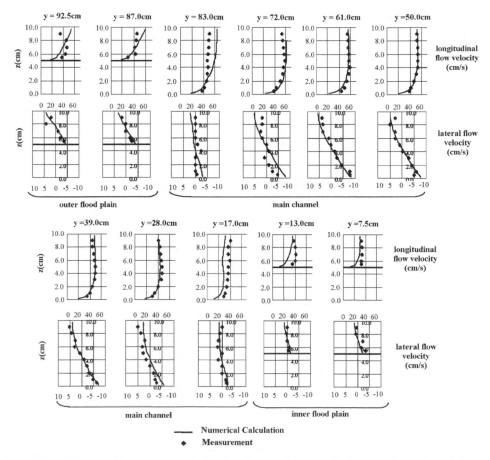

Fig. 2.54 Vertical distributions of the longitudinal flow velocities and the lateral flow velocities for various lateral locations (Run A3).

2.4.   Meandering Compound Channels

2.4.1.   Introduction

Less information is available concerning flow structure in meandering compound channels than for straight compound channels, due in part to the difficulties in measuring sufficient velocity and turbulence data over the entire spatial domain, i.e. typically one wavelength, and to the difficulties in visualizing large sets of results. Furthermore, the shape of the cross-sections of meandering river channels change along the course of the meander, making the range of conditions to be tested more extensive than straight channels with prismatic trapezoidal cross-sections. Early work by the US Army Corps of Engineers (1956) and Toebes & Sooky (1967) led to a better understanding of the influence of geometrical parameters on stage-discharge relationships and the additional energy losses present in such channels. Since then, studies have tended to be divided into two main categories: those with straight floodbanks, set well back from the river channel, and those with floodbanks that themselves meander or have variable planform, as may occur in natural rivers. Representative studies of those in the former category include: Ervine *et al.* (1993), Imamoto Ishigaki & Fujisawa (1982), James & Myers (2002), McKeogh & Kiely (1989), Morvan *et al.* (2002), Sellin, Ervine & Willetts (1993), Shiono & Muto (1998), Shiono *et al.* (1999) and Stein & Rouve (1988), and those in the second category include: Ervine & Macleod (1999), Fukuoka *et al.* (1996), Lambert & Sellin (1996), Shiono *et al.* (1994). A review of flow in natural meandering channels is given by Ikeda & Parker (1989).

2.4.2. Experiments in meandering compound channels

Shiono & Muto (1998) carried out detailed turbulence measurements in compound meandering channels using a 2 component fibre-optic LDA system (TSI). The experimental flume was made of perspex with a rectangular cross section, with dimensions 10.8 m long, 1.2 m wide and 0.35 m deep. The longitudinal slope (valley slope) of the flume was set at $0.001 \pm 0.8\%$ throughout the measurements. There were 5 meander waves constructed for the cases of $s = 1.370$ and 1.571 and 6 for the case of $s = 1.093$ where the sinuosity, s, is defined as the ratio of the meandering channel length to the meander wavelength. The configuration of the channel with $s = 1.37$ is shown in Fig. 2.55. The test section for measurements was set in the half meander wavelength of the 4th meander for the $s = 1.370$ and 1.571 cases and in the 5th meander for $s = 1.093$. The test section was divided into 13 or

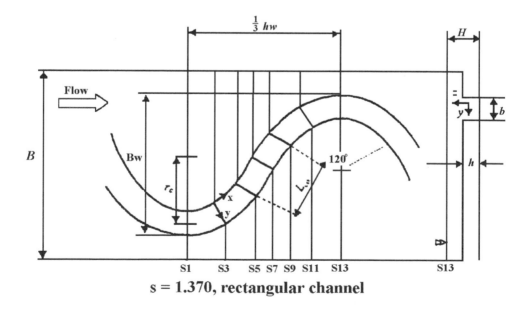

s = 1.370, rectangular channel

Fig. 2.55 Meandering channel configuration for test section.

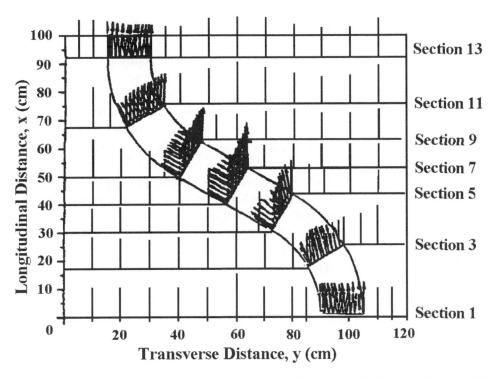

Fig. 2.56 Layer averaged velocity distributions in the rectangular channel for $s = 1.37$ and $Dr = 0.5$ (after Shiono & Muto, 1998).

9 cross sections and the measurements were undertaken at every other section, i.e. 7 or 5 cross sections respectively, as shown in Fig. 2.56.

Velocity measurements for compound meander channels have been conducted by: Toebes & Sooky (1967) with a Prandtl tube; Sellin *et al.* (1993) with a series of propeller current meters; Stein & Rouve (1988) and McKeogh & Kiely (1989) with a laser-Doppler anemometer (LDA). They observed that the primary velocity within the main channel below the bankfull level tends to follow the meander channel, whereas above the bankfull level it tends to follow in the valley direction. Toebes & Sooky (1967) reported that the rotating direction of the dominant secondary cell at the apex in the overbank case is opposite to that in the inbank case. Stein & Rouve (1988) measured the secondary flow along the meandering channel. Willetts & Hardwick (1990) illustrated, from their observations at the crossover section, that this dominant cell at the bend apex is created by the impinging floodplain flow from the upstream side as it flows over and into the main channel. They suggested that the differences are not only in the structure, but also in the originating mechanism of secondary flows between inbank and overbank cases. On the other hand, Imamoto *et al.* (1982) and Shiono *et al.* (1994) carried out experiments on a meander channel with meandering floodplain walls and reported that the dominant cell for the overbank case rotated in the same direction as that for the inbank case. This clearly indicated the effect of the phase of the meandering floodplain walls on the secondary flow structure.

A recent research programme on the Flood Channel Facility (FCF, Phase B) in the UK successfully illustrated the general structures of mean flow, secondary flow and shear flow in compound meander channels (e.g. Sellin *et al.*, 1993, 1996; Ervine *et al.*, 1993). Despite the achievements, the FCF Phase B programme was undertaken for limited ranges of experimental conditions, in particular there was a lack of detailed turbulence measurements along the meander channels. Shiono & Muto (1998) therefore carried out detailed turbulence measurements in compound meandering channels using a TSI 2 component fibre optic LDA system. Although the channel aspect ratio was small, the basic flow mechanisms in compound meandering channels for overbank flow were identified.

The strong vertical shear layer, shown in Figs. 2.56 and 2.57, generated by the floodplain flow crossing over the main channel flow, is controlled by the angle between the meandering channel and the floodplain wall and water depth, as shown schematically in Fig. 2.58. This is in contrast to flow in a straight compound channel, in which the shear is normally generated by the lateral velocity difference between the main channel and floodplain flows, shown in Figs. 2.1, 2.13 and 2.44, and schematically in Fig. 2.3. The shearing mechanisms in a straight compound channel and a meandering compound channel with straight floodplain banks with overbank flow are therefore significantly different. The intermediate case of a straight channel with skewed floodplains exhibits somewhat similar flow structure to the crossover region of a meandering channel, as indicated by Elliot & Sellin (1990).

The measured secondary flow vectors in Fig. 2.57 highlight one of the most important features of secondary flow structure in meandering channels. That is that the sense of rotation of the secondary flow cell at bend apices changes before and after inundation, and originates from different flow mechanisms occurring in overbank and inbank flows. In the studies of Shiono & Muto (1998), using a laboratory meandering channel with either uniform rectangular or trapezoidal cross-sections, the kind of data shown in Fig. 2.57 confirm this. For inbank flow, the vorticity analysis showed an increase in clockwise vorticity by stretching due to the centrifugal force along the bend, and a decrease in size over the crossover region after leaving the bend. For certain overbank flows, two clockwise vorticities were generated by the flood flow crossing over the main channel flow in the crossover region. One, which is driven by the centrifugal force, was located in the middle of the cross section and the other one appeared to stay at the inner edge of the channel within the crossover region. It is of interest that these two cells are not formed as a reaction to each other.

The turbulence data of Shiono & Muto (1998) also indicate the nature of the turbulent kinetic energy in the crossover region, as shown by the contours of turbulent kinetic energy production in Fig. 2.59 for Sections 5, 7 and 9. For inbank flow, the bed and wall generated turbulence is generally the most dominant feature, but for overbank flows the turbulent intensities just below the bankfull level

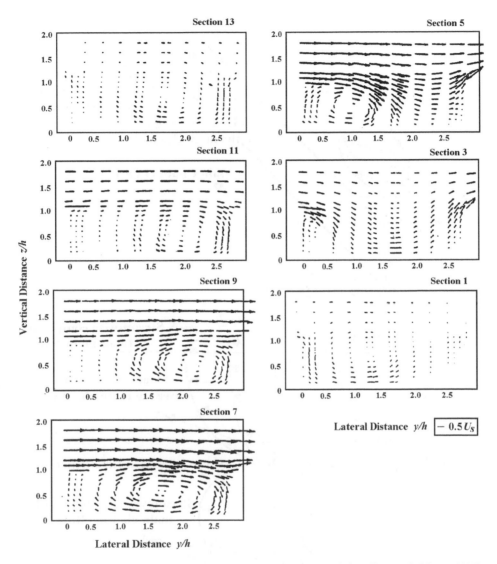

Fig. 2.57 Secondary flow vectors for $s = 1.37$ and $Dr = 0.5$ (after Shiono & Muto, 1998).

become more important. This is also supported by observations of $-\overline{uw}$ for overbank flows, which show that the turbulent shear generated at around the bankfull level is more dominant than the bed generated turbulence and strongly affects the flow structure. However the magnitude of the shear stress $-\overline{vw}$, induced by the strong secondary flows in the lower layer, can reach 3–4 times the bed generated turbulence.

The main contribution of the shear stresses to the turbulence production comes from the $-\overline{vw}$ term, generated by the secondary flows. In contrast, the terms $-\overline{uw}(\partial\overline{U}/\partial z)$ and $-\overline{uv}(\partial\overline{U}/\partial y)$ are main contributors to the turbulence production for a straight compound channel case.

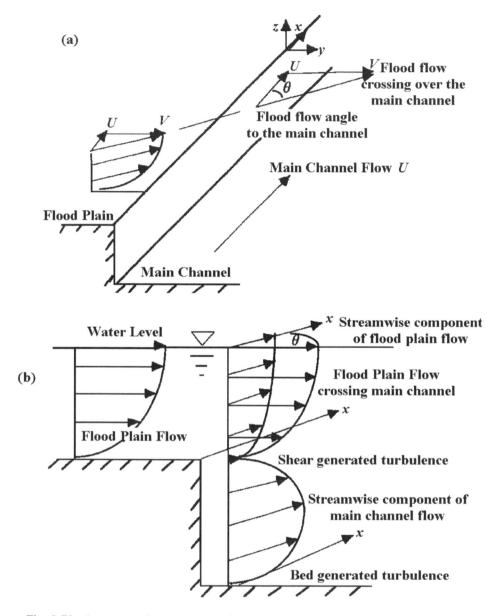

Fig. 2.58 Interaction between upper layer flood flow and lower main channel flow.

There are thus two turbulence flow characteristics which exist within the cross section in the crossover region. One is free turbulence flow that occurs above the bankfull level at the inner edge of the channel, and the other is the combined wall and shear-generated turbulence below the bankfull level, as shown schematically in Fig. 2.5. It should be noted that the eddy viscosity becomes larger when the flow possesses secondary flow.

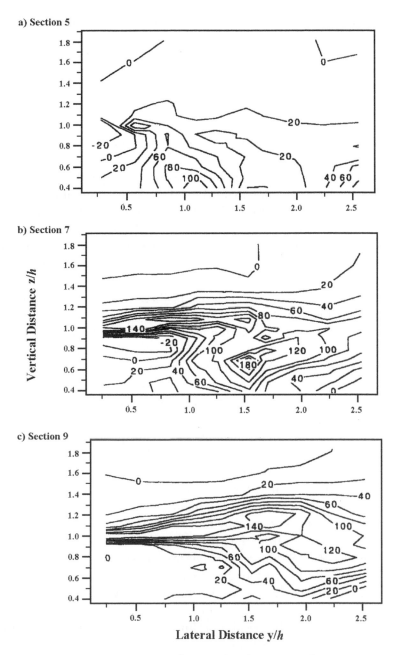

a) Section 5

b) Section 7

Vertical Distance *z/h*

c) Section 9

Lateral Distance *y/h*

Fig. 2.59 Turbulent kinetic energy production distributions in the cross-over region for $s = 1.37$ and $Dr = 0.5$.

2.4.3. Flow structure

For overbank flow in a compound meandering channel with meandering flood-plain walls, secondary flow of the 1st kind predominates in the main channel, similar to that in a curved channel mentioned in Section 2.3 (Imamoto *et al.*, 1982).

D. W. Knight *et al.*

However, the behaviour of flow on the floodplain is somewhat different from that in a uniformly-curved channels. The secondary flow from the main channel now creates large horizontal circulations on the floodplain behind every bend, depending on the sinuosity of the meandering floodplain walls and their phase relative to the meanders in the main channel (Shiono *et al.*, 1994; Fukuoka *et al.*, 1996 & 1997). They also demonstrated the effect of alignment of floodplain walls on the primary velocity and secondary flow, and showed the importance of the phase of the meandering floodplain walls for controlling sediment transport rates and conveyance.

For overbank flow in a compound meandering channel with straight floodplain walls, the flow behaviour in the main channel is quite different from that for the meandering floodplain wall case. The turbulent flow structure for this type of flow is described in detail elsewhere by Shiono & Muto (1998) and Stein & Rouve (1988). With straight floodplain walls, the floodplain flow tends to plunge into the main channel flow near the crossover region, as described by Willetts & Hardwick (1990), and shown in Figs. 2.4 and 2 5.

Figure 2.60 shows surface flow patterns for three relative depths, Dr, of 0.15 to 0.25, taken from Shiono & Muto (1998). At relatively shallow floodplain depths, i.e. for $Dr = 0.15$, the flow within the main river channel appears to dominate, such that the flow entering from the upstream floodplain is turned into the general direction of the flow in the streamwise direction of the flow in the lower main channel. As the depth of flow increases, i.e. for $Dr = 0.25$, the upper flow predominates, and only a small proportion of that flow is turned in the direction of the meandering river channel.

A secondary flow in the main channel flow is thus generated by the horizontal shear applied by the upper floodplain flow, which typically is in the valley slope direction, to the flow in the lower main channel, which is in the local direction of the meander channel. This may then be sufficiently strong to produce a complete reversal in the sense of rotation of the main secondary flow cell at the apex of the next bend downstream, especially for smooth floodplains. The overbank flow condition is thus different from the inbank condition, as illustrated elsewhere by Knight & Shiono (1996). Secondary flow vectors are shown in Fig. 2.57 for the case of $s = 1.37$, at $Dr = 0.50$, corresponding to the data shown in Fig. 2.56. The roll up of the lower flow in the main channel by the upper flow is shown most clearly in the crossover region (Section 7, Fig. 2.56).

Another key feature is the ejection of fluid from the lower main river channel onto the floodplain, just downstream from the apex of each bend. This causes high boundary shear stresses to occur in this region (Knight *et al.*, 1992). The distributions of boundary shear stress are greatly affected by these three-dimensional flow structures, and not only vary with depth, but also along the channel in the streamwise direction. These should be compared with distributions of boundary shear stress in straight channels, which although complex, do not vary in the streamwise direction (Knight, Yuen & Alhamid, 1994).

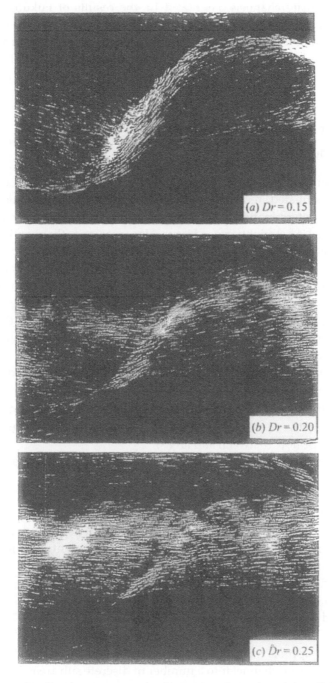

Fig. 2.60 Flow visualization on water surface painted sawdust (after Shiono & Muto, 1998).

For low sinuosity channels, i.e. $s < 1.10$, the results of Fukuoka *et al.* (2000) indicate that certain aspects of the flow field in compound meandering channels, e.g. the position of the filament of maximum velocity, may be regarded as similar to those in simple meandering channels, provided the aspect ratio of the main channel is large and the relative depth is small.

2.5. Effect of Flow Structure on Resistance

2.5.1. Introduction

One of the common tasks of a river engineer is to make estimates of water level based on an estimated, recorded or simulated flood discharge. For inbank flow the theoretical determination of the stage-discharge relationship at a given cross-section of a river is a straightforward issue. However, once the river is in flood, and flowing out-of-bank, then it becomes much more difficult due to the flow structure described earlier. The stage-discharge relationship (H v Q) is of great practical importance, as it not only links discharge with water level in flood routing models, but also it is frequently used to obtain estimates of water level at extreme flood discharges. Such discharges are typically higher than those that have ever been actually recorded, and therefore the corresponding water levels have to be obtained by extrapolation of the rating curve.

When overbank flow occurs, the issues that require special consideration are:

- use of hydraulic radius, R, in calculations (abrupt change at bankfull stage)
- interaction between main river and floodplain flows (lateral shear, etc.)
- proportion of flow (between sub areas)
- heterogeneous roughness (roughness differences between river and floodplains)
- vegetation zone (effect on flow, planform vortices)
- unusual resistance parameters (global, zonal and local)
- significant variation of resistance parameters (with depth and flow regime)
- distribution of boundary shear stresses (affects sediment, mixing and erosion)
- sediment transport (rate, equilibrium shape, deposition, etc.)
- flood routing parameters (wave speed and attenuation)
- critical flow (definition, control points)
- valley and channel slopes for meandering channels (sinuosity)

One traditional approach to the hydraulics of flows in complex cross sections has been to subdivide the channel into a number of discrete sub areas. The conveyance capacities are then calculated via a 'standard' resistance equation, such as the Manning (n) or Darcy-Weisbach (f) equations, using the appropriate sub areas values for A, P, R, f & n, in which A = area and P = wetted perimeter. The individual conveyances are then summed to give the total discharge for the whole channel. One example of the difficulty in applying this approach is shown in Fig. 2.61,

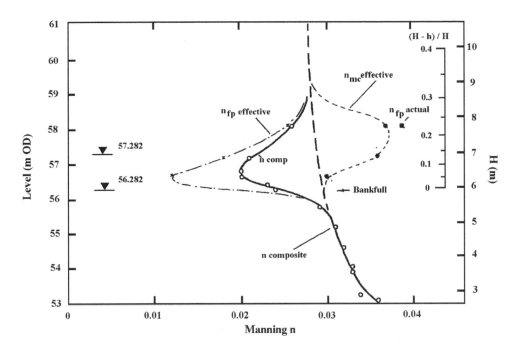

Fig. 2.61 Variation of Manning 'n' resistance coefficient for overbank flow at Montford, River Severn (after Knight *et al.*, 1989).

taken from the River Severn at Montford. The schematised cross section and section properties are as shown in Figs. 2.62 and 2.63.

Using measured friction slope and overbank velocity data in the field, Knight *et al.* (1989) calculated the conveyance in three sub areas and their corresponding roughness values. Figure 2.61 shows the variation of Manning's n with depth, and how n_{comp}, (composite or global value for the whole channel) decreases sharply just above bankfull level. This effect is entirely fictitious, due to abrupt changes in the hydraulic radius, R, and has nothing to do with any real change in channel roughness. Indeed, at the water margin near the bankfull stage one might expect the roughness to actually increase, due to bankside vegetation. Figure 2.61 also shows that in order to obtain the correct conveyance capacity for the main river channel, the sub area (zonal) resistance values for the main channel, n_{mc} (effective), had to be increased, due to the retarding effect of the two floodplains. The sub area resistance values for the floodplain, n_{fp} (effective), had to be correspondingly reduced, to levels well below their actual values for the floodplain based on textural roughness considerations, in order to obtain the correct sub-area conveyance. This example serves to show the care that is needed in distinguishing between different types of roughness coefficient, their strong dependence on depth, and the difficulties associated with predicting a stage-discharge curve for natural rivers when flowing overbank.

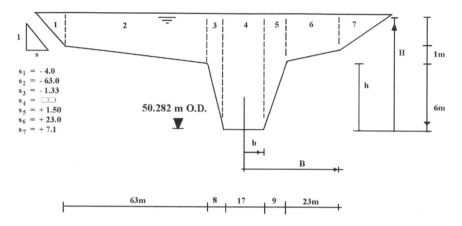

Fig. 2.62 Cross section of River Severn at Montford Bridge.

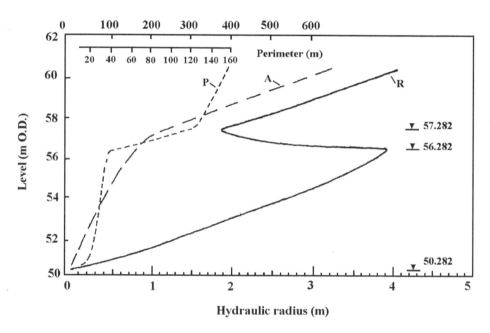

Fig. 2.63 Variation of hydraulic parameters with level.

2.5.2. Resistance to flow in 3-D, 2-D and 1-D simulations

The fundamental equation for the streamwise motion of a small element within the cross-section of an open channel, with a plane bed inclined in the streamwise direction with an angle θ, is given by Equation (2.25) in which $S_o = \sin\theta$. The driving gravity force is thus balanced by the two Reynolds stress terms, that control the vertical and lateral shearing processes, arising from friction forces on the channel bed and sides respectively, and the force term that is required in maintaining the secondary flows transverse to the mean streamwise direction of flow, with velocity components V and W. The large-scale lateral eddies shown in Figs. 2.1–2.3, 2.13 and 2.44 also require some energy input to maintain their structure. Equation (2.25) thus requires some turbulence closure equations with wall functions before it can be solved to give a 3-D simulation of the flow field, as discussed further in Chapter 4. It therefore also follows that in this 3-D approach, any resistance coefficient will be influenced considerably by the 3-D flow structure and these wall functions.

Since river engineers are often concerned with the parameters at the boundaries, (2.25) has to be integrated over the depth, width or area before being of much practical use. If lateral distributions are of importance, as they usually are in rivers, then integration over the depth is preferable, leading to a 2-D depth-averaged form of (2.25), as shown by (2.29) in Section 2.2.2.

Defining Γ as the right hand side of (2.29), as indicated in (2.30), then f, λ and Γ are the local friction factor, dimensionless eddy viscosity and secondary flow parameters respectively, defined by (2.28) and (2.30). Equation (2.29) thus becomes

$$\rho g H S_0 - \tau_b \left(1 + \frac{1}{s^2}\right)^{1/2} + \frac{\partial}{\partial y}\{H\bar{\tau}_{yx}\} = \Gamma \tag{2.36}$$

Equation (4.1) indicates that the boundary shear stress, τ_b, is influenced by the lateral shear and secondary flow. It therefore follows in this 2-D approach that any resistance coefficient will also be influenced by these particular flow structures.

2.5.2.1. Resistance coefficients

If instead of over integrating over the depth, (2.25) is now integrated over the cross-sectional area of the channel, it will lead to a 1-D equation, in which all flow structure effects are simply lumped into a single bulk-flow, i.e. discharge related, resistance parameter (Yen, 1995). The Manning (1891) and Darcy-Weisbach (1857) equations are well known examples of this approach, usually expressed as,

$$Q = (AR^{2/3}S_f^{1/2})/n \tag{2.37}$$

$$Q = (8g/f)^{1/2}AR^{1/2}S_f^{1/2} \tag{2.38}$$

in which n and f are resistance coefficients for the entire cross section, labeled here as a 'global' resistance coefficients. These coefficients effectively ignore the flow structure components arising from the individual terms in the 2-D or 3-D equations. The 'global' resistance coefficients are known to vary with depth or discharge in most channel flows as indicated by Knight (1981), Myers & Brennan (1990), Myers et al. (1999), Wallis & Knight (1984) and Yen (1995). The variation of the resistance coefficient, f, with Reynolds number and relative roughness is normally defined by the Colebrook-White equation. It should be remembered that this resistance law strictly only applies to flows in circular pipes, although it is frequently used for flows in simple prismatic open channels. It is shown later not to be very suitable for flows in open channels with complex cross sections.

It therefore follows that in any 1-D approach the shape of the cross section is not explicitly included, and has to be accounted for by some other means (Engelund, 1964; Kazemipour & Apelt, 1979; Keulagan, 1938; Knight, Yuen & Alhamid, 1994, Rhodes & Knight, 1994a, 1994b). Where the cross section changes in the stream-wise direction, then reach-average resistance coefficients may need to be specified, and some weighting factor applied in order to obtain the appropriate longitudinal energy slope, as suggested by Samuels (1989 & 1990). If the roughness distribution around the wetted perimeter is heterogeneous, as is often the case in natural channels, then additionally there has to be reliance on ancillary equations (Alhamid, 1991; Knight, Alhamid & Yuen, 1992; Task Force, ASCE, 1963). These topics are discussed further in Section 2.5.3. It should be clear from this brief introduction that the 1-D roughness coefficients, f or n, are essentially somewhat crude measures of the net effect of vertical shear, lateral shear and secondary flows, even in straight prismatic channels. In meandering channels, streamwise curvature further complicates the resistance to flow, which is therefore dealt with separately in Section 2.5.4.

2.5.3. Resistance to flow in straight compound channels

For the purposes of channel resistance, the mean boundary shear stress ($\tau_o = \rho g R S_f$) is also traditionally linked to the section-mean velocity, U_A, by the 'global' or overall friction factor, f, by the empirical, yet dimensionally valid, relationship (see Equation 2.39). This is the basis of the Darcy-Weisbach resistance law, given as Equation (4.3). For wide channels, defined here as $B/H > 20$ in order to give less than 5% error in mean boundary shear stress (Knight & Macdonald, 1979), then the local depth, H, may be used instead of the hydraulic radius, R, to give an approximate value for the mean boundary shear stress, τ_o. In depth-averaged models it is customary to define the local boundary shear stress, τ_b, in terms of the local depth and the depth-mean velocity, U_d, at that particular location. The local boundary shear stress, τ_b, and its relationship to the cross section-average, τ_o, is difficult to determine theoretically, even for simple prismatic channel shapes, as indicated in the review by Knight, Yuen & Alhamid (1994).

Care needs to be taken to distinguish between the depth-mean velocity, U_d, and the section-mean velocity, U_A, as well as between the local depth, H, and the hydraulic radius, R, when defining resistance. It is also very important to distinguish between 'global', 'zonal' and 'local' friction factors. The three friction factors commonly used in 1-D and 2-D river simulation models are therefore defined as,

$$\tau_o = \left(\frac{f}{8}\right)\rho U_A^2 \qquad \tau_z = \left(\frac{f_z}{8}\right)\rho U_z^2 \qquad \tau_b = \left(\frac{f_b}{8}\right)\rho U_d^2 \qquad (2.39)$$

$$\text{(global)} \qquad \text{(zonal/sub-area)} \quad \text{(local/depth-averaged)}$$

The zonal, or sub-area friction factor, f_Z, is now defined as one pertaining to a sub-area, and U_A is replaced by the sub-area mean velocity, U_Z, and τ_o is replaced by τ_Z, the mean zonal boundary shear stress. A distinction has therefore to be drawn between these three resistance coefficients, used in 1-D and 2-D models, and the viscous sub-layer or wall function relationships used in 3-D models. Whichever resistance factor is used, its specification should be understood, and its numerical value properly assigned in order to obtain the correct boundary shear stress from calculated or measured velocities. It should be noted that the sub-area and depth-averaged friction factors defined in Equation (4.62) implicitly include the effects of secondary flow/vorticity and lateral shear. This also implies that appropriate f, λ and Γ values must be specified if lateral distributions of both U_d and τ_o are to be determined accurately. Since depth-averaged models are commonly used in river engineering, this particular type of approach is dealt with next.

2.5.3.1. Resistance based on depth-averaged parameters

The Shiono & Knight method (SKM) is based on (2.29) with the three calibration coefficients, f, λ and Γ, concerned with bed friction, lateral shear (via depth-averaged eddy viscosity) and secondary flow (via the lateral gradient of $H(\rho UV)_d$) respectively, defined by,

$$\tau_b = \left(\frac{f}{8}\right)\rho U_d^2; \quad \overline{\tau}_{yx} = \rho\overline{\varepsilon}_{yx}\frac{\partial U_d}{\partial y}; \quad \overline{\varepsilon}yx = \lambda U_* H; \quad \frac{\partial\,(H\rho UV)_d}{\partial y} = \Gamma \quad (2.40)$$

In this approach the cross-section of a river is discretised by a finite number of linear elements, dividing the section into a number of separate domains, within which these coefficients may be regarded as sensibly constant. The number of domains should be kept relatively small if the analytical solution given in Section 2.2.2.2 is used, as shown by the example in Fig. 2.62 with seven domains. If a numerical solution to Equation (2.29) is attempted, then the number of domains may be increased to suit the particular geometric and hydraulic features of the cross-section. Once the three calibration parameters are prescribed for each domain then the analytical or numerical solution gives the local values of the depth-averaged velocity and boundary shear stress for each domain, and hence the lateral distributions of both parameters across the entire channel. See Abril (1995 & 1997), Knight & Abril (1996), Shiono & Knight (1988, 1990, 1991), Knight, Shiono & Pirt (1989), Knight & Shiono (1996), Knight & Yu (1995) and Yu & Knight (1998), Abril & Knight (2004), McGahey & Samuels (2003), McGahey (2006) and the website concerned with the Conveyance Estimation System (CES) developed at HR Wallingford [www.river-conveyance.net]. The PhD thesis of McGahey (2006) provides validation of the Conveyance Estimation System (CES) methods, based on SKM.

The lateral division method (LDM) is similar to Shiono & Knight method (SKM), except that the secondary flow term, Γ, is ignored (i.e. $\Gamma = 0$). In the conventional LDM, all the diffusion process is normally put into a single parameter, such as the depth-averaged lateral eddy viscosity, $\overline{\varepsilon}$, or its dimensionless value, λ. The LDM application described by Wark, Samuels & Ervine (1990) uses the discharge intensity, $q \, (= U_d H)$, as a variable, rather than U_d, and so the governing equation corresponding to Equation (4.59) is

$$\rho g H S_o - \frac{\beta f q^2}{8H^2} + \frac{\partial}{\partial y}\left[\varepsilon_t \frac{\partial q}{\partial y}\right] = 0 \tag{2.41}$$

where β is a factor relating stress on an inclined surface to that in a horizontal plane, and ε_t is used as a 'catch-all' parameter, similar to $\bar{\varepsilon}_{yx}$ in Equation (4.63).

However, in the LDM it could equally well be assumed that either the lateral eddy viscosity parameter, λ, or the 'local' friction factor, f_b, should be enhanced to compensate for assuming $\Gamma = 0$. In the former case, which is the conventional LDM, the corresponding modified SKM equation, taken from Shiono & Knight (1990) is

$$\rho g H S_o - \tau_b + \frac{\partial}{\partial y}\left[\rho H (\bar{\varepsilon}_s + \bar{\varepsilon}_t) \frac{\partial U_d}{\partial y}\right] = 0 \tag{2.42}$$

i.e. the turbulent eddy viscosity $\bar{\varepsilon}_t$, is modified by adding to it an amount equivalent to the effect of the secondary flow term (large scale vorticity) $\bar{\varepsilon}_s$. Alternatively, the normal value of $\bar{\varepsilon}_t$ may be maintained, and the local friction factor increased, which effectively accounts for the secondary flow term in the bed shear stress expression. The LDM equation is then written from Equation (2.29) as,

$$\rho g H S_o - \left(\tau_b + \frac{\partial}{\partial y} H (\rho U V)_d\right) + \frac{\partial}{\partial y}\left[\rho H \bar{\varepsilon}_t \frac{\partial U_d}{\partial y}\right] = 0 \tag{2.43}$$

Although the use of $\bar{\varepsilon}_t$ as a 'catch-all' parameter is known to be incorrect, especially when determining boundary shear stresses, the method can give reasonable lateral distributions of velocity, and hence discharge. Clearly particular care needs to be taken over the calibration procedure and selection of appropriate values for f and λ. For example, it is known from Knight & Shiono (1996), that the values of or λ which fit the velocity data, do not in general fit the boundary shear stress data, and are not in agreement with actual physical measurements. Although useful for stage-discharge calculations, the LDM may not be so useful as SKM for dealing with sediment transport or resistance phenomena, which are dependent on local boundary shear stress values. However, in the absence of any field data on secondary flows with which to calibrate Γ, this may be the only appropriate pragmatic approach.

Typical default values of λ are 0.07–0.50 for trapezoidal channels, with 'standard' values being 0.067 (boundary layers), 0.13 (rectangular open channels), 0.16 (trapezoidal data), 0.27 (FCF, smooth floodplains), 0.22 (FCF, rough floodplains), and higher. Detailed experimental measurements of λ have been undertaken by many researchers, e.g. Alhamid (1991), Haque (1959), Holley & Abraham (1973), Knight, Yuen & Alhamid, (1994), Nokes & Wood (1988), Yuen (1989). Their use in models has also been investigated by numerous researchers: e.g. Knight & Abril (1996), Ackers (1991), Liu *et al.* (1999), Shiono & Knight (1990), Wark, Samuels & Ervine (1990), Wormleaton (1988) and Younis (1996).

Wormleaton (1988, 1996) has proposed a variation on the LDM, known as the lateral velocity distribution method, which is based on the argument by Lean &

Weare (1979) that the eddy viscosity, ν_t, in the 2-D momentum equation could be better defined as,

$$\nu_t = \lambda_1 U_* H + \lambda_s U_s l_s \tag{2.44}$$

The first term on the right hand side of (2.44) is perceived to be the contribution from bed-generated turbulence, with λ_1 assumed to be a constant and equal to 0.16, whereas the second term is perceived to be the contribution from the lateral velocity gradient, with U_S and 1_S selected as characteristic velocity and length scales respectively. He found λ_S to be 0.013 from his experimental data. This approach was re-examined by McKeogh & Kiely (1989) and Wark et al. (1990) for practical use.

Shiono & Knight (1991) investigated the values of the depth-averaged parameters, f, λ and Γ, through a series of structured experiments in the FCF (Shiono & Knight, 1988, 1989, 1990; Knight & Shiono, 1990, 1996). Using the analogy of the Blasius equation in turbulent flow, they found that the ratio of the friction factor in the main channel to that on the floodplain was a function of $Dr^{3/7}$ as follows:

$$\frac{f_{fp}}{f_{mc}} = -0.1065 + 0.8893\, Dr^{-\frac{3}{7}} \tag{2.45}$$

The ratio of the eddy viscosity in the main channel to that on the floodplain, without the secondary flow term, was also studied. They obtained the best agreement with their data when employing values for λ as follows: $\lambda_1 = 0.45$ in the main channel, $\lambda_2 = \lambda_4 = 0.2$ on the side slopes and $\lambda_3 = \lambda_2/Dr^{1.5}$ on the floodplain. Their estimated values of λ, however, appear to be higher than those in the literature for smooth laboratory flumes, given earlier. Shiono & Knight (1990) explained this via theoretical considerations of (2.44) and (4.69). The λ values include all the effects of bed-generated turbulence, shear generated turbulence and secondary flow. From the result of their turbulence measurements, with the secondary flow term included, Shiono & Knight (1991) obtained the empirical formula: $\lambda_{fp}/\lambda_{mc} = (2Dr)^{-4}$ and values for the non-dimensional Γ values, $\beta(= \Gamma/(\rho g H S_o)$ of $\beta = 0.15$ in the main channel and $\beta = -0.25$ on the floodplain.

Knight & Abril (1996) further investigated the relationships for the variation f, λ and Γ with depth in overbank flow through a refined calibration study of the FCF data. They found the following empirical formulae for the ratio of main channel to floodplain values, as:

$$\frac{f_{fp}}{f_{mc}} = (1 - 0.331) + 0.331\, Dr^{-0.719} \tag{2.46}$$

$$\frac{\lambda_{fp}}{\lambda_{mc}} = (1 - 1.2) + 1.2\, Dr^{-1.44} \tag{2.47}$$

$$\frac{\Gamma_{fp}}{\Gamma_{mc}} = -0.4 + 0.810\, Dr^{-0.8} \tag{2.48}$$

It is seen from above that the flow parameters on the floodplain are strongly dependent on both those in the main channel and on the relative depth, Dr.

2.5.3.2. Resistance based on cross section-averaged parameters

For practical purposes, the 1-D cross-section averaged resistance coefficient, i.e. the 'global' n and f value, is often required, especially when dealing with issues concerning the stage-discharge relationship. Although the H v Q relationship is essentially a resistance relationship, it is good practice to actually plot out the variation of resistance coefficients with depth or Reynolds number for rivers, since the graphs may reveal shape or other effects. Such plots are usually required in any case in the calibration or validation stages of numerical model simulations.

Figure 2.61 has already indicated the way in which Manning's n typically varies with depth for a natural river with overbank flow. Figure 2.65 now shows the variation of the Darcy-Weisbach friction factor, f, with Reynolds number, Re, for some smooth rigid boundary laboratory compound channels, with different floodplain widths, giving rise to different B/b ratios. These data were obtained from the FCF, with channels whose cross sectional shape is as shown in Fig. 2.64.

Figure 2.65 shows that as the inbank flow increases, the global f values decreases until the bankfull stage is reached (point A), as would be expected from the Colebrook-White equation for a smooth surface. Once overbank flow begins, for a given floodplain width, the f value abruptly drops, as does the Reynolds number, producing a discontinuity in the diagram. The f values changes abruptly from point A to point B as flow goes overbank, thereafter gradually increasing as the depth on the floodplain increases and the compound channel reverts back to a simple channel at high relative depths. The non-standard nature of the overbank 'global' values is mirrored by a corresponding pattern for the 'local' f overbank values, obtained by integrating all the local friction factors along the wetted perimeters, and the two patterns may be combined to produce the herring-bone pattern shown. The global $f \sim Re$ relationship for the River Severn data at Montford, corresponding to data already presented in Fig. 2.61, is shown in Fig. 2.66, a similar format in Fig. 2.65, but with river discharges superimposed. Clearly, the data in Fig. 2.66 have a similar pattern to that in Fig. 2.65, with a discontinuity at

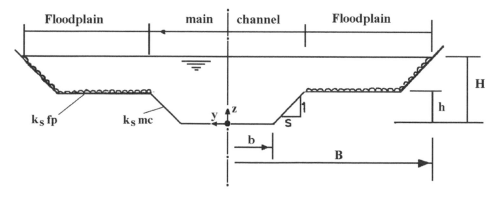

Fig. 2.64 Cross section of a compound channel with notation.

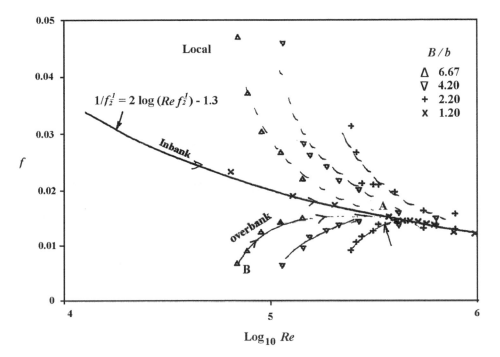

Fig. 2.65 Resistance relationship $(f - Re)$ obtained for a series of experiments in the Flood Channel Facility.

bankfull level, but would not be so easily understood without first appreciating the shape effects inherent in the cross-section averaged or 1-D laboratory resistance data.

The unusual variation of resistance coefficients due to shape effects is shown even more clearly in Fig. 2.67(a), where the same FCF data in Fig. 2.65 is shown replotted in the form of Manning's n versus depth. Whereas Fig. 2.61 showed the variation of Manning's n with depth for a natural river, Fig. 2.67(a) shows the variation of Manning's n for the cases already shown in Fig. 2.65. A significant discontinuity in the 'global' resistance is again apparent at the bankfull stage (points A & B), due to the discontinuity in the hydraulic radius as the flow goes overbank.

Similar cross-sectional shape effects may be seen in compound channels with mobile boundaries, as indicated in the data from the FCF (Phase C), shown in Figs. 2.67(b) and 2.68. In the Phase C experiments, the bed of the main river channel was moulded in a loose sand ($d_{35} = 0.8$ mm), and the floodplains were either kept smooth or roughened with dowels, as described elsewhere by Bousmar *et al.* (2006), Knight *et al.* (1999), Knight & Brown (2001), Tang & Knight (2001), Atabay *et al.* (2004 & 2005), Karamisheva *et al.* (2005), Tang & Knight (2006). Figure 2.67(b) shows that the global Manning's n values increase with depth, as the dunes develop on the river bed, from around 0.015 to 0.025 at the bankfull stage (point A).

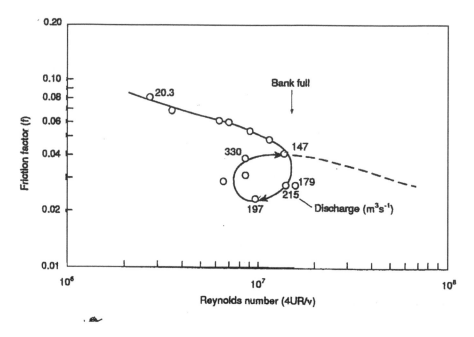

Fig. 2.66   Resistance relationship (*f* versus *Re* for River Severn at Montford).

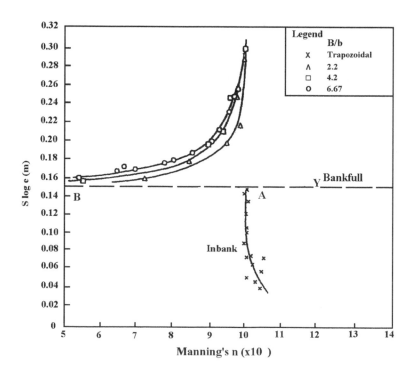

Fig. 2.67   (a) Resistance relationship in FCF Phase A experiments (*n* versus depth).

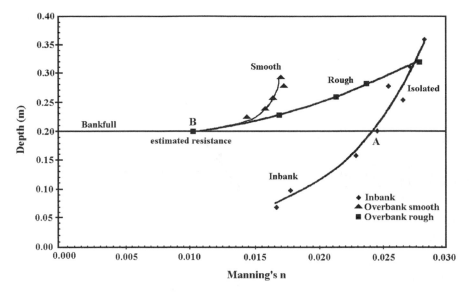

Fig. 2.67 (b) Resistance relationship in FCF Phase C mobile bed experiments (n versus depth).

At this stage the global n value suddenly appears to drop to around 0.010 (point B), before increasing again for both smooth and rough floodplain cases. Also shown are some data obtained by isolating both floodplains, in order to obtain the roughness of the main channel at high flows, without any main channel/floodplain interaction. The cross-sectional shape effect in Fig. 2.67(b) is therefore somewhat similar to that shown in Fig. 2.67(a), except for the effect of the sand dunes, which increase the bed roughness. The corresponding global $f \sim Re$ relationships for these same experiments are shown in Fig. 2.68. Although more complex, they still exhibit a similar trend to those shown in Fig. 2.65, with a discontinuity occurring between points A and B. In Fig. 2.68, the mobile bed data show that the inbank f values increase with discharge (or Re), due to the bedforms, whereas in Fig. 2.65 the smooth channel inbank data decrease, as might be expected from the Colebrook-White equation. In both sets of data, there is still the usual discontinuity at the bankfull stage, again due to the hydraulic radius effect. The overbank f values with smooth floodplains in Fig. 2.68 are more or less constant at around 0.04, whereas the roughened floodplain f values increase with Re, as would be expected with higher velocities over a floodplain roughened with dowels.

2.5.3.3. Resistance with a heterogeneous roughness distribution

The previous section has highlighted the effect of cross-sectional shape on resistance with homogeneous roughness distributions. However, most river channels

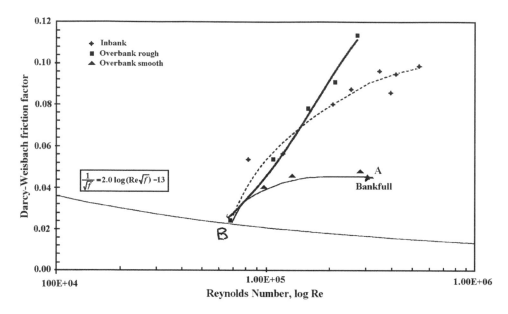

Fig. 2.68 Variation of the overall friction factor with Reynolds number for mobile bed experiments in the FCF (Phase C).

have a non-uniform distribution of roughness around the wetted perimeter. This further complicates the specification of an appropriate global resistance coefficient for the channel cross section. It is sometimes convenient to use a 'composite' roughness coefficient for the whole channel, based on aggregating values from individual sub-areas, rather than many individual values. Although the difficulties in using this approach for channels with complex variations in hydraulic radius have already been demonstrated, this method is still advocated in various textbooks and software. In this approach, certain simplifying assumptions still however have to be made in order to obtain a 'composite' roughness value. These have ranged from making pairs of complementary assumptions, such as:

(i) the total shear force equals the sum of the constituent sub area shear forces and that the sub area velocities vary in proportion to the depth to a one sixth power law (Pavlovskii, 1931),
(ii) the mean velocities are equal in each sub area and the friction slope is the same for all sub areas (Horton, 1933; Einstein, 1934),
(iii) the total discharge equals the sum of the constituent discharges and that the friction slope is the same for all sub areas (Lotter, 1933).

These are summarised in Chow (1959), examined in more detail by Yen (1992) and discussed in relation to flood channels by Knight, Demetriou & Hamed (1983, 1984a, 1984b). It is now recognised that the three methods given above are all flawed, as the assumptions listed are untenable, particularly for overbank flow.

Recourse should therefore be made either to using the local, zonal or global resistance coefficients as outlined previously, or alternatively, to using one of the more modern approaches, such as that based on the concept of 'coherence', as proposed by Ackers (1992b). Since this method is particularly suitable for flows in compound channels, and deals with the division of flow between domains, as well as boundary shear stress, it is dealt with more fully in Section 2.6.2.2.

Vegetation within or alongside a river is one example of heterogeneous roughness that greatly affects the resistance to flow in a river channel. Seasonal growth and decay, as well as managed weed cutting programmes, mean that instream and bankside vegetation typically vary spatially and temporally within most managed river systems. General guidance about the effects of vegetation on resistance is given in Klaassen & van der Zwaard (1974), Kouwen & Unny (1973), Kouwen & Li (1980), Kouwen, Li & Simons (1980) and Hasegawa et al. (1999). The vegetation on floodplains may be of a more varied and dense nature and requires special treatment (Technology Research Centre, Japan, 1994). The effect of vegetation on flow structure in compound channels has been described in Section 2.2.3.6.

An estimate of flow resistance for straight weedy channels may be obtained from the HR Wallingford equations, given by Fisher (1992) as

$$n_{\text{tot}} = n_c + 0.0239 K_w / Fr \tag{2.49}$$

$$(k_s)_{\text{tot}} = (n_{tot}/0.038)^6 \tag{2.50}$$

where n_c is for the channel only, K_w is the fractional surface area of weed cover, Fr is the Froude number, k_s is the Nikuradse equivalent roughness length and n_{tot} is the total roughness. For a detailed review of roughness due to different types of vegetation, see www.riverconveyance.net, which contains extensive documentation on a Roughness Advisor (RA) embedded in the 'Conveyance Estimation System' (CES) referred to earlier. The RA covers many types of submerged, emergent and floating types of aquatic vegetation, as well as land-based plants (Fisher & Dawson, 2003). For the combined effects of unsteady flow and vegetation, see Sellin & van Beesten (2004).

2.5.4. Resistance to flow in meandering compound channels

2.5.4.1. Introduction

Various methods have been formulated to account for the additional resistance to flow caused by streamwise curvature and the flow structure outlined in Figs. 2.3 and 2.4. See for example Chang (1983, 1984, 1988), Ervine & Ellis (1987), Ervine et al. (1993), Ikeda & Parker (1989), Leopold et al. (1960), McKeogh & Kiely (1989), Shiono, Al-Romaih & Knight (1999), Sellin & Willetts (1996), Shiono et al. (1999), Soil Conservation Service (1963), Toebes & Sooky (1967), Wark et al. (1994) and

Rameshwaran & Willetts (1999). One of the earliest and simplest empirical evidence, that Manning's n should be increased for meandering channels as follows

$$n_{mc}/n_{sc} = 0.43s + 0.57 \text{ for } s < 1.7$$

$$n_{mc}/n_{sc} = 1.30 \text{ for } s \geq 1.7$$

where n_{mc} refers to the meandering channel, n_{sc} to the equivalent straight channel and s = sinuosity. For overbank flow sinuosity is only one of a number of factors which influence resistance, as Figs. 2.3 and 2.4 have shown, and consequently energy losses are attributable to a variety of physical causes, notably boundary friction, secondary flow, turbulence, as well as flow expansion and contraction. All of these phenomena are depth dependent, and therefore will contribute to the overall resistance in different amounts as the discharge increases (Shiono *et al.* 1999a, 1999b).

A meandering channel may be divided into three sub-channels:

(1) the main channel below the bankfull stage,
(2) the floodplain within the meander belt width, and
(3) the floodplain outside the meander belt width.

The energy losses in each sub-channel then have to be determined individually and summed to give the total resistance. For example, the resistance to flow in sub-channel (1) might be conceived as being due to the shear stress on the horizontal plane at the bankfull level, the energy loss due to the secondary flow below the bankfull level, as well as the energy losses associated with bed and wall friction. The first two such losses are known to make a significant contribution to the total energy loss in the sub-channel below the bankfull stage. For sub-channel (2), within the meander belt width, additional resistance to flow arises from the energy losses due to contraction and expansion effects. These make a significant contribution to the total energy loss in the upper layer for high overbank flows, whereas the bed friction and streamwise shear at the bankfull level were found to be more significant for shallow overbank flows (Shiono & Muto, 1998; Shiono *et al.*, 1999).

The resistance to flow in meandering channels is further complicated by the changes in cross-sectional shape in the streamwise direction, induced by the secondary flow cells. In natural river bends, steep banks generally form at the outside and shallow zones of deposition at the inside (Bathurst *et al.*, 1977; Thorne *et al.*, 1997; Nelson & Smith, 1989). For inbank flows, the dominant secondary flow cell changes its direction of rotation from clockwise to anti-clockwise as the flow passes through left and right-handed bends respectively. In between bends there is a crossover region, where the depths are relatively small compared to depths at the adjacent bend apices. As a result of these changes in cross-section, there may be significant localised streamwise convective accelerations and decelerations, which must also be taken into account in the overall force balance (Dietrich & Whiting, 1989). For overbank flows, these dominant secondary flow cells may rotate in the opposite sense for inbank flow, due to the roll up of the lower main channel flow in the crossover region by the faster floodplain flow above it. The mechanism is illustrated in Figs. 2.4 and 2.5 and some experimental data for this phenomenon may be seen in Knight & Shiono (1996). The flow resistance in an open channel with an alternative riffle-pool arrangement is another example of where stream-wise changes in shape are important (Michioku *et al.*, 1999).

2.5.4.2. Energy loss coefficients

The bed shear stress is known to decrease once overbank flow occurs (Knight *et al.*, 1992) and the sectional mean velocity below the bankfull level also decreases at an early stage of overbank flow (Shiono *et al.*, 1993). Rameshwaran *et al.* (1999) observed the reduction of sediment transport rate in compound meander chan-nels with a mobile bed also at an early stage of overbank flow. These suggest that a substantial energy loss occurred below the bankfull level reducing the main chan-nel velocity, the bed shear stress and the sediment transport rate at the early stage of overbank flow. A better understanding of the energy expenditure in the lower layer is therefore the key to understanding the sediment transport mechanisms in the main channel.

Shiono *et al.* (1999) investigated mean energy losses due to boundary friction, secondary flow and turbulence at the bankfull level along the meander channel for overbank flows using the turbulent flow data. They also revisited the Ervine & Ellis (1987) method for stage-discharge estimation, based on the three channel sub-sections cited earlier, namely (1) the main channel below the bankfull level, (2) the floodplain within the meander belt and (3) the floodplain outside the meander belt.

In the Ervine & Ellis method (1987), for the main channel below the bankfull level the energy loss caused by the turbulent shear stress on the "horizontal" plane at the bankfull level is ignored. However, Shiono *et al.* (1999) found significant energy loss due to the turbulence at the bankfull level as a result of their turbulence measurements and introduced an extra shear stress term into the Ervine & Ellis equation, which should be re-written as:

$$K\frac{V_a^2}{2g} = K_{bf}\frac{V_a^2}{2g} + K_{sf}\frac{V_a^2}{2g} + K_{ts}\frac{V_a^2}{2g} = \frac{S_0}{s} \qquad (2.51)$$

where V_a is a sectional mean velocity below the bankfull level, K_{bf}, K_{sf} and K_{ts} are the dimensional energy loss coefficients (m^{-1}) due to the boundary friction, the secondary flows and the streamwise turbulent shear stresses at the bankfull level respectively, K is the sum of individual coefficients, s is sinuosity, and S_0 is the valley slope. The coefficients, K_{bf} and K_{ts} may be defined by using the Darcy-Weisbach equation:

$$K_{bf} = \frac{f}{4R} \quad \text{and} \quad K_{ts} = \frac{f''}{4R_{a(l)}} \qquad (2.52)$$

where f is a friction factor $(= 8gRS_f/U^2)$ and R is an appropriate hydraulic radius.

The friction factor f'' is related to the turbulence and $R_{a(l)}$ is the apparent hydraulic radius for the lower layer, defined with the cross sectional area at the bankfull level, A_{bf}, divided by the perimeter of an imaginary wall P_a, i.e. $R_{a(l)} = A_{bf}/P_a$. In compound meandering channels, P_a is taken to be equivalent to the top width of the main channel. Further assuming that f'' can be related to the boundary shear stress acting on the imaginary plane at the bankfull level, in a similar manner to the relation between the normal friction factor and the bed shear stress, gives:

$$f'' = \frac{8\tau a}{\rho Va^2} \tag{2.53}$$

where τ_a is the turbulent streamwise shear stress acting on the imaginary plane at the bank level. The friction factor, f, was obtained from experiments in a wide straight channel of the same bed material.

Relative magnitudes of K_{bf}, K_{ts} and K_{sf} were obtained and are shown in Tables 2.7 and 2.8. Table 2.7 gives values of K_{bf} and K_{sf} corresponding to Ervine and Ellis's original equation in which $K_{ts} = 0$, and Table 2.8 gives measured values based on Equation (2.51). The figures clearly show that the dominant mechanism for energy loss varies according to the relative depth and sinuosity, Dr and s.

The K_{sf} values based on Ervine & Ellis's equation (1987) deduced from Chang's (1983) formula, and based on an analytical consideration of inbank flow for a wide channel, are clearly different from the measured values using Equation (2.51).

Table 2.7 Energy loss coefficients for Equation (2.53), ignoring K_{ts} term.

s	Dr	K_{bf} (m^{-1})	K_{sf} (m^{-1})
1.093	0.15	0.2495	0.2248
1.093	0.5	0.2495	0.2248
1.37	0.15	0.2580	0.2281
1.37	0.5	0.2580	0.2281
1.571	0.15	0.2632	0.2301
1.571	0.5	0.2632	0.2301

Table 2.8 Energy loss coefficients for Equation (2.53), including K_{ts} term.

s	Dr	K_{bf} (m^{-1})	K_{sf} (m^{-1})	K_{ts} (m^{-1})	K (m^{-1})
1.093	0.15	0.2495	0.3344	0.0671	0.6509
1.093	0.5	0.2495	0.0413	−0.0272	0.2637
1.37	0.15	0.258	0.6016	0.2118	1.0714
1.37	0.5	0.258	0.0892	0.1112	0.4584
1.571	0.15	0.2632	0.5643	0.3432	1.1707
1.571	0.5	0.2632	0.0155	0.2312	0.5099

As mentioned previously, not only is the structure of the secondary flow for the inbank and overbank flows different, and develop from different originating mechanisms, but also their strength is quite different also. Hence the amount of energy expenditure should differ and, as a result, the formulation for Chang's K_{sf} was also reconsidered by Shiono *et al.* (1999). The secondary flow generation mechanism for overbank flow appears to be somewhat similar to that causing the flow to re-circulate behind a rearward facing step in an expansion, as first proposed by Ervine & Ellis (1987). Further examination of K_{sf} was carried out and the following relationship between K_{sf}, s and Dr obtained (see Fig. 2.69):

$$K_{sf} = \{-6.9367s^2 + 17.464s - 8.8983 - 2.634\,Dr\}K_{ex} \tag{2.54}$$

where K_{ex} is the total expansion loss along a meander wavelength estimated from

$$K_{ex} = \left(1 - \frac{H-h}{H}\right)^2 \frac{s(\sin\overline{\theta})^2}{Bw} \tag{2.55}$$

where Bw is the meander belt width, H is the main channel water depth, h the floodplain height, $\overline{\theta}$ is the mean angle of incidence averaged over the meander wavelength. The angle $\overline{\theta}$ is also used in the contraction loss estimation. It is clear from Equation (2.55) that both the relative depth and sinuosity, s, are approximately equally weighted and therefore both important.

The energy loss coefficient, K_{ts}, in the lower layer varies considerably with the sinuosity and relative depth, as shown by the data presented in Table 2.8. Some variation is to be expected because there is a velocity difference between the streamwise components of the lower main channel flow and the floodplain flow. The difference generates a turbulent shear stress on the horizontal plane at the bankfull level, and clearly varies with sinuosity and water depth. In this particular experimental study, there existed a clear relationship between the sinuosity and the relative depth, as shown in Fig. 2.70, expressed as

$$K_{ts} = 0.561\,s - 0.281\,Dr - 0.504 \tag{2.56}$$

This again shows the relative importance for Dr and s. In this test case, it is clear that sinuosity dominates the energy loss due to the turbulence at the bankfull level.

Similar analysis of the energy loss for the floodplain within the meander belt was carried out and the results shown in Table 2.9. The measured turbulent shear stresses at the bankfull level were again found to be comparatively large. It is therefore important that the resulting energy loss should be included in Ervine & Ellis's (1987) equation, which may be re-written as:

$$K_{bff}\frac{V_b^2}{2g} + (K_{ex} + K_{co})\frac{V_b^2}{2g} + K_{tsf}\frac{V_b^2}{2g} = S_o \tag{2.57}$$

where V_b is the sectional mean velocity above the bankfull level within the meander belt, and the coefficients K_{bff}, K_{ex}, K_{co} and K_{tsf} are the boundary frictional head loss coefficient per wetted area of the floodplain, the total expansion loss along a meander wavelength, the total contraction loss along a meander wavelength, and

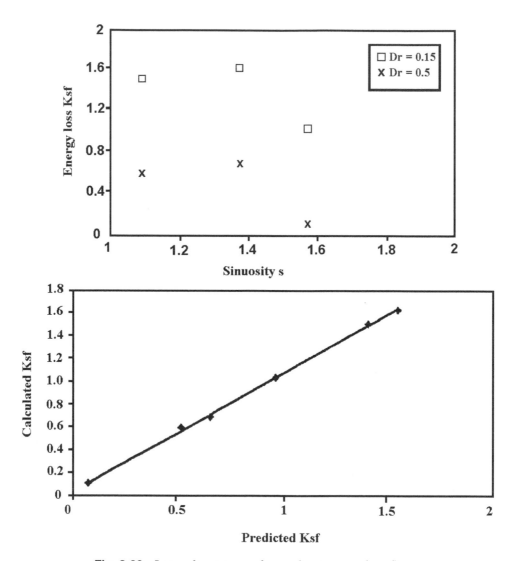

Fig. 2.69 Lower layer energy losses due to secondary flows.

the head loss coefficient due to the turbulence over the meander wavelength and width, respectively.

2.5.5. Effect of flow structure on boundary shear stress distributions

2.5.5.1. Resistance and boundary shear

Equation (4.63) has shown that in 1-D models, for steady uniform flow in a straight prismatic channel the resistance to flow is directly linked to the mean boundary shear stress. Furthermore, Equation (4.60) shows that the mean boundary shear stress may be readily determined for each flow rate, using given values of R and

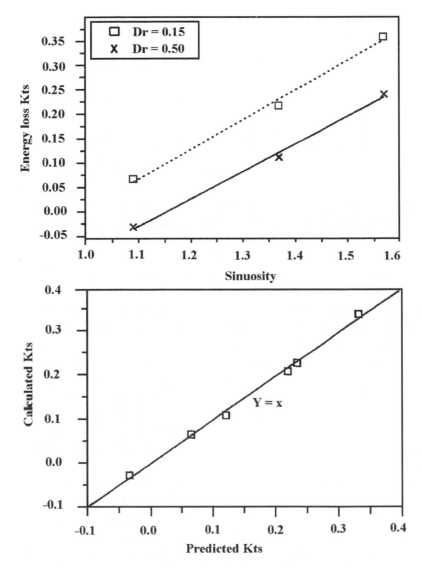

Fig. 2.70 Lower layer energy losses due to shear stress at bankfull level. (a) K_{ts} versus sinuosity, s; (b) calculated versus predicted K_{ts} values from Equation (2.58).

S_f ($= S_o$). However, once 2-D or 3-D models are adopted, then Equation (4.62) has shown that the zonal or local boundary shear stresses are linked to the certain velocities by different types of friction factor. This implies that in order to predict the local boundary shear stress from a local velocity, the correct value of f_b is required. For example, where the distribution of U_d is known, e.g. from a 2-D or 3-D model, then in order to obtain the correct lateral distribution of τ_b around the wetted perimeter of the channel, the lateral or spanwise distribution of f_b needs to be specified. The link between τ_b and U_d is now examined for flows in straight

Table 2.9 Energy loss coefficients for Equation (2.59) including K_{tsf} term.

s	Dr	K_{bff} (m^{-1})	K_{tsf} (m^{-1})	$K_{ex} + K_{co}$ (m^{-1})
1.093	0.15	0.6790	−0.0538	0.0555
1.093	0.5	0.0706	0.0315	0.1847
1.37	0.15	0.9260	0.2699	−0.1320
1.37	0.5	0.0908	0.0585	0.1356
1.571	0.15	0.9422	0.5530	−0.0328
1.571	0.5	0.0831	0.0713	0.1573

rigid boundary channels for which there are sufficient data. A general review of the lateral distributions of boundary shear stress in open channel flow is given elsewhere by Knight *et al.* (1994).

Figures 2.2, 2.3, 2.17, 2.24 and 2.30 have shown that for low floodplain depths the generally faster moving water in the main river channel and the slower moving water on the floodplain create a shear layer between the two regions and intense vorticity. The depth-mean velocity, U_d, will decrease from a higher value in the river to a lower value on the floodplain. The boundary shear stress likewise responds in a similar manner, as indicated by one set of FCF results for a floodplain width of 2.25 m (Series 2) shown in Fig. 2.71. These data correspond to those already shown in Fig. 2.9 in a different format. At low inundation depths on the floodplain, marked △ (Exp 020101), the normalised boundary shear stress, τ_b/τ_{fp}, where $\tau_{fp} = \rho g(H - h)S_o$, varies from above 3 in the river to around 0.2 on the floodplain (Fig. 2.72). At larger depths of inundation (Exp 020801), the distribution is much more uniform across the channel, with τ_b/τ_o values tending towards 1.0.

Figure 2.71 also shows that the shear layer between the floodplain and a river extends into both the river and the floodplain. It should be noted that these turbulent shear layer widths vary slightly according to whether they are based on τ_b or U_d (Rhodes & Knight, 1995b). The same issue arises in the basic definition of a boundary layer thickness, δ, which is usually defined in terms of the distance from the wall to a point in which the local velocity varies less than 1% from the free stream velocity, i.e. $u/U = 0.990$.

However, at this point in laminar boundary layers the local fluid shear stress is still some 5% of the wall or boundary shear stress, i.e. $\tau/\tau_W = 0.05$, leading to a 17% under prediction for δ. If the more stringent definition for δ is taken as being $\tau/\tau_w = 0.01$, then δ is larger and the Blasius (1908) equation gives $u/U = 0.999$. As indicated in Section 2.2, a turbulent shear layer is composed of two parts, one produced by the Reynolds stresses and the other by the secondary flows and vorticity. The extent of the shear layer is shown more clearly in Figs. 2.72 and 2.73, taken from other FCF data (Series 01 with 1:1 side slopes and 4.1 m wide floodplains and Series 10 with 1:2 side slopes and 2.0 m wide floodplains). The lateral gradient of boundary shear stress, $\partial\tau_b/\partial y$, is seen to be significantly different in

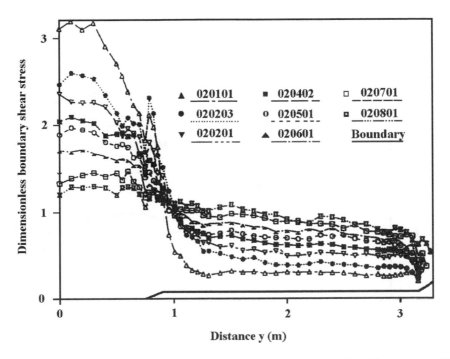

Fig. 2.71 Dimensionless boundary shear stress distribution, normalized by $\tau_{fp} = \rho g(H - h)S_o$. The data correspond to those shown in Fig. 2.9.

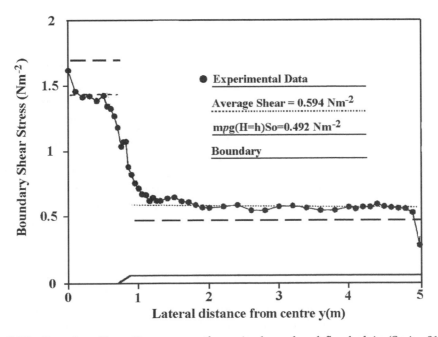

Fig. 2.72 Boundary Shear Stress across the main channel and flood plain (Series 01) for condition shown in the legend.

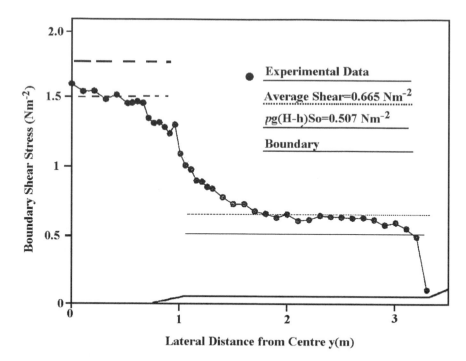

Fig. 2.73 Boundary shear stress distribution from the FCF experiments (Series 10) for condition shown in legend.

Figs. 2.71 and 2.73, indicating that the lateral spread is affected by the floodplain width, depth of flow as well as the side slope of the main river channel.

The lateral distributions of local boundary shear shown in Figs. 2.71 and 2.73 indicate that τ_b varies from $\sim 1.6\,\mathrm{Nm^{-2}}$ in the river channel to $\sim 0.6\,\mathrm{Nm^{-2}}$ on the floodplain. Also shown are the average values of the measured floodplain data, which are 0.594 and 0.665 $\mathrm{Nm^{-2}}$ respectively, some 21% and 31% higher than the calculated 'two-dimensional' values of 0.492 and 0.507 $\mathrm{Nm^{-2}}$, based on Equation (2.43) and the local depth, ($\tau_{fp} = \rho g(H - h)S_0$). The differences arise because the right-hand-side (RHS) of Equation (2.29) is non-zero, due to the influence of secondary flows and planform vortices shown in Figs. 2.13, 2.14 and 2.15. The flow is not in fact 'two-dimensional' in the usual sense, despite a constant plateau of values of both τ_b and U_d on the floodplain. This again emphasizes the importance of defining and using the correct 'local' friction factor, and including both lateral shear and secondary flow terms in the governing depth-averaged equation. The same effect may be observed for different shaped channels and ducts, with the difference from the 'two-dimensional' values being in some cases as much as 100% (e.g. see Fig. 5.11 in Knight & Shiono (1996). See also Holden & James, 1989; Knight & Cao (1994), Rhodes & Knight (1995a, 1995b), Shiono & Knight (1991) and Wormleaton (1996).

The corresponding main channel values of τ_b and U_d are likewise different from the values calculated from the local depth. In the cases shown in Figs. 2.71 and 2.73 the main channel bed shear stresses are 31% and 25% lower than those predicted using Equation (2.43) and the local depth ($\tau_b = \rho\, gHS_o$). This is clearly of importance when estimating sediment transport rates or dealing with associated phenomena. The lateral distribution of τ_b is of particular use in 2-D sediment models, where threshold criteria, active bed layer widths or the variation of transport rates across the river are required. Since the bedload sediment transport rate typically varies third power of the shear velocity, any errors in estimating the zonal or main channel boundary shear stress will have a significant effect on the zonal transport rate.

Ackers (1992a) has suggested that the zonal discharge adjustment factors may be used to obtain a preliminary estimate of the zonal boundary shear stresses, by multiplying the basic values by the appropriate Discharge Adjustment Factor squared (DISADF^2), based on the 'coherence' method (COHM), described later in Section 2.6.2.2. However the value so obtained, based on COHM, is not as accurate as that obtained using the Shiono & Knight method (SKM). Both these methods are described in detail in Section 2.6.2.

Although the distributions of τ_b shown in Figs. 2.71–2.73 are fairly uniform outside the shear layer, it should be appreciated that perturbations may occur where secondary flows are strong. This is usually the case in narrow channels (Bhowmik, 1982; Knight & Lai, 1985; Nezu et al., 1989), in corners (Knight & Cao, 1994; Knight et al., 1994), near re-entrant corners (Perkins, 1979) and for loose boundary channels at low Reynolds numbers (Ikeda, 1981; Nezu et al. 1988; Nezu & Nakagawa, 1989). Figures 2.74(a) and (b) illustrate some data from Yuen (1989) presented by Knight et al. (1994).

Despite these perturbations in τ_b (and U_d), the local friction factor, f_b, tends to be more or less constant within the shear layer. Evidence for this is shown in the studies leading to Equations (2.42) and (2.43) by Shiono & Knight (1991) and in studies on shear layers in closed conduits by Rhodes & Knight (1994a, 1994b, 1996). Figure 2.75 shows some of the very detailed duct data from Rhodes & Knight (1994b), indicating constancy in the lateral distributions of f_b. In these particular data the average value of the local friction factor is closely approximated by the corresponding zonal (sub-area) value. The difference between the average value in floodplain and that in the main channel increases with decreasing relative depth, Dr. It was this feature that in part led to the Shiono & Knight method (SKM) described in Sections 2.2.2.2 and 2.6.2.3, in which the cross-section is divided into zones or regions where the resistance factor, f, along with factors for turbulence and secondary flow (λ and Γ) may be treated as constants in numerical analysis.

The effect of increasing the main channel bed roughness is illustrated in Figs. 2.76 and 2.77, using data from a series of experiments by Atabay (2001) in which comparable experiments were performed in similar shaped compound channels, with and without sediment. See Atabay & Knight (1999), Knight et al. (1999) and Atabay et al. (2004 & 2005). The main channel bed was moulded either in loose sand or by

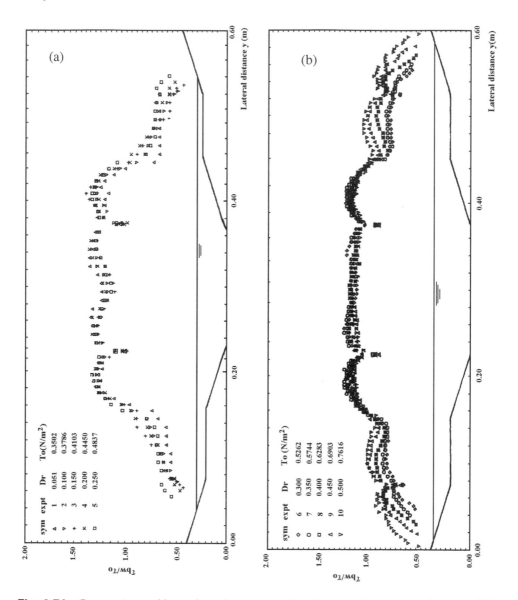

Fig. 2.74 Comparison of boundary shear stress distributions in compound trapezoidal channels for various relative depths (a) Dr ($0.051 < Dr < 0.250$) and (b) Dr ($0.3 < Dr < 0.5$).

a smooth rigid surface at the same mean bed elevation of the sand. Figure 2.76 shows how the sediment bed forms affect the lateral distribution of U_d within the main channel, and Fig. 2.77 shows how the boundary shear stresses are affected on the floodplain. Figure 2.76 shows that the increase in roughness in the river channel causes the main channel velocities to decrease and the floodplain velocities to increase, resulting in very little lateral shear at the main channel/floodplain interface as the velocities are then of a comparable magnitude at this particular

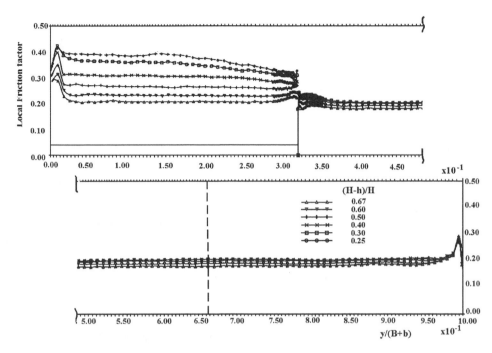

Fig. 2.75 Distributions of local friction factor, for $\theta = 90°$ and $Dr/H = 0.25$ to 0.67.

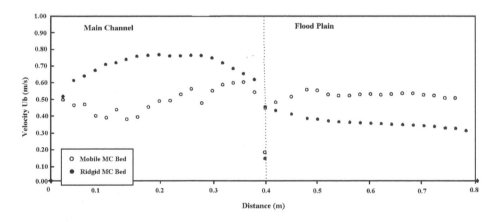

Fig. 2.76 Comparison of Velocity Measurement between rigid and mobile main channel bed, ($Q = 21\,\mathrm{l/s}$).

discharge. Knight (1999), Knight *et al.* (1999), Knight & Brown (2001), Karamshiva *et al.* (2005), Atabay *et al.* (2005), Tang & Knight (2006) and Bousmar *et al.* (2006) contain further details of sediment induced resistance effects.

Using data similar to that presented in Figs. 2.71–2.75, Knight & Demetriou (1983) have shown that integration of the local boundary shear stress values along

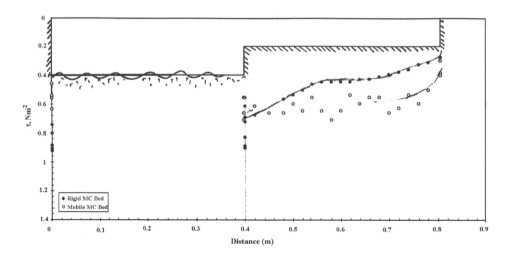

Fig. 2.77 Comparison of Boundary Shear Stress between rigid and mobile main channel bed, ($Q = 21\,\mathrm{l/s}$).

certain elements of the wetted perimeter will give the mean shear forces acting on those same elements, which may then be used for analysis purposes. Figures 2.78 and 2.79 show one example of this, using some data from the same series of experiments by Atabay (2001) referred to earlier in Figs. 2.76 and 2.77, with rigid and mobile bed boundaries. These figures also serve to illustrate the effect of both cross-sectional shape (symmetric and asymmetric floodplains) and heterogeneous

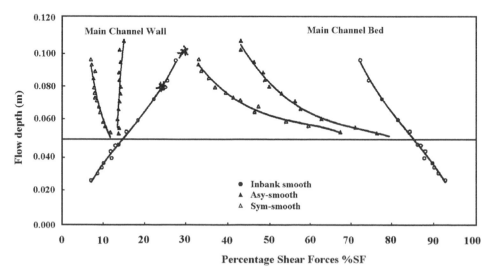

Fig. 2.78 Percentage shear forces on the main channel bed and walls for the rigid compound channel.

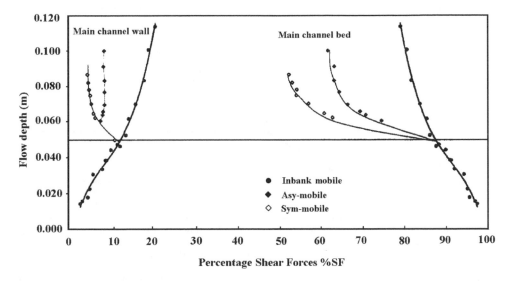

Fig. 2.79 Percentage shear forces on the main channel bed and walls for the mobile compound channel.

roughness (caused by the sediment bed forms). Clearly the percentage shear forces, $\%SF_i$, which act on the wall and bed elements are strongly dependent on the flow depth and so functional relationships may be readily determined. These can then be used to calculate the mean boundary shear stress acting on the same elements of the wetted perimeter for any depth. To illustrate this, Figs. 2.80 and 2.81 show the effect of the lateral transfer of momentum between the floodplain and the river, for both smooth rigid compound channels and ones in which the bed of the main river channel is composed of sediment. The effect of the momentum transfer is seen to decrease the bed shear stresses in Fig. 2.80, but to increase them in Fig. 2.81. Full details are given in Knight & Atabay (2001).

Attention has been drawn in Sections 2.3 and 2.4 to the variation in channel planform and cross-section shape that can occur in the streamwise direction in natural rivers. Given the complexity of the flow structure described in those Sections, there are inevitably very significant spatial variations in both boundary shear stress and resistance in natural channels. Very little boundary shear stress data exist for non-straight channels, especially for those with overbank flow, those with overbank flow, except for Bousmar (2002), Bousmar & Zech (2004), Bousmar *et al.* (2004) and Rezaei (2006). For data on laboratory meandering channels see Chang (1984), Knight *et al.* (1992) and Al-Romaih (1996), for data on natural rivers see Bathurst (1979), Dietrich & Whiting (1989), Nelson & Smith (1989) and, Odgaard (1986) and for other data on related quasi-uniform flows, see Ikeda *et al.* (1984) and Kironoto & Graf (1995).

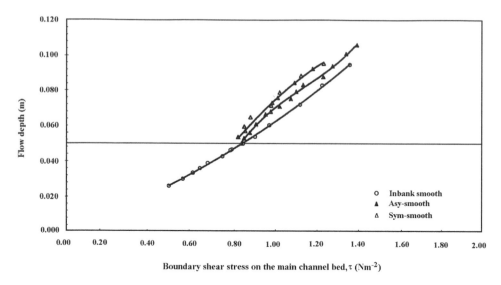

Fig. 2.80 Boundary shear stress on the main channel bed for the rigid compound channel.

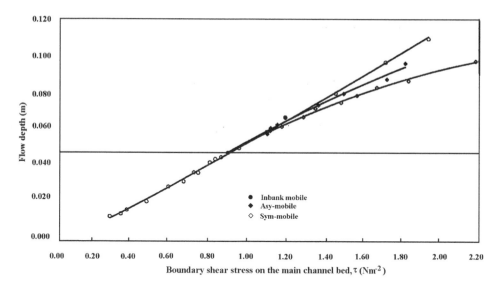

Fig. 2.81 Boundary shear stress on the main channel bed for the mobile compound channel.

2.5.6. Effect of flow structure on velocity distributions

2.5.6.1. Velocity distributions

A certain amount of velocity data have already been presented, some of it in the form of isovels (Figs. 2.28–2.30), secondary flows (Figs. 2.26, 2.27), vertical distributions of longitudinal velocity (Fig. 2.54), lateral distributions of vertical velocity

(Figs. 2.25, 2.26), lateral distributions of streamwise velocity at certain elevations (Figs. 2.24, 2.31) and depth-averaged velocities (Figs. 2.14, 2.29, 2.30, 2.38, 2.47, 2.53, 2.56). Since velocity and boundary shear stress are linked to resistance in Equation (2.39), it is important to appreciate the effect of flow structure has on isovels.

The non-uniformity in the flow field makes it essential that appropriate kinetic energy and momentum correction coefficients, α and β, are used in 1-D analysis of flow in compound channels, since their values can be much higher than those for simple prismatic channels (Knight, 1992). The variation of α and β with depth is also important in defining critical flow and control points in natural rivers with floodplains (Blalock & Sturm, 1981; Yuen, 1989), as well as for calculations of back-water and bridge afflux. The definition of critical flow for overbank flow is itself problematic, as indicated by Yuen & Knight (1990), Bhowmik & Demissie (1982), Demissie (1982) and Lee et al. (2002). Both laboratory and field data show that sub-critical and supercritical flow can occur simultaneously in different parts of the same cross section of a compound channel, thus requiring at least a 2-D level of analysis. The high velocity and low depths on the edge of the floodplain near the river can make conditions where $Fr > 1$, even though $Fr < 1$ in the main chan-nel, where Fr is the local Froude Number $(= U_d / \sqrt{\{gH\}})$. The flow is then locally supercritical, with large surface waves, on the edge of the floodplain adjacent to the river.

Another reason why the flow structure and isovels are important is because they indicate the division of flow within the cross-section. This is generally required in all 2-D analysis, in partitioning flow between a river and its floodplain, in valida-tion studies of numerical models, in eco-hydraulics where the velocity distribution may be important for habitats, and in flood routing studies.

Equation (2.29) has expressed the lateral variation of U_d with y, and (2.31) and (2.32) give the analytical solutions. The lateral distribution of U_d is similar to that for boundary shear stress, τ_b, described in Section 2.5.5. The rougher the flood-plain relative to the main river channel, the greater is the difference in U_d between these regions and the extent of the shear layer width. In the case of where the main channel is rougher than the floodplains, as in some FCF experiments with large sand dunes (Knight et al., 1999), or in natural rivers with grassed floodplains, then the lateral shear term will obviously be small, see Brown (2001) and Bous-mar et al. (2006). The shear layer width should strictly be judged by the secondary flow/vortex spread, and not just by where U_d values becoming constant, as men-tioned in Section 2.5 and shown by Shiono & Knight (1990) and Rhodes & Knight (1995).

2.5.6.2. Discharges in sub areas

The lateral distributions of U_d are also useful, as integration with y give sub-area or zonal discharges. These can then be used to give both the proportion of total flow that occurs on the floodplain, and in the limiting case the complete stage-discharge relationship, for those few cases in which it is possible to integrate Equation (2.31) directly.

The division of flow between the river and its floodplains is frequently required in modelling for validation. Some typical results from laboratory experiments in which the percentages of the total flow that occur in the main river channel or floodplains have been plotted against relative depth for different floodplain geometries, roughnesses and flow conditions are presented in Knight & Demetriou (1983), Knight, Demetriou & Hamed (1983), Myers & Lyness (1997) and Atabay & Knight (1999). Figure 2.82 shows some representative data for $\%Q_{mc}$ and $\%Q_{fp}$ for $B/b = 4$, taken from Knight & Demetriou (1983). It is clear that $\%Q_{mc}$ values differ significantly from those that would be obtained by simple area percentages, shown by the dotted line. Figure 2.83 illustrates how the $\%Q$ curves vary with floodplain width, and Fig. 2.84 how they vary with symmetry and sediment, under similar geometric and operating conditions. A complete listing of the $\%Q$ values for the FCF data may be found in Knight (1992) and and in www.flowdata.bham.ac.uk.

The proportion of flow in different sub-areas or zones should always be determined and plotted, as errors in these flows can affect the accuracy of any predicted H v Q relationship, as well as having a significant effect on calculations of sediment transport rates. Examples of typical errors in predicted total and zonal discharges are shown in Wormleaton *et al.* (1982 and 1985), Wormleaton & Merrett (1990) and Knight *et al.* (1984a, 1984b). Traditional sub-division methods that neglect the interaction process will typically over-estimate the discharge in the main channel and under-estimate the discharge on the floodplains.

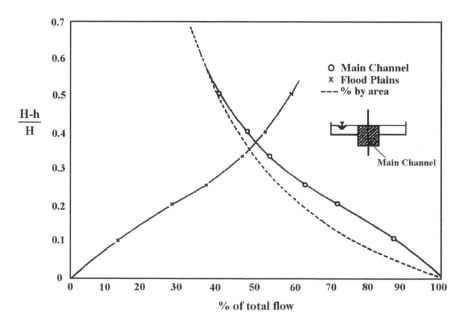

Fig. 2.82   Percentage of total flow in main channel and flood plains, $B/b = 4$.

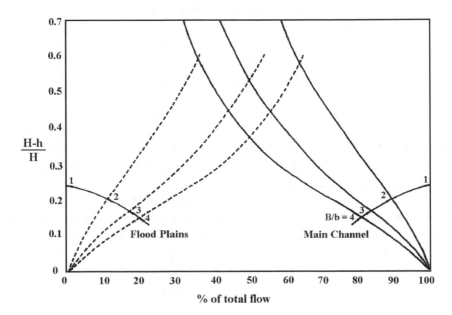

Fig. 2.83 Percentage of total flow in main channel and flood plains.

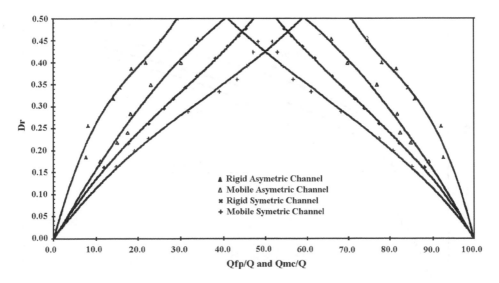

Fig. 2.84 Percentage of total flow in main channel and floodplains.

2.6. Effects of Flow Structure on Conveyance Capacity

2.6.1. Effect of flow structure on stage-discharge curves

The River Severn data in Fig. 2.61 have shown how the 'global' or single channel Manning's n values may decrease sharply just above the bankfull level, due to the abrupt change in hydraulic radius, and how the 'zonal' or sub-area resistance values for the main channel have to be effectively increased in order to obtain the correct sub-area discharges. Likewise, the sub-area resistance values for the floodplain have to be correspondingly reduced, to levels well below their actual roughness values. Since the global resistance coefficient (1-D analysis) is essentially the stage-discharge relationship in another form, the data already presented in Section 2.5.3 serve to show the difficulty in predicting this H v Q relationship. Similarly, the variation in the zonal resistance coefficient (2-D analysis) highlights the difficulty in determining the division of flow between a river and its floodplains, or between smaller domains, for say comparison with the results from a depth-averaged numerical model.

The stage-discharge curve, H v Q, is of particular interest to river engineers since it relates the two primary parameters in many flood problems, namely water level and discharge. Figure 2.85 shows a typical stage-discharge curve, and how the single and divided channel methods of computation produce different predictions. It also shows how discontinuities can occur at the bankfull level for compound channels, as already seen in the resistance relationships in Figs. 2.61, 2.65, 2.66 and

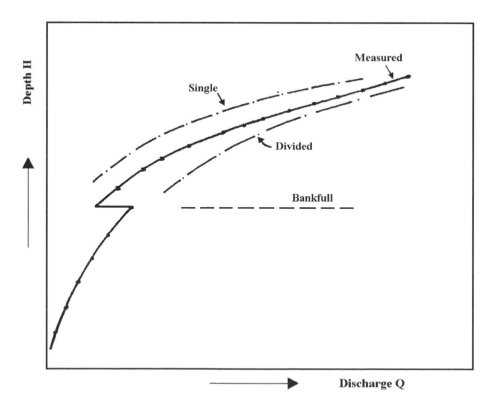

Fig. 2.85 Typical stage-discharge curves for overbank flow showing how single and divided channel methods compare with measured data.

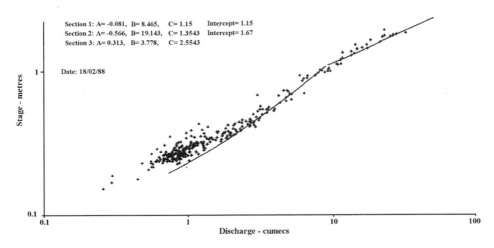

Fig. 2.86 Stage discharge relationship for River Penk at Penkridge.

2.67(a). Typical stage-discharge curves for a natural river and small-scale laboratory channels are shown in Figs. 2.86 and 2.87 respectively. Many other examples may be found in the literature (Atabay & Knight, 1999; Bousmar & Zech, 1999; Gessler *et al.* 1998; Myers & Lyness, 1997; Sellin *et al.*, 1993; Shiono *et al.*, 1999; Willetts & Hardwick, 1993). In general, as Q increases H increases, but under certain circumstances it should be noted that once the bankfull stage is reached, there may be an actual reduction in Q, despite a larger flow area. This is usually more apparent in laboratory channels than in nature (Knight *et al.*, 1983a, 1983b, 1984a, 1984b,

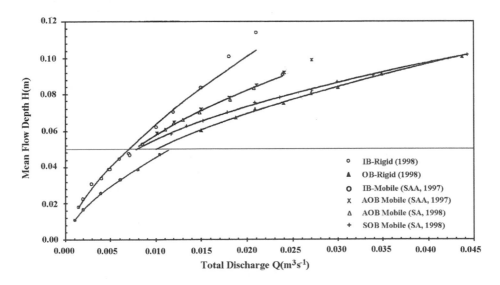

Fig. 2.87 Comparison of the stage-discharge relationships.

Knight, Samuels & Shiono, 1990), but even in the field very shallow inundation of the floodplains is problematic to analyse and model. However, once above this small floodplain depth region, Q increases significantly due to the increased flow area, with the slope of the H v Q curve decreasing as the width of the floodplain increases. There is therefore a functional relationship between H and Q, albeit complicated by the interaction process between the main river and the floodplains or by storage effects. For unsteady flow the stage-discharge curve is not unique and is further complicated by dynamic effects (Knight, 1989; Lai, Liu & Lin, 2000; Tominaga *et al.*, 1995; Knight, 2005).

The proportion of flow in different sub-areas or zones should always be determined and plotted, as errors in these flows can affect the accuracy of any predicted H v Q relationship, as well as having a significant effect on calculations of sediment transport rates. Examples of typical errors in predicted total and zonal discharges are shown in Wormleaton *et al.* (1982 and 1985), Wormleaton & Merrett (1990) and Knight *et al.* (1984a, 1984b). Traditional sub-division methods that neglect the interaction process will typically over-estimate the discharge in the main channel and under-estimate the discharge on the floodplains.

2.6.2. Prediction methods for stage-discharge relationships

2.6.2.1. An overview of various divided channel methods (DCM)

The simple sub-division and composite roughness methods given in Chow (1959) are not appropriate for compound channels, for the reasons given in Section 2.5.3.4. In the light of the knowledge gained about flow structure in compound channels a number of suggestions have been made as to how these divided channel methods (DCM) might be modified to simulate the interaction process in straight compound channels more accurately (Lambert & Myers, 1998). The methods fall into five categories and are briefly described as follows:

The first approach is based on altering the sub-area wetted perimeters, typically excluding the length of the vertical interface at the river/floodplain in the calculation of P for the floodplain, but including it in the value of P for the main channel. This is intended to have the effect of retarding the flow in the main channel sub area and enhancing it in the floodplain sub-area. However, since the length of the division line is small at low depths of submergence on the floodplain, when the interaction effect is high, this approach will inevitably fail (Knight, Demetriou & Hamed, 1983 & 1984; Wormleaton, Allen & Hadjipanos, 1985). Although many variants on this kind of approach have been tried, it is now recognised that it is better to adjust the discharges in each sub-area by some appropriate method, as a separate step in the calculation procedure. This is the basis of the second approach, called the 'coherence' method (COHM), which is discussed in more detail in Section 2.6.2.2.

The third approach is based on quantifying the apparent shear stresses (ASS's) or apparent shear forces (ASF's) on the sub-area division lines. This requires knowledge of the depth-averaged Reynolds stresses and vorticity terms, described in detail in Section 2.2.3. These inter-facial forces can then be included in a 1-D analysis to give the effective shear force, or resistance, for each sub-area, and hence the correct sub-area conveyance capacity. This in turn gives the correct division of flow within the cross-section. Many authors have attempted to develop empirical equations for these ASS or ASF quantities on specific division lines. Most of these equations include the ratio of the floodplain depth to main channel depth, $(Dr = (H - h)/H)$, as a primary variable, and the floodplain width and relative roughness as subsidiary variables (Knight & Hamed, 1984; Wormleaton & Merrett, 1990). Although most of the equations may fit particular experimental data sets well, they are not generally applicable.

A fourth approach, which follows on from the third, is to specify the division lines between sub-areas along lines of zero shear stress. However, as the isovels in Figs. 2.23–2.30 have indicated, the 3-D nature of the velocity field makes it extremely difficult to generalize the position of these division lines for all types of channel shape, flow depth and roughness configuration (Knight & Hamed, 1984). Moreover, it is known from three dimensional turbulence considerations, that orthogonal lines to the isovels do not necessarily imply lines of zero shear stress (Melling & Whitelaw, 1976; Tracy, 1965).

A fifth approach is to combine the divided channel methods using vertical and horizontal interfaces, together with a weighting factor, as suggested by Lambert & Myers (1998). In this approach a single weighting parameter is applied to the component velocities predicted by the vertical and horizontal division method to produce an intermediate velocity which more closely represents the observed velocity in both the main channel and the floodplain. Values of this weighting factor vary from 0.5 for homogeneously roughened channels to 0.2 for channels where the floodplain is rougher than the main channel.

It should again be noted that even though a particular approach performs well in predicting the total conveyance capacity for a given stage, it does not necessarily imply that the method is soundly based. It is always imperative to check that the division of flow between sub-areas is also correct, as the two adjustment procedures for adjusting the main channel discharge downwards and the floodplain discharge upwards, are clearly self compensating. As many authors have shown (e.g. Ackers, 1992b; Knight & Hamed, 1984; Wormleaton, Allen & Hadjipanos, 1982; Cassells et al. 2001; Lyness, Myers & Wark, 1997), it is quite possible to have a moderately successful H v Q prediction method based on division lines. However it may lead to gross errors in the internal distribution of the flow, let alone the distribution of boundary shear stress (Knight & Patel, 1985; Knight, Yuen & Alhamid, 1994).

D. W. Knight *et al.*

2.6.2.2. The coherence method (COHM)

The coherence method (COHM) of Ackers (1991, 1992b, 1993a, 1993b) is now well established as one of the better 1-D approaches for dealing with overbank flow and the related problems of heterogeneous roughness and shape effects. The 'coherence', COH, is defined as the ratio of the basic conveyance, calculated by treating the channel as a single unit with perimeter weighting of the friction factor, to that calculated by summing the basic conveyances of the separate zones. Thus

$$COH = \frac{\sum\limits_{i=1}^{i=n} A_i \sqrt{\left[\sum\limits_{i=1}^{i=n} A_i / \sum\limits_{i=1}^{i=n} (f_i P_i) \right]}}{\sum\limits_{i=1}^{i=n} \left[A_i \sqrt{(A_i / (f_i P_i))} \right]} \qquad (2.58)$$

where i identifies each of the n flow zones, and A is the sub-area, P the wetted perimeter and f the Darcy-Weisbach friction factor. The closer to unity the COH approaches, the more appropriate it is to treat the channel as a single unit, using the overall geometry. Some COH values are shown in Fig. 2.88 for a variety of river channels, indicating that in extreme cases COH may be as low as 0.5.

Where the coherence is much less than unity then discharge adjustment factors are required in order to correct the individual discharges in each sub-area.

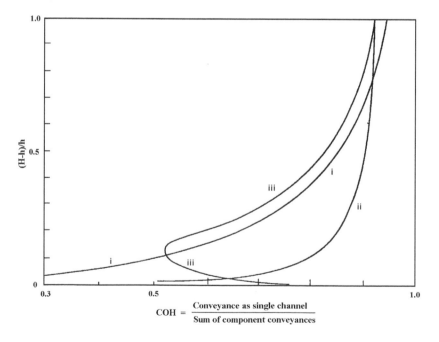

$$COH = \frac{\text{Conveyance as single channel}}{\text{Sum of component conveyances}}$$

Fig. 2.88 Channel coherence (COH) as a function of ratio of floodplain depth to main channel depth; (i) wide horizontal berms, (ii) narrow horizontal berms, (iii) natural river with sloping floodplains (after Ackers, 1993a).

Experimental studies of overbank flow in the FCF (Ackers, 1993a) suggest that different discharge adjustment factors (DISADF) are required in at least four distinct regions of depth, as indicated in Fig. 2.89. The experimental evidence shows (Ackers, 1993b) that

$$\text{COH} < \text{DISADF} < 1.0 \quad \text{i.e.} \quad \frac{Q_{\text{single}}}{Q_{\text{zones}}} \leqslant \frac{Q_{\text{actual}}}{Q_{\text{zones}}} \leq 1 \tag{2.59}$$

This implies that when overbank flow occurs, for a given stage or depth the actual discharge is always less than the basic value calculated on the basis of summing the discharge in different zones, but greater than the value based on treating the channel as a single unit, i.e.

$$Q_{\text{single}} \leq Q_{\text{actual}} \leq Q_{\text{zones}} \tag{2.60}$$

It also means that for a given discharge the actual stage is higher than that predicted by zonal summation but lower than that predicted on the basis of treating the channel as a single unit. This is illustrated schematically in the H v Q curves shown in Fig. 2.85.

The conveyance or discharge capacity of a channel, Q, is related to the energy slope by the geometric and roughness parameters. Based on Equations (2.37) and (2.38), Chow (1959) defined the conveyance, K, as

$$Q = KS_f^{1/2} \tag{2.61}$$

Ackers (1992a) introduced a modified conveyance parameter, K_D, in order to make it more suitable for use in overbank flow analysis, using (2.38) and defined K_D as

$$K_D = Q/\sqrt{(8gS)} = A\sqrt{\{A/(f\,P)\}} \tag{2.62}$$

Thus for a typical compound channel that is divided into three sub-areas, the main river channel and two symmetric floodplains, then the basic conveyance, K_{DB}, (before allowing for any interaction effects) is given by the sum of the individual conveyances for each sub-area as

$$K_{DB} = A_C\sqrt{\{A_C/(f_C\,P_C)\}} + 2A_F\sqrt{\{A_F/(f_F\,P_F)\}} \tag{2.63}$$

where the subscripts 'C' and 'F' refer to main channel and floodplain respectively. The actual discharge is then obtained from Equation. (4.4), by multiplying the basic conveyance by a 'discharge adjustment factor', DISADF, to give the correct discharge, allowing for any interaction effects. Thus

$$K_D = DISADF\ K_{DB} \tag{2.64}$$

As already shown by Equations (2.61) and (2.62), numerical values of COH are generally less than one, and typically the discharge calculated by assuming a single channel is less than the discharge calculated by summing the zonal values. The actual discharge is usually somewhere between these two values, as shown by Fig. 2.89.

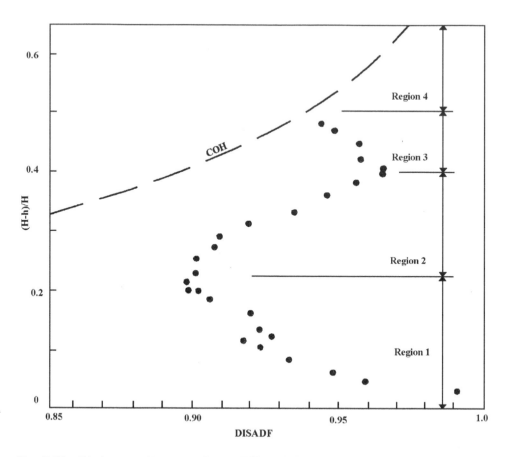

Fig. 2.89 Discharge adjustment factor, DISADF, for FCF tests, Series 2 (after Ackers, 1993a).

Vertical division lines should be used between zones, or sub-areas, and not used in the wetted perimeters for any of the zones. The 'basic' zonal discharges are calculated from standard resistance equations, i.e. Equations (2.37) and (2.38) and added together to obtain the 'basic' discharge, which is then adjusted to account for the effects of the interaction between the main channel and the floodplain flows. The adjustment required depends on the characteristics of the channel and also varies with stage. Ackers (1992b) provided a different adjustment function for each of the four regions of depth, and a logical procedure for selecting the correct value from those calculated assuming each adjustment factor in turn. He also provided additional corrections to account for the effect of deviations of up to 10° between alignment of the main channel and the floodplains, and a procedure for dividing the computed total discharge at any stage into main channel and floodplain components, based on experiments by Elliot & Sellin (1990). Ackers (1992a) also suggested that the square of the discharge adjustment factor could be used to give the mean boundary shear stress in each sub-area.

2.6.2.3. The Shiono and Knight method (SKM)

The Shiono and Knight method (SKM) is a logical development of the 1-D coherence method (COHM) described previously. Four examples of the use of this model are shown in Knight & Shiono (1996). In each case, experimental data from the FCF was chosen to test a finite element model, and comparisons made for two channels and three methods of calibration. Selecting appropriate individual values for f, λ, and Γ in the main channel, together with Equations (2.46)–(2.48), gave good agreement between the experimental data and the model. The same data was also calibrated with constant λ values ($\lambda = 0.13$) and variable Γ values. A third method of calibration was tried in which λ again held constant at 0.13, but different f values were used, with Γ values set to zero. In this case, the velocity distribution could be simulated well, but the lateral variation of boundary shear stress could not. This demonstrates the importance of the Γ term in Equation (2.36) and the importance of using the correct friction factor (i.e. the local friction factor) when calibrating a model, to give the boundary shear stress.

Attempts have also been made to determine the H v Q relationships theoretically for different types of compound channel based on the SKM (Liao & Knight, 2006). Integration of U_d over the whole channel width provides a complete specification of the overall discharge, but this it is only possible to integrate Equation (2.29) in a closed form for certain compound shapes. Further work is in progress to generalize these particular solutions, which include the effects of bed friction, secondary flow and lateral shear, as they are of value in benchmarking numerical simulations and in making predictions about water levels in channels with generalized shapes. See McGahey & Samuels (2003), Knight (2005a&b), Knight & Chlebek (2006), Omran & Knight (2006), McGahey (2006) and the CES website for further details see (www.river-conveyance.net).

2.7. Concluding Remarks

The previous sections have shown the complexity of flow structure in compound channels, that makes analysis even at a 1-D or 2-D level difficult. They have pointed to the significant advances that have been made over the last 25 years, partly due to careful and well-focussed experiments, but also due to better measurement techniques becoming available for both laboratory and field work. The insights gained by such work forms the basis of different modelling approaches, described further in Chapter 4 and applied in Chapter 5.

The advances that have been made through small-scale experiments must never be a substitute for seeking to acquire data at large-scale through fieldwork. However, it must be recognised that measuring sufficient parameters (e.g. velocity distributions, turbulence, coherent structures, boundary shear stress distributions, sediment fluxes, etc.) in natural channels is simply not possible in sufficient spatial and temporal detail under flood flow conditions. This partly explains the rationale behind the well-focussed laboratory experiments described earlier, aimed at understanding flow structure and quantifying particular phenomena. To date most of these experiments have been conducted in rigid boundary channels in order to simplify these tasks. However, it is recognised that most natural river channels are formed in sediments, and therefore any fluid/sediment interaction will also affect the flow structure. Issues related to sediment transport and morphology with overbank flow are considered next in Chapter 3.

Chapter 3

Sediment Processes

E. M. Valentine, S. Ikeda, D. W. Knight, I. K. McEwan, W. R. C. Myers,
G. Pender, T. Tsujimoto, B. B. Willetts and P. R. Wormleaton

3.1.   Introduction

Chapter 2 has dealt with the fluid mechanics of compound channel flows, assuming that the flow boundary (comprising the bed and banks of the inner channel, and the floodplain) does not change. In the field, as opposed to the laboratory, such fixed boundaries could only occur by virtue of extensive and careful engineering of the river system and are very rare indeed. Generally, the bed and banks are erodible and the river flow transports sediment, so degradation and aggradation occur at different places in the system at different times. As a result, the geometry and hydraulic behaviour of the river system are modified to an extent that is often important to river engineering and management.

As reported in Chapter 2, work on the flow structures in compound channels began in the 1960's and has continued to the present. However, research into sediment processes in compound channels is much less well developed for two reasons. First, an understanding of the flow mechanisms in the compound channels is a pre-requisite to understanding the sediment behaviour. Therefore the description of flow presented in Chapter 2 is a pre-requisite for progress in understanding sediment processes in compound channels. Second, there are many unresolved issues concerning sediment processes in simple open channel uniform flows and so it is unrealistic to expect a resolution of these issues in the context of compound channels before they are resolved for the simpler case.

3.1.1.   Sediment transport in compound channels

As has been shown in Sections 2.1 & 2.2, the flow structures in compound open channels are generally more complex than those in simple prismatic channels. These complex flow structures modify the 3-dimensional flow fields compared with those found in open channels of simpler cross-section, and in particular the near-bed flow conditions, which have a profound influence on sediment mobility. When dealing with sediment transport phenomena near-bed conditions are typically quantified through the use of parameters such as local shear stress or shear velocity, sometimes associated explicitly with turbulence characteristics. The different near-bed conditions associated with overbank flows therefore can have a significant influence on sediment phenomena, such as transport rate, surface texture and grading, bank erosion, floodplain deposition, equilibrium cross-section shape, bedforms and channel resistance. In order to understand this influence better, several series of experiments have recently been undertaken in fixed-planform compound channels, with erodible sediment beds. Fixed-planform channels are different from free-formed and natural channels in having non-erodible banks. The flow field in such channels is assumed to be the same as that in a free-formed channel of the same geometry. This device is often adopted in empirical studies, as a first step towards understanding the complexities of sediment behaviour in compound channels. (It mirrors many numerical geomorphological models, in which a river system is treated as quasi-steady, with flow and sediment being uncoupled for convenience in analysis.)

The account of flow in compound channels given in Chapter 2 revealed processes of momentum exchange between channel and floodplain and within each of these sub-units of the cross section. The processes differ greatly between the case of a low sinuosity channel with parallel floodplain/s and that of a meandering channel with a floodplain of non-uniform width. However, in both cases the momentum transfer is associated with internal circulations in the flow that exchange fluid in planes normal to the direction of dominant discharge. In the first case the exchange is dominated by vorticity (with near-vertical axis) along the banks of the inner channel; in the second by spillage of water from the inner channel on to one floodplain near each meander bend apex and a compensatory return from the other floodplain. In both cases, the lateral transfer of water is accompanied by lateral transfer of any sediment (and other dissolved and suspended material), transported in the water. When the transporting capacity of the water is reduced in the new environment, some or its entire load will be deposited. Floodplains, therefore, exhibit sediment deposits after the recession of inundating floods (Walling *et al.*, 1996).

Channel non-uniformity or flow unsteadiness perturbs the equilibrium between the flow and the sediment response to it. At different places and/or different times, the flow exceeds, or falls short of, the threshold of motion by different amounts and the equilibrium rates of bedload and suspended load transport also differ. Moreover, each change with position or time involves a transitional zone or period in which a new equilibrium is approached. When there is a transported sediment load, i.e. above the threshold of motion, this adjustment can be accompanied by net erosion or deposition of sediment at the flow boundary and hence a change in the geometry of the boundary. For example, flow escaping from a channel onto a floodplain (or returning to the channel from the floodplain) encounters an abrupt non-uniformity of depth, prompting an adjustment of its sediment load (which is likely to be substantial in flood conditions in a natural river). Sediment may be deposited from the shallower flow on the floodplain, and water returned to the channel from the floodplain with a reduced sediment load.

A further consequence of overbank flow is the redistribution of shear stress round the perimeter of the inner channel, in both the straight and meandering channel cases. The local dislodgement and transport rates of sediment would be expected to respond to this redistribution of boundary shear stress. Furthermore, the interaction between channel and floodplain flow also modifies the secondary circulation patterns within the channel, very radically in the case of the meandering channel. The combination of changed distribution of the boundary shear stress and changed secondary flows inevitably leads to changes in the distribution of the moving sediment in the cross section, laterally for the bedload and both laterally and vertically for the suspended load.

Among the consequences of this redistribution of active sediment is different equilibrium channel morphology in overbank flow compared with that when flow is inbank (see Chapter 5). Because overbank flow is rarely of long persistence, achievement of the associated equilibrium channel geometry is probably infrequent. During floods the common channel form is transitional, between the inbank and overbank types. Field evidence about this is unavailable because of the difficulty of bathymetric observation when rivers are in flood, but the evidence from laboratory experiments is secure, as will be reported later in Section 3.5.

Vegetation has a marked influence on the flow, both in-channel and on floodplains. Where part only of the flow cross-section is vegetated, the additional flow resistance of the vegetation often produces a redistribution of the water discharge within the cross-section. This is likely to produce sluggish conditions among the plants, leading to preferential deposition of sediment there, and enhanced flow velocities in the plant-free part of the cross-section, which may therefore be more prone to scour than they would in the absence of any vegetation. Thus the distribution and management of flora in the channel and floodplain have a pronounced effect on the flow resistance and the morphology of the river system as well as on the habitat diversity in it. Flow efficiency and habitat diversity suggest conflicting management strategies. The selection of an effective compromise strategy for each situation calls for careful and objective evaluation of both the fluid mechanics and the ecology in each case. Current research is directed towards provision of the knowledge base needed for such evaluation. Some of it will be reported in Section 3.6.

3.2. Straight Channels with Fixed Planforms

3.2.1. Introduction

In this section we will review recent progress in the understanding of the transport of sediment in straight channels and demonstrate changes in sediment concentrations and transport rates with relative depth. To achieve this we will report on a series of experiments conducted in laboratory flumes involving uniform and graded sediments. The treatment will be restricted to the transport of bed-load in the main channel for both inbank and overbank flows.

3.2.2. Observations of uniform sediment behaviour

Experiments were conducted in a 1.2 m wide, 18 m long flume with a bed slope of 2.04×10^{-3}, as shown in Fig. 3.1. Symmetric floodplains were formed either side of a 400 mm wide by 50 mm deep rectangular main channel, containing sand ($d_{50} = 0.8$ mm), and additional sidewalls were used to create an asymmetric channel with just one floodplain when required. Sediment was re-circulated, and equilibrium conditions established prior to undertaking detailed velocity, boundary shear stress and other measurements. Operating procedures and further details

Fig. 3.1 Photograph showing a general view of the water surface level measurement using vertical tubes attached to the left hand side wall of the main channel.

are given in Atabay & Knight (1999) and Atabay (2000). A comparable set of experiments were undertaken with the mobile bed removed and replaced by a smooth rigid surface, thus enabling comparative studies to be made in channels with exactly the same mean geometry under identical operating conditions, with and without sediment in the main channel.

3.2.2.1. Sediment transport data

Figure 3.2 shows data for both symmetric and asymmetric geometry in terms of X v Q_t, and Fig. 3.3 the same data in terms of X v Q_{mc} where X is the concentration in ppm (by mass), calculated using either the main channel discharge, Q_{mc}, or the total discharge, Q_t. Several experiments were repeated for both geometries, at different times by different operators and the repeatability was found to be excellent in all cases, indicating that the setting up procedures were carefully controlled. It is clear from Fig. 3.2 that X reaches a maximum value near the bankfull discharge, which in these experiments was approximately $0.078 \, \mathrm{m}^3 \mathrm{s}^{-1}$. This confirms the trend postulated by Ackers (1992) in Fig. 3.4, which was based on rigid boundary data with constant main channel and floodplain roughness values. Figure 3.3 shows that when the main channel discharge is used instead of the total discharge, the differences between the various experiments are small. This implies that provided $\%Q_{mc}$ may be determined, as is now possible through either the coherence (COHM) or Shiono & Knight (SKM) methods, described in Section 2.7.2, then the sediment transport rate may be calculated, since the interaction effect between the floodplain and the main channel is already included in the analysis via the discharge adjustment factors, DISADF, described in Section 2.2.7.2, or via the three

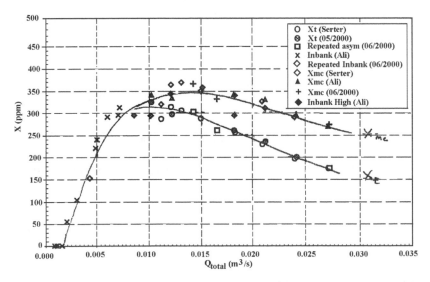

Fig. 3.2 Data from experiments conducted in a 1.2 m wide, 18 m long flume with a uniform sand ($d_{50} = 0.8$ mm).

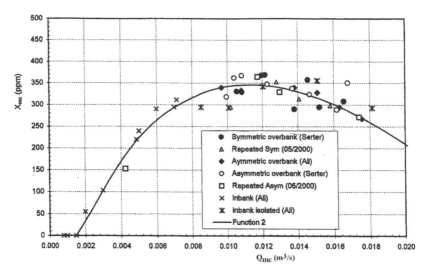

Fig. 3.3 Shows the same data as Fig. 3.2 this time plotted against main channel discharge.

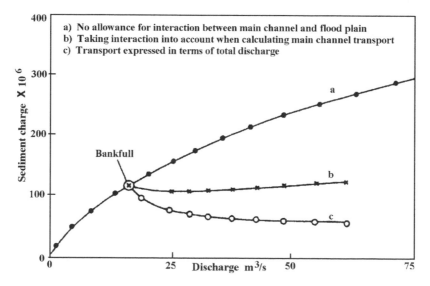

Fig. 3.4 Postulated sediment concentration (by mass in ppm) in a compound river section with a 0.25 mm sand bed (Ackers, 1992).

factors, described in Section 2.2.7.3. Whether the main channel discharge, mean velocity, or mean shear velocity should be used as the most appropriate parameter for calculating the sediment transport rate is still an open question. However, Fig. 3.3 indicates the efficacy of using the main channel discharge.

Larger scale experiments were undertaken in the Flood Channel Facility (see Section 2.4.2), using the same sediment size (d_{50} = 0.8 mm). In this series, a 40 m long trapezoidal main channel was constructed, with side slopes of 1:1, a depth of 200 mm, a top width of 2.0 m, and was flanked by two symmetrical 3.0 m wide floodplains set at a bed slope of 1.834×10^{-3} (Knight *et al.*, 1999). For this case the bankfull discharge was approximately $0.175 \, \text{m}^3\text{s}^{-1}$. Figures 3.5 and 3.6 show the sediment transport results for both smooth and roughened floodplain cases, plotted together with data for the isolated main channel case (trapezoidal). The overbank data are seen to follow somewhat similar trends to those already shown in Figs. 3.2 and 3.3, except that the peak concentration occurs at a discharge much lower than the bankfull discharge.

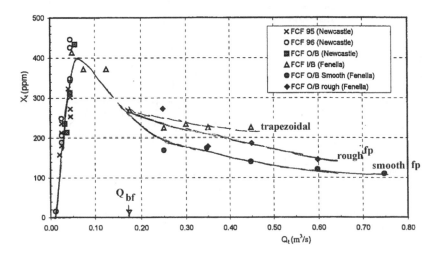

Fig. 3.5 Sediment concentration versus total discharge (FCF).

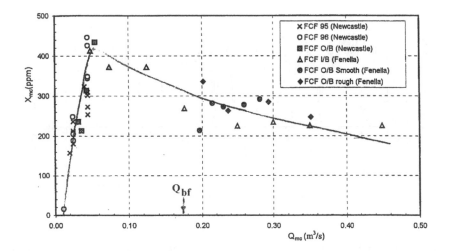

Fig. 3.6 Sediment concentration versus main channel discharge (FCF).

One of the most significant features of these experiments is the very large temporal variation in bed load transport rate. This is illustrated in Figs. 3.7 and 3.8, where data from other representative tests show similar variations in water surface slope, bed level and sediment transport rate. Figure 3.7 in particular, indicates the care that needs to be taken in measuring water surface slope or energy gradient, as S_f features prominently in calculations of the friction velocity, u_*, which

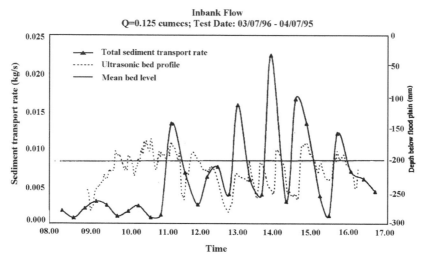

Fig. 3.7 Relationship between temporal variation in bed profile and manually measured sediment transport rate (after Knight & Brown, 2000).

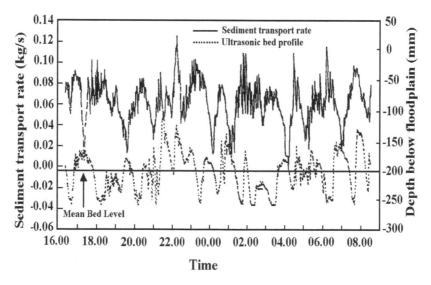

Fig. 3.8 Relationship between temporal variation in bed profile and automatically measured sediment transport rate (after Knight *et al.*, 1999).

is needed in most sediment transport equations. Poor estimates of overall or local friction velocity may partly explain why 70% of sediment transport predictions are still only within a factor of 2, as illustrated in the reviews by Chang (1988) and Ackers & White (1973).

A further series of experiments was undertaken in a 1/4 scale replica flume of the FCF which is a 1.89 m wide 16 m long flume set at the same bed slope as the FCF of 1.834×10^{-3}. Full details are given in Cassells (1998) and Myers *et al.* (2000).

3.2.2.2. Bed topography

Typical bedforms occurring in the main channel sections of experimental compound straight channels are shown in Figs. 3.9 and 3.10. Although these bedforms

Fig. 3.9 Sand dunes for low overbank flow with rough floodplains, $Q = 0.350 \, \text{m}^3 \text{s}^{-1}$.

Fig. 3.10 Sand dunes for bankfull flow, $Q = 0.175 \, \text{m}^3\text{s}^{-1}$.

are influenced by the main channel/floodplain interaction effect, they can be analysed by standard methods (Nordin, 1971), provided the sub-region velocity or boundary shear stress is known. Limited amounts of bedform data in channels with overbank flow are now becoming available through work on the FCF, and the other experiments in smaller flumes reported here. Atabay (2000), Ayyoubzadeh (1997), Brown (1997) and Cassells (1998) contain further details. There are a number of equations and methods for dealing with alluvial channel resistance. See, for example, the review by Chang (1988), and the methods of van Rijn (1984), and White, Paris & Bettess (1980). Their application to predicting zonal resistance coefficients (i.e. in the main channel part of a compound channel), and then using that via one of the methods described in Section 2.7 for the stage-discharge relationship is clearly difficult. Furthermore most alluvial channel resistance equations have been derived on the basis of a 'wide' river with simple cross sections. They may therefore not be applicable for compound channels where the aspect of the main river channel is less than 20. Both shape and composite roughness considerations now have to be taken into any calculation method.

3.3. Meandering Channels with Fixed Planforms

3.3.1. Introduction

There are few experimental data sets concerning meandering two-stage channels with mobile bed material, and consequently this section is centred upon recent sets of experiments performed using the UK Flood Channel Facility (FCF) at HR Wallingford. These were all carried out under equilibrium conditions in uniform flow. Valuable data were obtained upon adjustments of the bed profiles and also the response of sediment transport rate to overbank flow conditions.

3.3.2. Observations of uniform sediment behaviour in meandering channels

A sine-generated meandering channel with sinuosity of 1.34 and wavelength of 14.96 m was used in all of these mobile bed meandering tests on the FCF. The adopted planform incorporated two repeated meander wavelengths sandwiched between entry and exit transition sections. This sequence of uniform meanders, although different from conditions found in nature, improves the reliability of results and simulates experimental set-ups used by others (Willetts & Hardwick, 1993; Kiely, 1990). The main channel was formed with a mobile bed, having a top width of 1600 mm and 45° sloping concrete sides. In the graded bed tests, the bank was gradually brought to vertical around the outside of the meander apex, thus modelling more realistically the situation in natural rivers with cohesive banks. The main channel bed was initially screeded flat to a depth which roughly complied with regime conditions. The floodplains had a longitudinal slope of 1.86×10^{-3} with no cross-fall. Experiments were carried out with smooth concrete floodplains and also with the floodplains artificially roughened. Further details of the flume operation and data collection processes are described by Lyness *et al.* (1998).

The sand was a closely graded, uniform sediment with a d_{50} value of 0.835 mm. The maximum available floodplain width on the FCF was 10 m but tests were carried out with a reduced width of 8 m by installing temporary longitudinal walls on the floodplain. This enabled an increased floodplain depth to be obtained for a given discharge. These walls had a vertical slope of 45° giving the upper channel a trapezoidal cross-section.

The overbank test programme consisted of seven experiments with typical durations of 70–80 hours to allow flume conditions to reach dynamic equilibrium. Four experiments were undertaken with roughened floodplains to simulate vegetation effects. This was achieved using a single configuration of vertical rods, 0.025 m in diameter, arranged in a regular rhomboidal pattern and held in place by frames, which were above the water level. Roughness elements were surface penetrating for all flow depths and covered the floodplain to a density of 12 rods/m². This system of roughening is not entirely representative of a natural floodplain where vegetation generally consists of a combination of surface penetrating and submerged elements that would usually become progressively flattened with rising overbank water depth. However, it allowed modelling of more severe growth conditions that result in greater water surface elevations along a river reach.

Discharges of 0.175 m³/s, 0.250 m³/s, 0.350 m³/s and 0.600 m³/s, resulting in overbank depths ranging from 0.038 m–0.178 m, were tested on roughened flood plains. Identical discharges, with the exception of the 0.175 m³/s discharge were tested on smooth flood plains and resulted in overbank flow depths that ranged from 0.042 m–0.092 m. An outline of the tests completed is given in Table 3.1, and further details can be found in Lyness *et al.* (1998).

3.3.2.1. Sediment transport data

The variation of dimensionless total load transport with relative depth, **Dr** is shown in Fig. 3.11 for both floodplain configurations examined at overbank conditions. Sediment rates were observed to increase with discharge and depth for the inbank phase until a maximum value is reached at the bankfull level. As can be seen in Fig. 3.11, immediately above bankfull, a sudden decrease in sediment rate is observed in both the rough and smooth floodplain overbank flows. Although not well defined, further increasing total channel discharge, above **Dr**>~ 0.2, creates a flow condition where diminishing trends in sediment rates are reversed. Increases in discharge and depth above this, result in increasing transport rates, with those for the smooth floodplain being greater than those for the roughened floodplain.

In the case of the smooth straight floodplain, Rameshwaran *et al.* (1999) carried out a set of complementary experiments on a 1/4 scale model of the FCF using a

Table 3.1 Flow data and boundary characteristics for FCF Phase C meandering channel tests.

Test No.	Q (m³/s)	H (m)	s	S_{fp}	Main channel boundary	Floodplain boundary
3	0.175	0.238	1.34	1.86×10^{-3}	0.835 mm sand	Rod roughness
4	0.250	0.265	1.34	1.86×10^{-3}	0.835 mm sand	Rod roughness
5	0.350	0.300	1.34	1.86×10^{-3}	0.835 mm sand	Rod roughness
6	0.600	0.378	1.34	1.86×10^{-3}	0.835 mm sand	Rod roughness
7	0.250	0.243	1.34	1.86×10^{-3}	0.835 mm sand	Smooth concrete
8	0.350	0.259	1.34	1.86×10^{-3}	0.835 mm sand	Smooth concrete
9	0.600	0.292	1.34	1.86×10^{-3}	0.835 mm sand	Smooth concrete

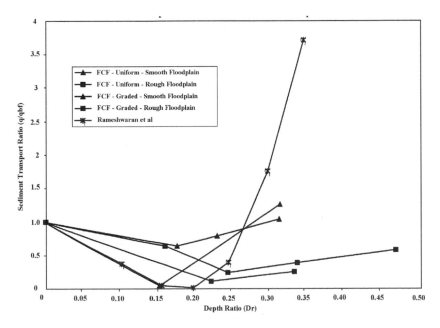

Fig. 3.11 Variation in dimensionless sediment transport ratio (q/qbf) with depth ratio.

uniform sand with $d_{50} = 0.85$ mm. These data are also shown in Fig. 3.23 for comparison. The smooth floodplain results for both sets of tests indicate a reduction in sediment transport rate to a minimum at depth ratio of around 0.2.

Observations of sediment behaviour in meandering two-stage channels indicate that movement follows the path of the main channel, and hence the shear stresses inducing motion are generated by the flow in the main channel. It follows that the observed decrease in sediment transport at low overbank flows is due to a reduction in these main channel shear stresses. This is caused by interaction, mainly in the region of the cross-over, between the main channel and floodplain flows which reduces the main channel discharges, velocities and hence boundary shear stresses below those at bankfull. Knight's (1992) direct boundary shear stress measurements for a rigid bed meandering channel with smooth floodplains show that τ_0 initially reduces at low overbank levels, which accords with these observations, although topographic bed changes may produce τ_0 different from these earlier observations.

The variation of sediment concentration with depth ratio, is shown in Fig. 3.12, where the sediment concentration is non-dimensionalised in terms of its value at bankfull. As sediment transport is normally confined to the main channel of a compound section, the resulting concentration is usually expressed in terms of the main channel discharge. However, as bed load movement occurred over the floodplains during the overbank tests with roughened floodplains, the determination of sediment transport rates in this manner is not appropriate and consequently, sediment transport rates for these tests relates to total channel discharges.

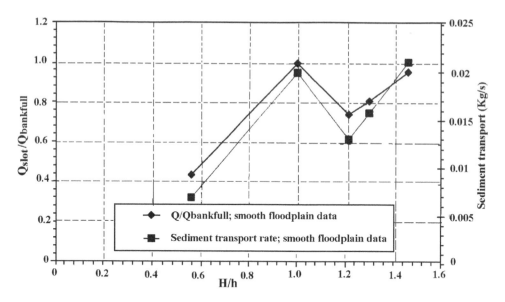

Fig. 3.12 Variation of $Q_{slot}/Q_{bankfull}$ and sediment transport rate with H/h for smooth floodplain test case.

The concentration initially rises to a peak at the bankfull discharge. As flow enters the overbank phase of the channel, the extra discharge dilutes the sediment concentration and so it is reduced for higher flows. Note that the transport rate for the roughened floodplain test series tends toward a steady value fort $Dr > 0.25$ (Fig. 3.13).

Associated with the migration of sediment along a channel is the development of undulations on the main channel sand bed. For flows confined to the main channel, the channel bed gets progressively deeper towards the outer extreme of the apex and exhibits the typical deposition and erosion patterns associated with meander bends. Deep scour around apex sections is usually compensated by deposition and a resultant mean bed level above that of the screeded condition (i.e. 0.2 m below bankfull) at cross-over sections. Final bed features for overbank flows differ from those for the inbank tests. For the smooth floodplain test, bedforms occur that are regular in nature, and can be described as a series of repeating dunes with amplitudes in the range of 40–80 mm. These are due predominantly to localised shear stresses. Although the frequency of consecutive bedforms was difficult to specify, crest to crest distances were approximately 450–700 mm. Bed features that arose during roughened floodplain tests had a much more irregular and variable pattern and, rather than being described as bedforms, appear to take the form of a scour-deposition morphology governed primarily by secondary re-circulation currents. The structure of these bed formations was influenced by the sediment movement in the channel which moved from the main channel at cross-over sections, onto the floodplain and back into the main channel at the next cross-over in a direction along the longitudinal centre-line. The

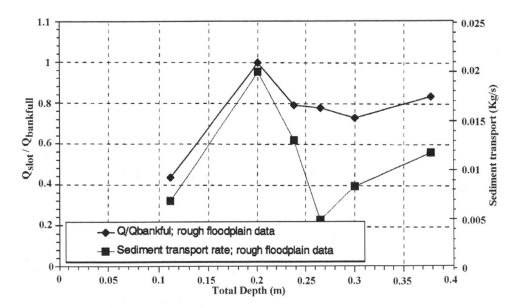

Fig. 3.13 Variation of $Q_{slot}/Q_{bankfull}$ and sediment transport rate with Total Depth for rough floodplain test case.

quantity of sediment migrating along the channel in this manner increased with increasing discharge and flow depth and reached a maximum for a total discharge of $0.6 \, m^3/s$. Furthermore, this sand movement along the floodplain contributed to an increased bed level at cross-over sections along the channel. The difference, in the range of 150 mm to 200 mm, between the high and low points of the channel bed for the roughened floodplain tests, was greater than those observed for smooth floodplains.

3.3.3. Observations of graded sediment behaviour

The previous section described experiments on meandering channels with mobile beds consisting of single-size sediment and having rigid embankments. This section takes one more step towards reality by considering the effect of graded bed material upon the flow and sediment transport processes. The vast majority of natural river sediments are graded and, in many cases, broadly graded. This can promote behaviour, such as sorting and armouring, that is not found in single-sized material. These processes can significantly modify the hydraulic and morphological characteristics of the channel.

The grading of the bed material was chosen in order to avoid the formation of over-large bed features, and so that it was only partially mobile under overbank test flows, thus enabling sorting processes to occur. The material had a d_{50} of 1.33 mm and a gradation coefficient of 3.07. The material is weakly bimodal.

In the first series of tests the floodplain was straight, 10 m wide with no lateral slope. Tests were carried out with a smooth concrete floodplain and also with an artificially roughened floodplain. The floodplain was roughened for these tests using expanded metal strips bent into inverted 90° 'V' shapes and laid across the floodplain at 500 mm centres. This gave a greater roughness than the dowels used in the earlier tests, commensurate with the greater main channel roughness due to the larger grain size of the bed material. In order to assess the effects of floodplain planform, experiments were also carried out with a floodplain having a sinuosity of 1.1 synchronous with the main channel, and finally with flood banks parallel to the main channel. An outline of the tests completed is given in Table 3.2.

3.3.3.1. Sediment transport data

The variations of the relative total sediment transport rate with depth ratio and discharge ratio for the bankfull and overbank flows are shown in Fig. 3.10. The graded sediment results have broadly similar features to the earlier uniform sediment results in meandering channels. Both show an initial reduction in sediment transport at low overbank depths followed by an increase as the floodplain depth becomes greater. They indicate that the minimum sediment transport for smooth floodplains occurs when the depth ratio is in the range 0.15–0.20 and it is higher at 0.22–0.24 for rougher floodplains. However, there were significant differences between the uniform and graded experiments, which make detailed comparison unrealistic. The floodplain widths were different in the two cases (8 m and 10 m) and also the floodplain roughness in the graded experiments was substantially greater than for the uniform series. More importantly, however, the substantial

Table 3.2 Flow data and boundary characteristics for FCF Phase C meandering channel tests using graded bed material.

Run No.	Run Name	Description	Flow (l/s)	Floodplain Depth (mm)
1	BFMB	Bankfull — mobile bed	97	−6.4
2	BFEN	Bankfull — roughened banks	90	−1.6
3	BFFZ	Bankfull — frozen bed	97	−12.6
4	BFRF	Bankfull — frozen bed, riffles removed	97	−11.1
5	LOSW	Low overbank — smooth straight floodplain	163	26.8
6	HOSW	High overbank — smooth straight floodplain	449	68.9
7	LORW	Low overbank — rough straight floodplain	120	42.5
8	HORW	High overbank — rough straight floodplain	164	75.7
9	LORS	Low overbank — rough sinuous floodplain	121	43.5
10	HORS	High overbank — rough sinuous floodplain	143	75.8
11	HORN	High overbank — rough narrow floodplain	150	73.9

bedforms observed in the uniform sediment tests were not present in the graded tests probably due to the greater d_{50} value of the latter.

The rough straight floodplain results show a reduction to a minimum in the region of $Dr = 0.2$ but then a much slower rise than for the smooth floodplain case. This reflects the greater level of interaction between the main channel and roughened floodplains, which further reduces the discharge in the main channel sub-section. Indeed, with the smooth floodplains, as the depth increases the mean floodplain velocity exceeds that in the main channel.

It is apparent that, except in the extreme (and somewhat unrealistic) case of high flow over the smooth floodplain, the sediment transport rate is greatest for bankfull. This lends credence to the concept of bankfull as being the dominant channel forming discharge as proposed by Wolman & Miller (1960). It reinforces the use of bankfull as a practical flow upon which to base channel morphological design, as recommended by Hey (1978), since this flow is also relatively easy to identify in the field.

3.3.3.2. Lateral distribution of bed load

The lateral distributions of bed load for all of the FCF experiments are shown in Fig. 3.14. It is apparent that the bed load transport rate is not uniformly distributed laterally and, in many cases, it is predominantly concentrated within a narrow band. Normalised (to section mean) distributions of depth, depth mean velocity and bed load discharge are given in Fig. 3.15(a–e) for all cases. Also shown in this figure is a nominal local bed grain shear stress value, τ', calculated from local depths, D, and depth averaged velocities, V, using the van Rijn (1984) relationship:

$$\tau' = \frac{\rho g V^2}{\left(18 \log \dfrac{12D}{3d_{90}}\right)} \tag{3.1}$$

This equation, which assumes a logarithmic flow profile, is strictly applicable to wide uniform channels and is unlikely to provide accurate values in highly three-dimensional meandering channel flows. It is provided here simply as a reference by which deviations from the uniform case, and hence secondary flow effects, may be inferred and highlighted.

In the bankfull case the sediment is seen to move in a fairly narrow band which follows the inside of the apex and then moves across the channel from upstream to downstream in the region of the cross-over. This accords well with the Knight et al. (1992) measurements of bed shear stress in a rigid bed meandering channel, which shows high values around the inner bank of the apex at bankfull flow. A contributory factor to this is super-elevation of water surface around the apex. This results in a high longitudinal water surface slope on the inside bank going into the apex, which then shifts to the outside bank downstream of the apex. The increased surface slope translates to greater boundary shear stresses and hence sediment transport in these regions. Coupled with this, the centrifugal secondary circulation will also act to impel bed load transport towards the inner bank around the apex as

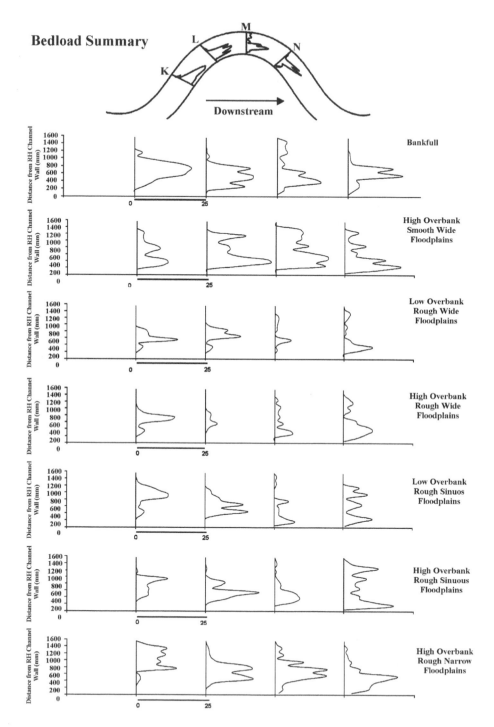

Fig. 3.14 Lateral bedload distributions for graded bed experiments in the Flood Channel Facility.

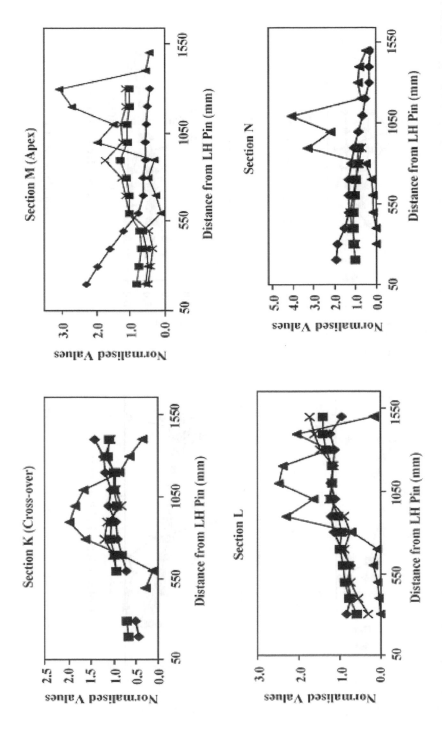

Fig. 3.15(a) Normalised (to section mean) distributions of depth (◆), depth mean velocity (■) and bed load discharge (▲) for Case BFMD (see Table 3.2). Also shown in this figure is a nominal local bed grain shear stress value, τ' (x) calculated from local depths, D, and depth averaged velocities, V.

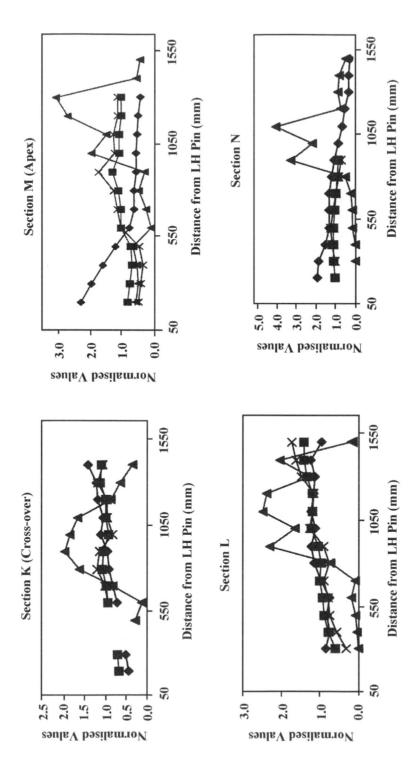

Fig. 3.15(b)   Distributions of depth, depth mean velocity, bed load discharge and τ' for Run HOSW.

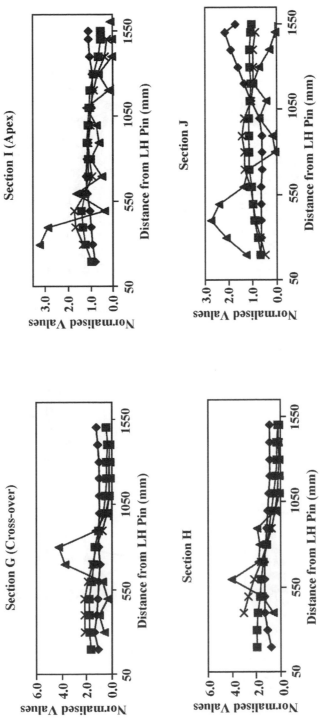

Fig. 3.15(c) Distributions of depth, depth mean velocity, bed load discharge and τ' for Run HORW.

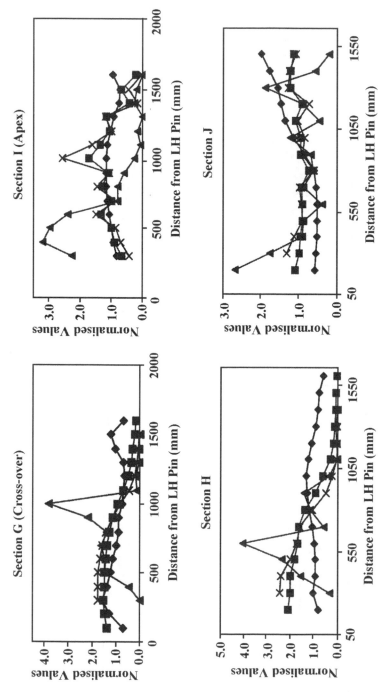

Fig. 3.15(d) Lateral distributions of depth, depth mean velocity, bed load discharge and τ' for Run HORS.

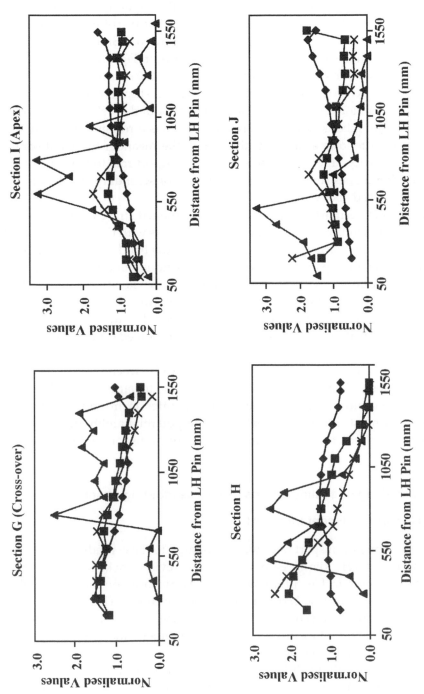

Fig. 3.15(e) Distributions of depth, depth mean velocity, bed load discharge and τ' for Run HORW.

seen in Fig. 3.15(a). The effect of this circulation, particularly on the smaller particles, is reflected in an analysis of the bed load particle fractions, which shows the larger particles to be following the thalweg rather than the line of maximum bed load transport. The observed patterns of bed load distribution show similar characteristics to the Muddy Creek field data of Dietrich & Smith (1984).

The high overbank smooth floodplain flow results, in Fig. 3.15(b), show the bed load to be somewhat more evenly distributed across the section than for bankfull flow. This again accords with the Knight *et al.* (1992) shear stress measurements, which indicate a more even lateral distribution of boundary shear stress in overbank rather than inbank flows. Nevertheless, there is a recognisable peak in the bed load, which follows a similar path through the meander to that at bankfull. In this case, however, the maximum sediment transport rate follows the thalweg through the bend apex due to a general reduction of secondary circulation in this region as discussed above.

Interaction between main channel and the roughened straight floodplain tends to shift the point bar downstream of the bend apex. As with the bankfull flows the sediment transport is largely concentrated in a narrow band, shown in Fig. 3.15(c). At the cross-over and upstream of the apex, this band is shifted towards the channel centre-line due to secondary circulation from the floodplain flows. It then tracks across the inside of the apex, following the thalweg. It remains on the point bar at the inner bank downstream of the apex, possibly due to the strong secondary circulation at this point resulting from the combined centrifugal and floodplain induced flows.

In all cases there appears to be little relationship between the calculated local grain shear stress, τ', and local bed load. This points towards a strong secondary circulation effect upon sediment transportation rates in meandering overbank flows. Doubts must therefore be cast upon the potential ability of one- or two-dimensional methods to accurately model sediment transport in these situations.

3.3.4. Modelling of sediment transport in meandering channels

3.3.4.1. Uniform sediment transport

Measured sediment transport rates were compared to those predicted by the total load formulae of van Rijn (1984 a,b,c) and the revised Ackers & White (1990) approach are shown in Fig. 3.16. Calculations were based on main channel geometrical parameters for the bed in its screeded condition. This was taken to be representative of the mean cross-sectional data along the length of the channel.

Sediment transport rates predicted by the sediment theories underestimate observed values by considerable margins so that for the lower discharge tests with both rough and smooth floodplain configurations, no transport is predicted. Such large errors are somewhat surprising because these formulae use a single characteristic sediment size to represent the range sizes in the bed material. The uniform grain size used in the FCF, although different to what would be expected

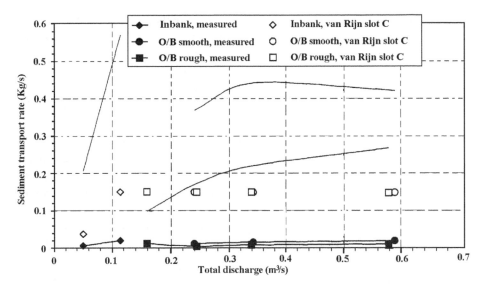

Fig. 3.16 Variation of actual, van Rijn and Ackers and White predicted sediment transport rates with total channel discharge.

in a natural river where sediment comprises small to large sediment sizes, would be expected to produce reasonable results for these theories. It is expected that errors for channels incorporating mixed or graded sediments could be significantly larger.

By assessing the development of the van Rijn theory, it is reasonable that, despite it being valid for straight channels, the logarithmic relationship between the bed shear velocity and the flow resistance attributable to the grains is invalid for those with a sinuous planform. This relationship is summarised as:

$$u'_* = \frac{\sqrt{g}V}{18 \log\left(\frac{12R_b}{3d_{90}}\right)} = \left[\frac{\sqrt{g}V}{C'}\right] \tag{3.2}$$

By performing a zero error back calculation with the measured sediment transport data, the actual values of u'_* required to generate the recorded sediment motion were obtained. As anticipated, the value of this parameter calculated by the van Rijn method is substantially lower than that required to produce the observed transport rates.

Following the failure of the van Rijn method to predict accurately sediment transport rates in meandering channels, two attempts were made to re-calculate the Chezy coefficient so that the theory's performance for meandering channels could be improved. Firstly, point velocity measurements taken using a Nortek ADV probe were re-integrated in the area of the main channel contained between the interfaces of zero shear (A_b) as specified in the Vanoni & Brooks' (1957) side wall correction method. With this discharge the velocity in the area A_b was determined and hence a revised Chezy coefficient for the main channel. Sediment transport rates were recalculated using this revised value of Chezy's C in Equation (3.2). These new calculated sediment discharges grossly overestimate the observed rates. The application of the Vanoni and Brooks' sidewall correction results in values of wetted perimeter, P_b, that are considerably less than those measured on the FCF, whilst the cross-section areas, A_b, are relatively unaffected. This leads to unrealistically low Chezy values, resulting in over-prediction of transport rates.

The second method involves omitting the sidewall correction, and determining a value of C, for the main channel from the observed geometry. This value of C is termed the slot C. On the basis that the hydraulic radius of the main channel is not compromised in any way, this value of slot C will be greater than that of the revised C above and transport rates will be reduced. The transport rates predicted by using this value of slot C in the van Rijn method are plotted against the total channel discharge in Fig. 3.17. The transport rates predicted using the revised C are also included as plain lines in Fig. 3.17, as is the measured data. Although the use of the slot C improves the sediment transport estimations, considerable errors remain.

The results of this analysis highlight the difficulties associated with the complex issue of sediment transport in meandering cannels. Although the limitations of the van Rijn theory for channels of sinuous planform have been identified and the logarithmic relationship between the grain resistance, C', and the granular bed shear velocity, u_*', suggested as possible reasons for the errors, the attempts to re-calculate a more appropriate value of C have been unsuccessful.

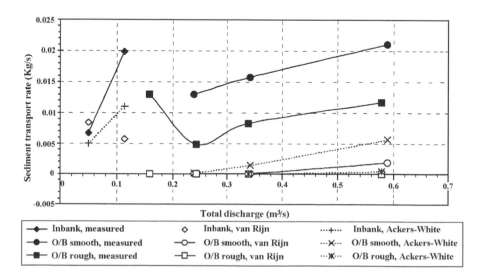

Fig. 3.17 Variation with total channel discharge of sediment transport rates predicted by the van Rijn method with a slot C.

3.3.4.2. Graded sediment transport

The observed sediment transport rate was compared with the values of bed material load calculated from several well-known transport formulae. The formulae used were the total load formulae of Engelund & Hansen (1967), Ackers & White (1973, 1990), van Rijn (1984a), Yang (1973), Karim & Kennedy (1990). Figure 3.18 shows the comparison between these formulae at the cross-over and apex where

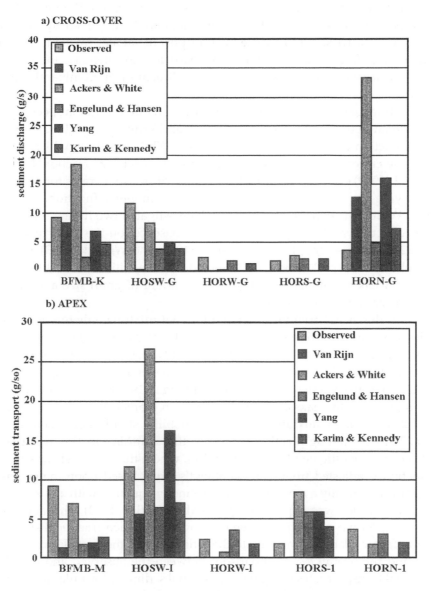

Fig. 3.18 Comparison between Observed and Calculated sediment transport rates assuming mean D_{50} sediment size (a) at the cross-over (b) at the apex.

the calculation is based upon a single representative sediment size taken generally as d_{50} (d_{35} for Ackers-White). Clearly there is a wide range of values for the various methods, however one significant pattern does emerge, which is that for all overbank cases, except the narrow floodplain, the methods give a higher discharge at the apex than at the cross-over. For the bankfull and narrow floodplain cases, the opposite is true.

One clear source of error in these equations is the use of a single characteristic sediment size to represent the range sizes in the bed material. In calculating sediment transport by fraction, it is necessary to recognise the interaction of different particle sizes in the bed material. This may be achieved by the use of hiding functions. These act to reduce the transport of smaller fractions, recognising the hiding effect on them of the larger particles, and increase the transport of larger particles due to their exposure in a sediment of predominantly smaller material.

These hiding functions are generally empirical and often equation specific. Three are considered in this investigation. Day (1980) and White & Day (1982) proposed a hiding function to be applied to the Ackers & White equation, which in effect adjusted the critical stress from its Shield's value for each size fraction. Proffitt & Sutherland (1983) extended this to a wider range of transport rates for each grain size, applying the hiding factor adjustment to the mobility function, F_{gr}, in the Ackers & White equation. Pender & Li (1995) investigated two hiding functions to be applied to the van Rijn equations. One was based upon the work of White & Day (1982) and the second, following Parker (1991) and Wilcock (1993), upon the equal mobility concept. Wilcock (1993) argued that the critical shear stress in unimodal and weakly bimodal mixtures is independent of grain size and depends upon the mean grain size in the mixture.

However, this is not true for strongly bi-modal mixtures where critical shear stress increases with particle size, possibly due to lateral segregation of grains in the bed. Wilcock provides a measure of bi-modality, B, based upon the size of the grains at the two modes and the proportion of grains in the modes. Equal mobility applies to mixtures where B is less than 1.7. In Fig. 3.19, the bed material used is clearly bi-modal with peaks at 0.4 mm and 2.7 mm diameters, however the value of B is 1.2, signalling a mixture where equal mobility should apply according to Wilcock. Thus the hiding function applied to the van Rijn function was the second of the two investigated by Pender and Li, where the critical shear stress for each fraction is adjusted from a single Shield's stress based upon the geometric mean grain size. Finally a modified Karim-Kennedy equation with adjustment for particle size fraction was applied as proposed by Karim (1998). This adjustment was based upon the proportion of the bed covered by each fractional size and a sheltering factor related empirically to the ratio between the particle size for each fraction and the overall d_{50}. Comparisons between the observed and calculated bed material transport rates using these fractional sediment formulae are given in Fig. 3.20 for the cross-over and apex sections. Generally (except in the case of the van Rijn formula) the calculation by fraction provides marginally better agreement

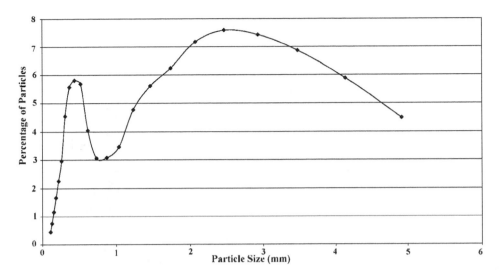

Fig. 3.19 Bed material Grain Size Distribution for the graded sediment experiment performed in the Flood Channel Facility.

with the observed values than those using single grain size, although the spread is still large.

All calculations to date have been based upon mean cross-section data, therefore a further refinement to the calculations can be made by integrating across the section using measured depth and depth averaged velocities. The results of these calculations are shown in Fig. 3.21. Overall the depth averaged integrated results show little, if any, improvement on those using the mean section values. This points to the strong influence of secondary circulation in determining bed load.

3.4. Free Planform Channels

3.4.1. Introduction

Hydraulic engineers are often concerned with the design and operation of alluvial channels and the determination of their geometrical parameters in a dynamic equilibrium condition. The problem is to determine the shape, cross-sectional area and the slope of the channel which will carry a given discharge of water and sediment, while flowing over an erodible bed of known characteristics.

3.4.2. The equilibrium channel

Methods of determining the stable cross-section fall into two categories. The empirical approach which relies on data collected from stable natural and artificial channels and the so-called rational approach which attempts to solve the geometrical parameters from relationships which describe the dominant processes such as flow resistance, sediment transport and bank stability.

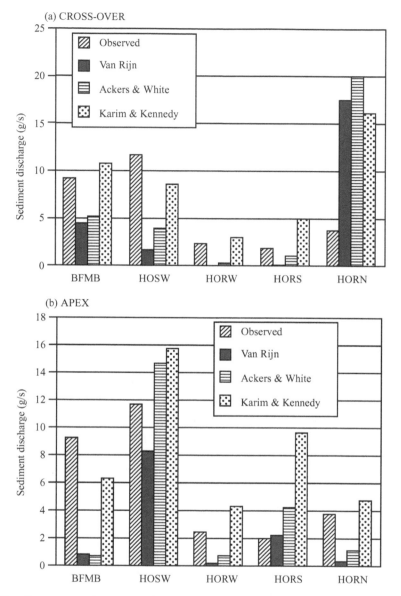

Fig. 3.20 Comparison between Observed and Calculated sediment transport rates by fractions with mean section parameters (a) at the cross-over (b) at the apex (see Table 3.2).

The design of 'stable' or 'equilibrium' free-formed channels may be based on either the rational equilibrium approach (White, Bettess & Paris, 1982) or on an empirical regime approach using observations on natural rivers and irrigation canals (Ackers, 1992a). Although various extremal hypotheses exist (e.g. Bettess & White, 1987), most are based on the principle that "for an alluvial channel the necessary and sufficient condition of equilibrium occurs when the stream power per unit length of channel, $\rho g Q S_f$, is a minimum, subject to certain constraints"

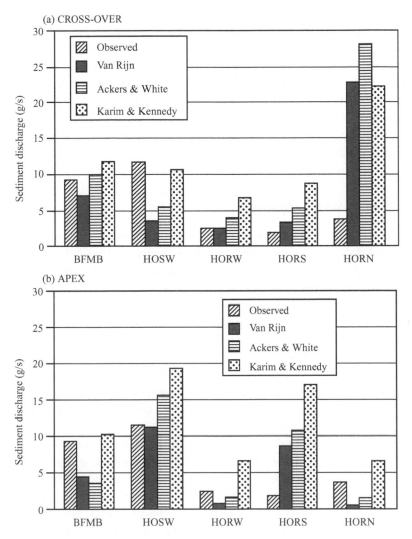

Fig. 3.21 Comparison between observed and calculated sediment transport rates by fractions with integrated section parameters at (a) the cross-over and (b) the apex.

(Chang, 1980). The implications of this have been illustrated elsewhere, e.g. by White, Bettess & Paris (1982), who show that for a given discharge and sediment size, the slope will reach a minimum at the equilibrium breadth that produces the maximum sediment discharge rate, X, for that slope, S. An alluvial channel in 'equilibrium' is thus hypothesized as one that maximises the sediment carrying capacity for given external constraints. Any regime channel so formed is thus shaped by its bankfull discharge, which in this context is regarded as an inbank flow.

However, once overbank flow takes place, then existing rational or empirical regime theories are no longer applicable and new design methods have to be developed. One such new method is based on the 'coherence' concept, described in Section 2.7.2.2, and illustrates how X is related to Q for both inbank and over-bank flows. The method was originally developed by Ackers (1992a), based on data from the 'Phase A' series of experiments (Knight *et al.*, 1989, 1992) carried out in the Flood Channel Facility (FCF), using a fixed-planform channel with rigid boundaries. The method COHM takes into account the floodplain/main chan-nel interaction process. Subsequently experiments with a mobile bed condition, (Series C) were undertaken (Knight *et al.*, 1999, 2001 and Valentine *et al.*, 2001), and these are also described herein.

3.4.3. Channel Width Adjustment

For overbank flows in straight and meandering channels, considerably fewer experimental data exist with regard to both velocity and boundary shear stress. This is discussed in Chapter 2. In general terms, for straight channels with flood-plains the boundary shear stresses under overbank flows vary in a more complex way than for in-bank flows, with stresses in the main channel decreasing because of the influence of the slower floodplain flows. Conversely, the floodplain bound-ary shear stresses are increased above the expected 2-D values due to the effect of the faster main channel flow.

The redistribution of boundary shear stress within the cross-section during over-bank flow has a profound effect on sediment transport rate (Ackers, 1992; Knight & Abril, 1996). Consequently, there is also an effect on bank adjustments that occur when a river flows in an overbank condition.

Width adjustment may take place as part of the natural evolution, by disrup-tion of the long-term equilibrium due to an extreme event, or by a morpholog-ical response to river engineering or management. It is therefore necessary for civil engineers to be aware of the geomorphic context of width adjustment. In general, boundary shear stress distributions in channels with non-uniform cross-sections, bank advance and retreat mechanisms are poorly understood. Specifi-cally, an improved understanding of the effects of over-bank flows on river width adjustment processes is required (Valentine *et al.*, 2001).

3.4.3.1. Channel response to changes in boundary conditions

A compound channel is composed of (i) the main channel, (ii) the side slope (at the boundary of the main channel and floodplain), and (iii) the floodplain. The flow in the floodplain (iii) is characterized by slower velocity and small submergence in general. Thus sediment transport is less active, and suspended sediment is dom-inant there. The sediment transport in the side slope (ii) is very much affected by transverse bed slope. The bed-load motion moves toward the main channel, and thus the widening of the channel and rising of the bed of the main channel take place. Along the side slope, the shear stress increases with depth. When it exceeds the critical tractive force, $\tau_{*c}(\theta_y)$, affected by the lateral slope θ_y, the bed-load transport occurs and it has the lateral components toward the main channel. Then, the side-slope erosion or the widening of the main channel takes place. The critical tractive force on the slope composed of sand (cohesionless materials) is given by Equation (3.3), and the lateral bed-load transport rate is calculated from the total bed-load transport rate as shown in Equation (3.4).

$$\tau_{*c}(\theta_y) = \tau_{*c} \cos\theta_y \sqrt{1 - \frac{\tan^2\theta_y}{\tan^2\phi_r}} \tag{3.3}$$

$$q_{By^*} = q_{B^*} \sin\phi_B; \quad \tan\phi_B = \tan\gamma - \sqrt{\frac{\tau_{*c}}{\mu_s\mu_d}} \tan\theta_y \tag{3.4}$$

where q_{B^*} is $q_B / [(\sigma/\rho - 1)gd^3]^{1/2}$; q_B is bed-load transport rate; σ and ρ are, respectively, the mass densities of sand and water; d is sand diameter; g is gravity acceleration; ϕ_B is bed-load direction; γ is flow direction deflected from the longitudinal axis; ϕ_r is angle of repose of sand; and μ_s are μ_d the static and dynamic friction coefficients of sand, respectively. Nakagawa *et al.* (1986) deduced Equation (3.4) by linear analysis of the equation of motion of single bed-load particle. When the materials are cohesive, Equation (3.3) should be modified under consideration on the cohesion. The bed-deformation including side slope is analyzed by the continuity equation of sediment as follows:

$$\frac{\partial z_b}{\partial t} = -\frac{1}{1 - \rho_0} \text{div}\mathbf{q}_B \tag{3.5}$$

where z_b is elevation of bed-surface; t is time; ρ_0 is porosity of sand; and $\mathbf{q}_B$ is vector of bed-load flux. If the bed-slope becomes steeper than the angle of repose, the slope failure occurs and the collapsing material is distributed on the main-channel bed. If the sediment on the slope is in motion, the erosion of the slope continues (non-equilibrium). With expanding of the main-channel width, the stage and the discharge per unit width decreases gradually, and when the condition of the slope becomes under the critical, the process stops (static equilibrium). Such a static equilibrium cross section was studied by Parker (1978b) and Ikeda (1981). If the secondary current exists to transport sediment against gravity, there is a dynamic equilibrium where the direction of sediment transport coincides the longitudinal axis (for example an equilibrium cross section geometry at river bend, (Engelund, 1974; Kikkawa, Ikeda & Kitagawa, 1976). Without secondary motion or sediment diffusion toward the floodplain, the dynamic equilibrium geometry cannot be expected for compound channel. If the sediment is transported in suspension, the lateral flux appears by horizontal turbulent mixing toward the side bank (see Section 3.7), and the balance of the lateral fluxes of sediment due to turbulent diffusion (for suspended sediment) and due to gravity (for bed load) brings a dynamic equilibrium cross section (Parker, 1978a; Izumi & Parker, 1997).

When materials composing the slope are in motion, side bank erosion occurs along the floodplain. Collapse occurs with bank erosion of the floodplain in a compound channel to form an angle of repose. If the flow is concentrated into the main channel, side-bank erosion accompanies a collapse of the part of the floodplain above the water surface which also falls down into the water. When the material is composed of sand, the side slope after collapse is around an angle of repose, while the side bank is very steep because of the capillary forces (see Fig. 3.22).

When the sediment supply to an equilibrium channel is stopped, the channel responds to reach a new equilibrium. This is termed a "non-equilibrium process" here. When the sediment supply to a channel is cut off, bed-degradation takes place in a straight channel with constant width (rigid side walls or revetment) and propagates downstream (Gessler, 1971). When the channel has a mobile flood-plain, degradation accompanies side bank erosion (widening of a main channel). As mentioned, widening may occur without bed degradation due to the cessation of sediment supply, but bed-degradation promotes widening. According to flume experiments (Goto *et al.*, 2000), the widening process takes place first, and then the bed degradation follows. In other words, the decrease of sediment supply causes bed degradation in a channel with a single cross section, but it promotes the erosion of floodplain or widening of the main channel in a compound channel. Goto *et al.* (2000) showed that 2D numerical analysis of flow and bed deformation can describe such degradation with widening by bank erosion.

When the water stage at the downstream drops suddenly in a stream with constant width (with rigid side walls), the bed-degradation propagates toward upstream (Ashida, 1971). In the case of a compound channel, a sudden drop of

Fig. 3.22 Laboratory experiment for bank erosion.

water elevation at the downstream end causes bed degradation and side-bank erosion of the floodplain. Because of the widening of the main channel, the upward migration of degradation is suppressed relative to conditions without any widening of the main channel. Goto *et al.* (2000) showed some examples from flume experiments, and they could explain the process by using 2D numerical analysis of flow and bed deformation (see Fig. 3.23).

3.4.4. Regime channels and overbank flow

In regime approaches, no explicit account has been taken of floodplain interaction with the main channel. Therefore, there is a need to develop prediction techniques which can be applied to two-stage, loose-boundary channel design, allowing for the effects of overbank flows (Valentine *et al.*, 2001).

In the context of compound channels, it is necessary to be aware of the underlying stability of the channel throughout the range of discharges which the channel

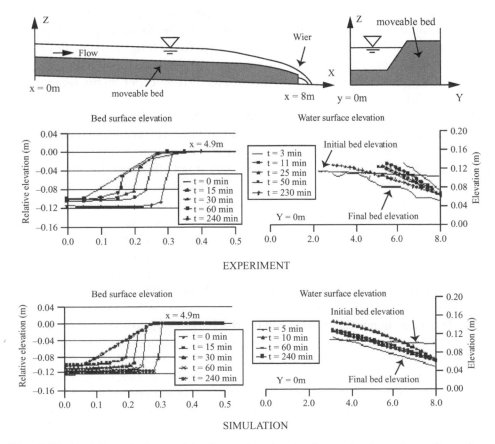

Fig. 3.23 Bed-degradation and bank erosion due to change in downstream boundary condition.

experiences. This section deals, therefore, with issues of stability relating to channels with mobile boundaries. Our knowledge is limited, however, both in terms of flow range and in terms of bed material type and size.

The limitations regarding flow rate concern the fact that most work has been carried out to test channel regime relationships where bankfull flow is viewed as the channel-forming discharge, and to laboratory conditions. Only recently have overbank flows been examined in the laboratory and it must be stated that there is a great need for further research, both in extending experimental work and in taking the investigation out into the field in well-structured studies which can overcome the problems of the complexities of field conditions.

This section considers primarily straight planforms, and is limited to information on uniform materials, subject to inbank/bankfull and overbank flows. There are very limited data on free planform meandering channels or in channels with graded sediment.

Ikeda (1981) performed half-channel asymmetrical experiments where the channel was effectively divided by the flume wall which suppressed meandering tendencies. This pioneering work demonstrated that a stable channel exists which allows bed-load transport in the central bed region. Experience of operating experiments at small scales (e.g. Valentine & Shakir, 1992a, 1992b; Babaeyan-Koopaei & Valentine, 1995) has shown that mobile equilibrium straight channels can be achieved in the laboratory but they are consistently wider and shallower than predicted, albeit with a similar cross-sectional area. These experiments represent a limited particle size and flow range.

The starting point for these experiments was the establishment of an equilibrium or near equilibrium channel with bankfull flow. The White, Paris & Bettess (1982) [WPB] method was first used to calculate the initial channel dimensions, given particle size, bed slope and discharge. A small allowance for the angle of repose of loose sediment in the banks was made when screeding the initial geometry, producing a trapezoidal channel cross-section of equal area and width to the prediction. In the above experiments, the uniform bed material sizes (1.3 mm and 0.9 mm respectively) were chosen such that bedforms were suppressed. Preliminary tests in the EPSRC Flood Channel Facility [FCF] (see Section 2.4.2) produced the same result. Initially prepared channel sections with higher width/depth ratios than predicted by the WPB method gave greater bank stability, at least in the range of bankfull discharges investigated.

The FCF experiments have extended the discharge range to 45 l/s. The previous experiments reported for stable, straight, free-form channels in uniform course sand were restricted to less than 10 l/s (Wolman & Brush, 1961; Ikeda, 1981; Shakir, 1992; Babaeyan-Koopaei & Valentine, 1995; Ayyoubzadeh, 1997). The small scale experiments are in the plane bed region but the FCF experiments had a smaller particle size ($d_{50} = 0.8$ mm) and exhibited two-dimensional sand waves, shown in Fig. 3.24.

Fig. 3.24 A general view of the FCF main channel showing typical developed two-dimensional sand waves and flow parallel ridges.

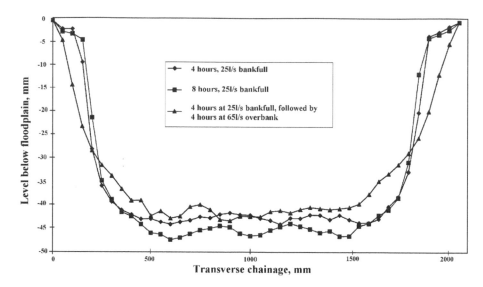

Fig. 3.25 Three final cross-sections for experiments with the 25 l/s bankfull channel with the same initial screed.

Figure 3.25 shows three final cross-sections. Noting the vertical exaggeration of the scales and the presence of dunes on the bed, the difference between the bankfull cases is insignificant despite the different experiment durations. The trapezoidal mean cross-section is also evident.

The bankfull, near-regime channels were again found to be wider and shallower than predicted by the WPB rational regime approach. This is consistent with experiments in smaller scale channels. The friction data for these mobile channels (Fig. 3.26) are consistent with theory, but it was found that sediment transport theory consistently underpredicts the measured rates (Fig. 3.27). A possible practical reason for this, which concerns the effect of the translation of the cross-sectional parameters from rectangular to trapezoidal, has recently been proposed by Valentine & Haidera (2001). These observations, of course, influence the rational regime predictions upon which the initial channel sections were based.

Thus the extremal hypothesis of the maximisation of sediment transport rate could not be tested unambiguously. Nevertheless, straight regime channel conditions were predicted reasonably well, although it was necessary to allow for bedform effects when assessing cross-sectional stability. The observed bedforms are consistent with the Simons & Richardson (1961) diagram of bedform domains. For these experiments they are located just above the plain bed domain and consist of two-dimensional sand waves.

Even when considering bankfull channels, topographic variability exists in the shape of individual cross-sections and also in the streamwise variation of both cross-sectional shape and planform geometry. The effects of these channel features may change markedly when the stream flows out of bank at times of flood. The high discharges associated with overbank flow may be particularly influential in

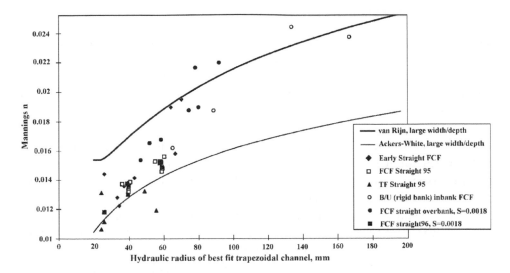

Fig. 3.26 Manning's *n* versus hydraulic radius, overview of Series C data.

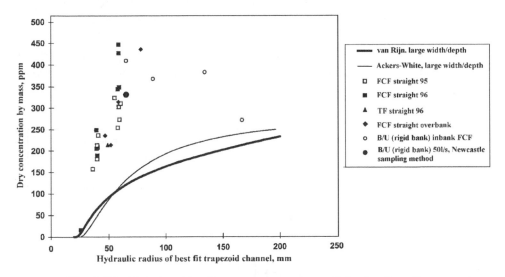

Fig. 3.27 Variation of sediment concentration with hydraulic radius.

producing geomorphological change due to higher, or, in some cases, lower than usual velocities and boundary shear stresses in the channel. This may lead to patterns of erosion and deposition that differ radically from those associated with in-bank flows.

As previously noted, the difficulties of taking measurements at field scale and the limited relevance of most, typically rather small-scale, laboratory experiments means there is a lack of data on overbank flows in mobile channels. However, limited data have been collected in two classes of experiments.

3.4.4.1. Fixed floodplain in the flood channel facility

These experiments were carried out to collect experimental data relevant to the regime behaviour of a straight, loose boundary (bed and banks) channel in a floodplain, to investigate the stability of the channel for overbank flow and bankfull discharges and to assess the performance of a rational regime method in predicting a stable channel geometry.

This work considerably extended the range of laboratory data, in terms of channel scale, discharge and width/depth ratio and represents the establishment of a methodology for overbank flow studies in loose-boundary channels. Considerable care was taken to ensure uniform quasi-steady flow conditions, as a basic requirement for assessing regime theory.

Figure 3.28 shows the variation of top width for channels of two different bankfull discharges and for overbank flows on previously formed bankfull channels. This demonstrates the consistent character of increasing channel width. In prototype conditions this is unlikely to occur due to the relatively short duration of a flood, but provides an indication of the instabilities caused by overbank flows.

The bank erosion rate was obtained by measuring width change over time and obtaining channel area from the automatic bed profiler at the end of the experiments. Bank erosion rate per unit length of channel, normalised with sediment transport rate per unit channel width, is plotted against width/depth ratio in Fig. 3.29, indicating that bank stability is greater in channels of higher width/depth ratio. The non-dimensional term provides a fairer comparison of stability between

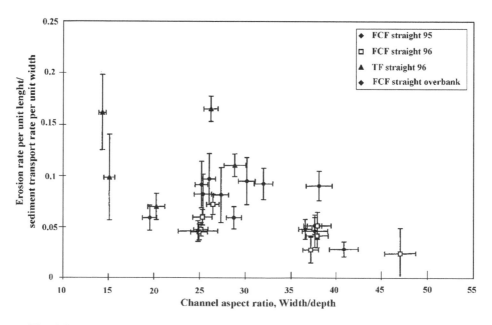

Fig. 3.28 Variation in top width for channels of bankfull discharge 25 and 45 l/s.

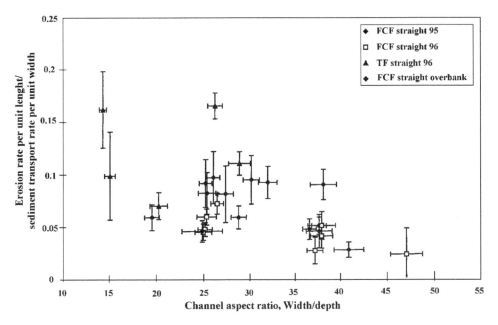

Fig. 3.29 Bank erosion rate versus width/depth ratio.

channels with different transport rates; a straight channel with a low transport rate would be expected to have a high degree of bank stability (Yalin, 1997). Experimental error is shown for 60% confidence limits resulting from propagation of errors in the input terms. Erosion rate along one bank per unit length is given as a function of stream power per unit area, non-dimensionalised by acceleration due to gravity, in Fig. 3.30.

This approach was prompted by Nanson & Hicken's (1986) analysis of field data on the expansion of meandering channel cutbanks. That analysis found a stronger link between erosion rate and stream power per unit length than stream power per unit area but the opposite was found for this study. It is important to note that the erodibility of the bank sediment has been ignored but, however this may be defined, it should be constant as the experiments were all performed with the same loose sediment. Of particular interest, erosion rates during overbank flows are similar to those for bankfull flows of the same stream power per unit area.

The scatter of data points in Figs. 3.29 and 3.30 is larger than can be accounted for by the estimated experimental error. This implies that width/depth ratio and stream power per unit area are not the only important factors affecting bank erosion rate. Stream power is dissipated mainly within the flow, rather than as work done on the bed and banks, and the proportion of this power expended on erosion or sediment transport is likely to depend in a complex way on such factors as flow symmetry and Reynolds Number.

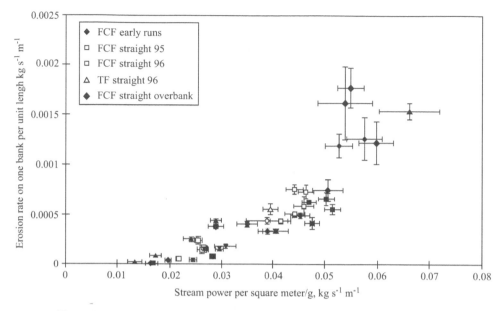

Fig. 3.30 Bank erosion rate as a function of stream power per unit area.

This is a weakness of using stream power as an independent variable affecting the boundary. However, tests using the alternative variables of shear stress, shear velocity and sediment transport per unit width (as predicted by the Ackers & White method) showed even greater scatter.

Overall, the results confirmed a good ability to predict stable channels for inbank flows but indicate that revised approaches are necessary for two-stage loose-boundary channels subject to significant overbank flow. Overbank flow stability was found to be poorer than the bankfull case. Rates of bank erosion were measured and constitute unique widening data for the main channel under overbank flow conditions. They show a much greater rate of widening than for the bankfull flow channels. These data may provide a basis for estimates of channel bank erosion in prototype channels, although it remains to be established how this can be expressed in the context of the prototype.

3.4.4.2. Floodplain flow velocities below tractive shear stress

In a subsequent limited set of small scale experiments (Haidera & Valentine, 1999) with a uniform sand of $d_{35} = 0.9$ mm, it was demonstrated that stable bank full flow channels, when subjected to overbank flows, adjusted to a new stable cross-section. It was observed from the rates of change and flow calculations that an overbank alluvial channel (compound channel) will adjust to reduce the interaction effect between the main channel and the flood plain by increasing its width, decreasing its depth and increasing side slope as it approaches single unit

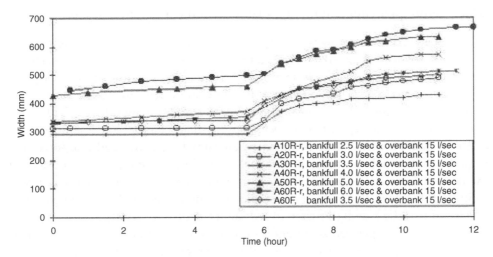

Fig. 3.31 Rates of widening during bankfull and overbank flow.

behaviour. The channel widening rates are shown in Fig. 3.31. The flow friction and sediment transport rates are affected by the cross-section adjustment.

3.5. Floodplain Processes

3.5.1. Introduction

Overbank deposition of sediment from river channels has important implications for floodplain development, deposition of contaminants adsorbed to sediment particles and the creation of future sediment sources for the river channel (e.g. Macklin & Klimek, 1992; Walling *et al.*, 1996). Given that the transfer of suspended sediment to, and its deposition on, the floodplain are affected by the interaction of channel and overbank flows, and that such interaction varies with channel planform, the deposition pattern may similarly be expected to vary with planform. It may, therefore, be altered by channel engineering, for example through channel straightening or through returning previously straightened channels to a more natural meandering state.

Field study has provided valuable information on rates of overbank deposition at both the event and longer time scales. Simm (1995) summarises nearly 40 such studies. New and imaginative techniques are also providing information on deposition patterns on a case study basis (e.g. Macklin & Dowsett, 1989; Walling *et al.*, 1992; Asselman & Middelkoop, 1995; Simm, 1995). However, the patterns so derived are typically site-specific, depending on factors such as micro-topography, inundation sequence and floodplain vegetation type and density (e.g. Nicholas & Walling, 1997; Gomez *et al.*, 1997). The patterns may also vary with the characteristics of each flood event.

3.5.2. Floodplain deposition due to interaction with main channel

When the main channel flow is accompanied with flow over the floodplain or flow in a vegetated zone, intensive turbulent mixing in a lateral direction causes lateral net transport of suspended sediment from the main-channel flow to the floodplain or the vegetated zone. Thus, even in a straight channel, sediment deposits on the floodplain or vegetated zone as a ridge. Basically, this is a lateral turbulent diffusion due to the velocity difference between the main channel and the floodplain or the vegetated zone. More sediment is entrained from the bed and/or transported longitudinally in the main channel, while less sediment is transported on the floodplain or the vegetated zone. Thus, there is a difference in sediment concentration between the main channel and the floodplain or the vegetated zone, and it reduces the lateral diffusion of sediment (Ikeda *et al.*, 1991).

Behaviour of suspended sediment is described by a diffusion-convection equation. In general, suspended sediment with moderate concentration scarcely influences the turbulent structure, and the flow-sediment system can be analyzed by a one-way calculation: The flow is analyzed, and then the sediment behaviour is analyzed. For example, the diffusion coefficient of suspended sediment is evaluated from the kinematic eddy viscosity. On the other hand, the bedload behaviour is governed by the bed shear stress and bed morphology because its space is limited to the near-bed layer.

In a compound channel, an interaction between main channel and floodplain is complicated and three-dimensional. Recent refined measurement and numerical analysis have clarified the characteristics of flow there. The time-averaged structure is characterized by an oblique upward motion from the edge of the floodplain to the main channel, which forms a cell with a longitudinal axis (Tominaga & Nezu, 1991). Such a time-averaged flow 3D structure is fairly well described by numerical analysis using a turbulence model such as an algebraic stress model (Naot *et al.*, 1993). It is not steady, but an instantaneous structure is not the same as the time-averaged one. Some researchers have tried to clarify a series of instantaneous structures by using a technique of LES (Thomas & Williams, 1995).

The non-submerged vegetation makes flow horizontally two-dimensional. For example, if the floodplain is covered by tall vegetation, the flow even in a compound channel is horizontally two-dimensional, and depth-averaged analysis becomes applicable. Even if the vegetation on the floodplain is partial but it covers the zone near the floodplain edge, such an effect is marked.

The flow in a channel with vegetation zone is rather two-dimensional, but it accompanies organized fluctuations with a low frequency, which is developed by instability of the horizontal shear flow. Such a low frequency fluctuation is still significant near the bed and it causes a fluctuation in bed-load direction. Since bed-load concentration in the main channel is higher than that in the vegetated zone, the fluctuation causes the net lateral transport of bed-load toward the vegetated zone, and a longitudinal sand ridge is formed near the vegetated zone (Tsujimoto, 1998). In the case of graded sediment, longitudinal sorting (a fine sand ridge) is developed along the edge of the vegetated zone. When the water stage is low, such a longitudinal ridge near a shoreline is often observed.

3.5.3. Floodplain deposition around vegetation

Since the floodplain is likely to be dry because of low frequency of submergence, vegetation often forms colonies (Fig. 3.32). When an isolated vegetated area is studied, the overbank flow is retarded around it then accelerates to recover velocity downstream. Thus during a flood, isolated vegetation captures fine sediment involved in the overbank flow, and it is deposited inside and downstream of the vegetated area. Figure 3.33 shows a fundamental flume experiment, where isolated vegetation is simulated by a cube made of entwined fibres. After the flood, deposition of fine sediment in general includes fertile soil suitable for the vegetation to invade and the vegetated area is enlarged. Instead of natural growth, additional units of the simulated vegetation were set in the experimental flume to investigate the growing process of the vegetated area (Tsujimoto, 1999). The study clarified that the isolated vegetated area on the floodplain spread downstream because of repeated flood and low-water stages. Though the bed material of the floodplain is rarely transported by overbank flow, morphological change takes place during major flood with intense overbank flow. The bed load deposits

Fig. 3.32 Vegetation colony on a floodplain.

upstream of the vegetated area (with some local scour just in front of the vegetation) and the sides of the vegetated area are eroded. This results in the vegetated area resembling a mound. The vegetation then becomes more firmly established and the vegetated mound is enlarged. A vegetation colony in the floodplain is often orientated longitudinally, and a sufficiently long vegetated zone creates lateral exchange of momentum of flow and suspended sediment. As a result, sand is deposited along the vegetated zone, and it grows gradually with fine and fertile soil being deposited.

Vegetation on flood plains retards the flow, which produces a difference of fluid velocities between the main channel and the flood plains. This difference yields instability of flow to generate lateral organized vortices at the boundary of the main channel and the flood plains (see Chapter 2). The situation also yields lateral transport of suspended sediment from the main channel toward the flood plains. A smaller bed shear stress at the flood plains induces deposition of fine sediment on the flood plains.

Vegetation on the flood plains in two-stage channels is typically seen in mangrove swamps. Such mangroves exist in Ryukyu Islands Chains which locate in the South-West part of Japan. A series of observations has been performed in a mangrove swamp which is located at the river mouth of the Nagura River in Ishigaki Island in 1999 and 2000 (see Fig. 3.33). The river has a typical width of the main channel of 30 m, and the total width of the flood plains is about 100 m. The

Fig. 3.33   Fine sediment deposited downstream of modelled vegetation.

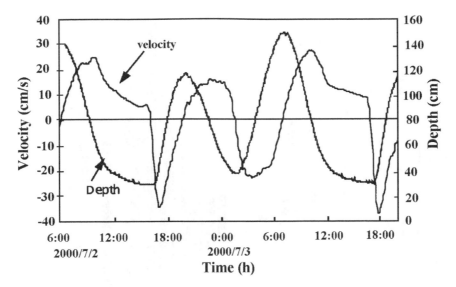

Fig. 3.34 Temporal variations of the longitudinal flow velocity component and the depth in Station B (at spring tide).

difference between the height of the bed elevation and that of flood plains ranges from 100 to 150 cm. The area is affected by tide, and the flood plains are inundated at high tide and the water disappears at low tide.

The flow rates associated with the tides were measured by using an electro-magnetic velocimeter at a station B in the main channel shown in Fig. 3.35. The

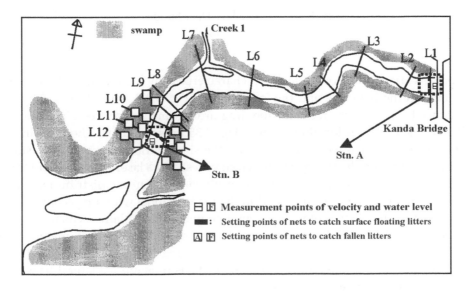

Fig. 3.35 Map of the Observation Area.

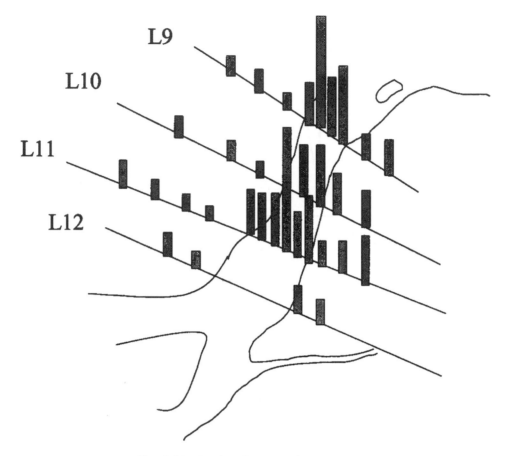

Fig. 3.36 Median Size Distributions.

temporal variation of flow velocity is characterized by a pronounced asymmetry between the high tide and the low tide (see Fig. 3.33). This is caused by the resistance to flow associated with the drag around mangrove trees. The surface soils were sampled at locations shown Fig. 3.35 by solid lines. The soils in the surface layer of 30 cm were sampled, and the size distributions were measured after drying the soils by using traditional sieve test and a laser measurement technique in tandem. Figure 3.35 shows the results of median size distributions. It is clear that the sediment is coarse sand in the main channel with a typical value of $d_{50} = 1.2$ mm. The sediment is, however, fine in the flood plains, which shows a typical value of 0.5 mm. The observation by eye shows that the bed materials in the flood plains include a lot of muddy soils ranging from clay to silt. The shallower depth and the vegetation in the flood plains thus induce lateral transport and lateral sorting of bed materials.

3.5.4. Floodplain processes by overbank flow

A compound channel is often characterized by the existence of a main channel and floodplain between levees. In the floodplain, there is much interesting micromorphology: secondary channels, side pools, dead zones, and so on. These are sometimes associated with vegetation, and vegetation influences fluvial processes related to these morphological features (Tsujimoto, 1999). Furthermore, these provide favourable habitats for many organisms which contribute to the fluvial ecosystem. In particular, the space in a compound channel plays the ecological role of the original floodplain of a river without levees. A characteristic of such morphology, in terms of habitat, is that they are sometimes submerged and sometimes dry, and that they are sometimes developed and sometimes destroyed as part of the riverine processes.

When overbank flow returns to the main-channel flow with some level-gap, gully head-cutting takes place, particularly in a meandering compound channel or a straight channel with alternate bars (Gay *et al.*, 1998). Gully head-cut is erosion migrating upstream (retrogressive erosion). In incised channels, head-cut processes have been focused on by many researchers (e.g. Bennett *et al.*, 2000), but head-cut is also one of elementary events in floodplain process (Fig. 3.37). Into the head of the head-cut (gully erosion), the flow is concentrated to a head, and an impinging jet makes a scour hole. Then, the upstream slope falls down into

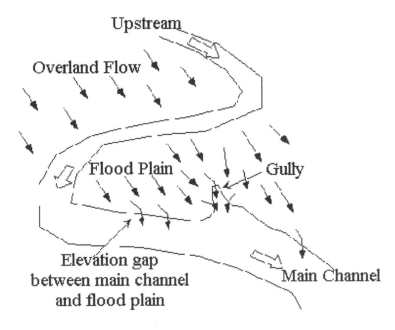

Fig. 3.37 Gully Headcut in Flood Plain.

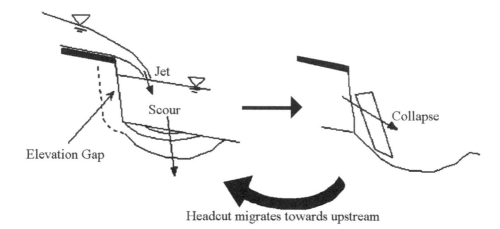

Fig. 3.38 Headcut Erosion.

the scour hole, and the head-cut head migrates upstream (Fig. 3.38). There is little work on two-dimensional head-cut (Izumi, 1993; Dey *et al.*, 2000).

3.6. Concluding Remarks

The flow structure of overbank flow in meandering river channels is determined to a large degree by secondary circulation due to interaction between channel and floodplain in the cross-over zone and centrifugal circulation around the bend. This significantly alters the distributions of velocity and bed shear in the channel. If the channel bed is mobile, then it will react to these changes by erosion or deposition, which will itself feed back into further re-distribution of shear stresses. Moreover, if the bed material is graded then selective fractional movement will lead to sorting and possibly armouring which further complicates the situation.

There is a wealth of laboratory and field data on inbank and bankfull flows in mobile bed meandering channels. However, data on these channels under overbank flow conditions are much rarer. The difficulty of collecting such information under field conditions, makes laboratory experiments more central to these investigations.

Experiments on rigid planform meandering channels have shown that flow structures typical of real rivers, such as riffles and point bars, can be successfully reproduced in the laboratory. These experiments confirm that overbank flow will significantly change the bed morphology in graded sediments and the gross flow structures causing this are largely understood. These complex flow structures have distinct affects upon sediment transport rates which, in turn, are beginning to be revealed if only qualitatively. Certainly it has been confirmed that sediment transport rate is a maximum at bankfull. It then falls significantly as the channel goes overbank and reaches a minimum at depth ratio of around 0.2 before increasing at higher depths. However, the complexity of the flow makes the quantitative determination of sediment transport rate very difficult, since local bed shear stress is affected significantly by secondary flows and cannot be determined simply from primary velocities. Current bedload and total load formulae are inadequate under these circumstances. It is conceivable that two or three dimensional numerical models may be required in order to capture the variations in bed shear that are necessary for accurate determination of sediment transport rate. Alternatively some empirical or semi-empirical adaptation of existing sediment transport formulae may be developed as a practical design tool.

Most of the experiments carried out to date use fixed planform channels with rigid banks. The banks of natural rivers are of course erodible, although often to a lesser degree than the bed material. Meandering rivers that are free to adjust their width, bank slope and planform have been little studied under overbank flow conditions, and it is inevitable that these extra degrees of freedom will alter to some degree the flow patterns observed for fixed planforms.

Perhaps the most significant gap in our current knowledge is the rate at which changes due to overbank flows evolve. Experimental measurements to date have largely concentrated upon equilibrium flow conditions. It must be recognised, however, that most rivers only come out of banks for relatively short and infrequent periods. These may not be long enough for equilibrium to be achieved. Particularly in graded bed rivers, morphological adjustments may depend upon the current state of the bed surface and sub-surface. This will have been built up over a history of flow events. Thus the frequency and magnitude of overbank events set against the adjustment and recovery rate of the river could well be the determining factors in the development of river morphology.

Chapter 4

Computer Simulation

Y. Kawahara, S. Fukuoka, S. Ikeda, G. Pender and Y. Shimizu

4.1.   Introduction

The use of computer simulation in river engineering is so popular that it has become an essential tool in the discipline, complementing the other analysis methods tools of experiment and theory. This is due to the relatively recent development of numerical methods and the rapid advances in computer technology.

One of the most important aspects of computer simulation is its impact on design. This is related to the decreasing cost of computations compared to the increasing cost of physical modelling. Indeed, decisions for many types of river engineering work are now based exclusively on results from computer model predictions. The role of computer modelling in design has a corollary in basic research. Assuming that a given computer model contains all the important physics then it can be used to undertake numerical experiments to help study the fundamental characteristics of the flow.

Inherent in the above observation is the assumption that computer simulations are accurate as well as cost effective; otherwise replacing physical modelling by computer simulation would be foolish. The results of computer simulations are only as good as the physical laws incorporated in the governing equations and boundary conditions. From the discussions in the preceding chapters it is clear that these continue to require improvement.

In the following, important points relating to computer simulation methods are briefly reviewed before more detailed information is presented in later sections. Computer methods consist of [1] mathematical models (the governing equations) [2] discretization methods (finite difference, finite volume and finite element methods), [3] co-ordinate and basis vector systems (cartesian and curvilinear non-orthogonal coordinate), [4] the numerical grid (structured, block-structured and non-structured), [5] the solution method (methods for solving large systems of equations), [6] convergence criteria (criteria for the convergence of iterative methods). These components all impact on the cost and accuracy of the simulations. Hence, users of computer codes and researchers developing new codes have to be aware of their overall structure and the inherent advantages and disadvantages of the different methods.

In this chapter, some of the mathematical models used to simulate flow in one-, two- and three-dimensions are described, this is followed by a discussion of selected discretization methods, finally several recent example applications are presented.

4.2. Flow Modelling

The most general mathematical expression of the physical laws (conservation of mass and momentum) governing river flow are the continuity equation and the Reynolds Averaged Navier–Stokes (or RANS) equations. These can be written as:
Continuity Equation:

$$\frac{\partial U_j}{\partial x_j} = 0 \tag{4.1}$$

Conservation of Momentum in X_i Direction:

$$\frac{\partial U_i}{\partial t} + \frac{\partial U_i U_j}{\partial x_j} = g_i - \frac{1}{\rho}\frac{\partial P}{\partial x_i} + \frac{\partial}{\partial x_i}\left(\nu\frac{\partial U_i}{\partial x_j} - \overline{u_i u_j}\right) \tag{4.2}$$

Here, U_i and u_i are the mean and fluctuating velocity, P the mean pressure, ρ the density, ν the kinematic viscosity, g_i the component of gravitational acceleration, $\overline{u_i u_j}$ the Reynolds stress.

Techniques for obtaining a full three-dimensional solution to these equations are discussed in Section 4.2.4; however, they are often employed in river engineering in a simplified form. The commonest practice is that the vertical momentum equation is reduced to a statement of hydrostatic pressure, i.e. negligible vertical motions. Computer models that employ equations based on a hydrostatic pressure assumption are termed shallow water models. There are three alternatives for modelling shallow water waves: area averaged models, depth averaged models and quasi-three-dimensional models. Each of these will be discussed in more detail in the following subsections.

4.2.1. Area averaged flow models

These models are used widely for the simulation of flood wave propagation over relatively long lengths of rivers. There are several computer packages available to undertake this type of simulation. In addition to the hydrostatic pressure assumption, the models assume that only the streamwise component of velocity is important so that the conservation of mass and momentum can be described by two equations applied along the river's longitudinal axis. Yen (1973) provides a formal derivation of the physical laws often referred to as the St Venant equations. They can be written as

Conservation of Mass:

$$B\frac{\partial \eta}{\partial t} + \frac{\partial Q}{\partial x} = q \qquad (4.3)$$

Conservation of Momentum:

$$\frac{\partial Q}{\partial t} + \frac{\partial}{\partial x}\left(\frac{\beta Q^2}{A}\right) + gA\frac{\partial \eta}{\partial x} + \frac{A\bar{\tau}_b}{\rho R} = 0 \qquad (4.4)$$

where Q is the river discharge, η is the water elevation, B is the flow top breadth, q is the lateral inflow per unit length, A is the flow area, R is the hydraulic radius and $\bar{\tau}_b$ and β are empirical parameters which must be specified. $\bar{\tau}_b$ is the mean boundary shear stress, but perhaps should be renamed the mean effective boundary shear stress, because in practice it is rarely determined by the value of bed roughness alone. It is normally replaced in the above equation by some functional relationship, which relates it to the river discharge. β is the momentum correction factor, which accounts for the fact that the velocity distribution over the cross-section is non-uniform. Together $\bar{\tau}_b$ and β represent the effects of turbulence in a lumped parameter sense.

Boundary conditions usually consist of specified flow hydrographs at upstream boundaries and specified water stages or rating curves at downstream boundaries, although alternative boundary conditions are possible. These equations when furnished with suitable initial conditions may be solved by a numerical method to give flows and stages throughout the modelled river reach, Cunge, Holly & Verwey (1980) give a fuller description.

These models are well proven and work well for calculating the flows and water levels during the propagation of floods through the river system. They do not, however, give much detailed information at local points on the river. At certain river sections, such as bridges, weirs, local expansions or contractions, Equations (4.3) and (4.4) do not apply and empirical laws must be used.

4.2.2. Depth averaged flow models

When both horizontal velocity components are important and strong vertical mixing is promoted by bed roughness, a depth averaged form of the RANS equations is appropriate. The depth averaged equations may be written in the Cartesian coordinate system as:

Conservation of Mass:

$$\frac{\partial \eta}{\partial t} + \frac{\partial \bar{U}h}{\partial x} + \frac{\partial \bar{V}h}{\partial y} = 0 \qquad (4.5)$$

Conservation of Momentum in X-Direction:

$$\frac{\partial \bar{U}h}{\partial t} + \beta\left[\frac{\partial}{\partial x}\left(\bar{U}\bar{U}h\right) + \frac{\partial}{\partial y}\left(\bar{U}\bar{V}h\right)\right] = F_x - gh\frac{\partial \eta}{\partial x} - \frac{\tau_{bx}}{\rho} + \frac{1}{\rho}\left[\frac{\partial}{\partial x}T_{xx} + \frac{\partial}{\partial y}T_{xy}\right] \qquad (4.6a)$$

Conservation of Momentum in Y-Direction:

$$\frac{\partial \bar{V}h}{\partial t} + \beta \left[\frac{\partial}{\partial x} \left(\bar{V}\bar{U}h \right) + \frac{\partial}{\partial y} \left(\bar{V}\bar{V}h \right) \right] = F_y - gh\frac{\partial \eta}{\partial y} - \frac{\tau_{by}}{\rho} + \frac{1}{\rho} \left[\frac{\partial}{\partial x} T_{yx} + \frac{\partial}{\partial y} T_{yy} \right]$$

(4.6b)

Here, η is the water elevation, $\bar{U}$ and $\bar{V}$ are depth-averaged velocities in the horizontal directions, β is the momentum correction factor, h is the water depth, τ_{bx} and τ_{by} are the bed shear stresses in the horizontal directions, F_x and F_y represent external body forces and T_{xx}, T_{xy}, T_{yx} and T_{yy} are the horizontal stresses.

Appropriate boundary conditions for the solution to the above equations are the specification of flow or water elevation, some combination of the two or some relationship describing the flow at water/land interfaces (zero flow normal to the land and some relationship between flow and stress tangential to the land).

The computational effort to carry out such modelling is significantly higher than in one dimension, but resulting simulations are able to represent flow events such as the arrival and extent of a flood, dam break and flow separation (recirculation and dead zones). The first models of this type used finite difference techniques (Leendertsee, 1967; Vreugdenhil & Wijbenga, 1982) and the method of characteristics (Townson, 1974). However, these methods suffered from the need to apply them on a Cartesian computational grid where river bends require to be approximated using a staircase like boundary. With the development of the finite element method (Brebbia *et al.*, 1978; Wang *et al.*, 1989) and associated numerical techniques (Brookes & Hughes, 1982; Hervouet, 1991), alternative codes were produced, such as RMS-2 (King & Norton, 1978) in the early 1970s and TELEMAC-2D in the late 1980s (Hervouet, 1991). The finite element method proved useful in representing complex geometries; however it is a demanding method to implement numerically and possesses mass-conservation difficulties. A hybrid of both finite difference and finite element methods the finite volume method therefore emerged. Initially, this was used to solve the full Navier-Stokes equations (Patankar & Spalding, 1972; Demirdzic *et al.*, 1987; Karki & Pantakar, 1988). It is conservative, numerically accurate and relatively simple to apply. It has been applied to model two and quasi three-dimensional flow problems (Lai & Yen, 1992). Recent two-dimensional simulation research has extended its application to flood events and dam break problems, in particular considerable work has been undertaken related to the treatment of wetting-and-drying and adaptive meshing (Lynch & Gray, 1980; Akanbi & Katopodes, 1988; Molinaro & Natale, 1994; Tchamen & Kawahita, 1988). In particular, the work by Bates *et al.* (1993, 1996) has illustrated the capability of the finite element code TELEMAC-2D to reproduce the dynamic behaviour of a flood wave and the corresponding inundation envelop. Such codes can now support river models of up to 60 km (Bates *et al.* 1996). Other researchers such as Wijbenga (1985), Akanbi & Katopodes (1988), King & Roig (1991), Paquier & Farissier (1997), Sleigh *et al.* (1998), Markhanov *et al.* (1999) have also reported the use of two-dimensional modelling to flood simulation. These papers have mainly reported on the dynamic flooding in the plan view.

Depth averaged flow models work well where the flow is nearly horizontal and vertical motions are negligible.

4.2.3. Quasi-three-dimensional flow models

The equations used in these models are the continuity equation, the momentum equations in x and y directions and the hydrostatic pressure relationship. Most applications of these models have been for water bodies with large horizontal dimensions compared to the depth, such as estuaries and lakes. However, some attempts have been made to simulate velocity fields in rivers (Benqué et al., 1982; Blumberg et al., 1990) and even flood flows. Laverdrine (1996, 1997) attempted to simulate a Flood Channel Facility experiment, but only very brief comparisons were presented. In cases where sediment transport was investigated, the limitation of the hydrostatic pressure assumption in failing to reproduce pressure driven recirculations becomes clear (Shimizu et al., 1990) as does the techniques inability to predict the flow distribution at the inflection point of a bend in a flooded channel with a high effective depth ratio (Fukuoka & Wanatabe, 2000).

4.2.4. Three-dimensional flow models

The equations upon which this type of model is based have been presented previously as Equations (4.1) and (4.2). In these non-hydrostatic pressure models the vertical momentum equation is retained in its entirety. This introduces two problems:

- How to solve for pressure?
- How to model variations in the location of the free surface?

Concerning the first problem, various algorithms for the pressure solution have been proposed. Patankar & Spalding (1972) and Patankar (1980) describe what has come to be known as the SIMPLE algorithm. The algorithm has several offshoots (SIMPLER, SIMPLEC, PISO) but they all work on approximately the same principle, described by Fletcher (1988). Using a guessed pressure field the momentum equations are solved to estimate the velocity field. The amount by which this velocity field deviates from satisfying the continuity equation is then used as a source for calculating a correction to the guessed pressure field, which in turn is used to make a better attempt at solving the momentum equations. This procedure is repeated until the solution converges.

There are two commonly used treatments of the free surface. The first approach is the introduction of a rigid lid approximation in which the free surface is replaced by an artificial plane surface parallel to the horizontal axes at some defined equilibrium water elevation. A non-zero pressure is allowed to develop on the surface and this in some way represents the height of water that would be present if the surface was completely free. Thus, the effect of the free surface on the internal flow is captured. However, it will be appreciated that there is some amount of fluid above the rigid lid that is omitted. McGuirk & Rodi (1977) and Leschziner & Rodi (1979) suggest that the rigid lid approximation is valid as long as the super-elevated regions are not greater than 10% of the total channel depth, but clearly more research is required in this area to ascertain the limitations of this method for various flow configurations. In this regard the work of Morvan *et al.* (2000) is noteworthy.

The second commonly used alternative to this is a surface tracking procedure, which has been proposed in various forms and is computationally much more demanding. Most free surface tracking models are related in some way to the original work of Harlow & Welch (1965). In this method, which became known as the Marker & Cell (MAC) method, they suggested that the fluid could contain marker particles whose co-ordinate position is recorded. These particles can then be moved according to local fluid velocity components. Using the new co-ordinates of these particles a new fluid region can be assigned in terms of the Eulerian computational mesh. This procedure was found to be too computationally demanding and subsequently was replaced by the Volume of Fluid (VOF) method (Nichols, Hirt & Hotchkiss, 1980). Instead of marker particles the VOF method assumes each computational cell to contain some fraction of fluid, F. Thus full cells have F equal to one and empty cells have F equal to zero. Cells with some fluid have F less than one but greater than zero. The F variable can then be advected as a Lagrangian invariant according to,

$$\frac{\partial F}{\partial t} + u\frac{\partial F}{\partial x} + v\frac{\partial F}{\partial y} + w\frac{\partial F}{\partial y} = 0 \tag{4.7}$$

where u, v, w are the local fluid velocities. Once F is known throughout the domain then the mesh can be reconfigured. The advantage of the MAC and VOF methods is that they allow arbitrary unrestricted deformation of the fluid including the case of wave breaking. Davis & Deutsch (1980) have applied the VOF technique to the simulation of free surface flow through a Parshall flume.

If the simulation of wave breaking is not required then a simpler procedure is the height function method (HF) by Hirt & Nichols (1981), in which a new free surface is computed as the height, H, above some datum by the following expression,

$$\frac{\partial H}{\partial t} + u\frac{\partial H}{\partial x} + v\frac{\partial H}{\partial y} = w \tag{4.8}$$

at each time step followed by reconfiguring of the mesh. Surface tracking approaches require fine numerical grids, which leads to high computer costs. Their application in river modelling is relatively rare.

Simulation of natural open channels using fully three-dimensional models has been recently discussed by Knight & Shiono (1996), Neary *et al.* (1999) and Bettess & Fisher (1999). It has long been known that flow features such as turbulence or the flow dynamics at bends or in flooded rivers are three-dimensional. Recent results regarding sediment transport have also indicated the importance of pressure (Fukuoka & Wanatabe, 2000). Yet little research work has been undertaken regarding the implementation of fully three-dimensional modelling techniques to river flows (Sinha *et al.*, 1998), and most flood flow models have involved simple prismatic channels (Krishnappan & Lau, 1996; Morvan *et al.*, 2000). Probably because there was no demand for such detailed investigation from the river engineering community, it is very expensive, to implement and such refined modelling is in contradiction with the uncertainty in defining other parameters necessary to describe a river's behaviour, such as total flow. Recently, Morvan, (2001), simulated flood flows in the River Severn, UK, using the three-dimensional code CFX and validated his results against available field data. Some recent reports from industry (Bettess & Fisher, 1999) and academia (Swindale, 1999) indicate that three-dimensional modelling could be of practical use in river restoration projects to aid with biological and morphological investigations.

4.2.5. Fully developed flow models

Fully developed flow models are models in which there are no changes in the longitudinal direction. Therefore, they compute the velocity, turbulence and shear stress distribution for a given cross-sectional shape, boundary roughness and uniform bed slope. These models have found application in river engineering providing practical predictions as well as enlightenment on the flow and turbulence mechanisms in natural and man-made river channels. They are strictly only applicable to fully developed flow conditions when the channel is long and straight enough to allow a longitudinal equilibrium (uniform flow) to be reached. They provide much better predictions of gross flow properties for two-stage river channels than traditional uniform flow formulae. It should be noted, however, that their range of application is strictly no greater than uniform flow formulae. It should be possible to introduce ad-hoc correction factors to take account of non-uniform flow as can be done for the traditional formulae, however, this should be done cautiously. These models are based on simplified forms of either the three-dimensional equations or the two-dimensional depth averaged equations. They are simplified to the case of fully developed steady flow in a prismatic channel by assuming that all velocity derivatives in the x-direction are zero as are all temporal derivatives. If the secondary motions are also ignored (usually in depth averaged models) then the lateral and vertical velocities can be taken to be zero. Haque (1959) gives a formal discussion of the fully developed depth averaged flow problem while Gosman & Rapley (1980) give an excellent account of the general principles of the fully developed three-dimensional flow problem.

This type of model has been applied to compound channel flow by Krishnappan & Lau (1986), Kawahara & Tami (1989), Prinos (1990), Naot, Nezu & Nakagawa (1993), Lin & Shiono (1992), Prinos (1992) and Cokljat (1993) among others. It has also been applied to simple channels by Naot & Rodi (1982) and Younis & Abdellatif (1989) who considered its effect on sediment transport predictions. Nearly all applications of this type of model have used the SIMPLE algorithm suitably modified for the parabolic flow case.

4.2.6. Flow resistance due to vegetation

Trees and other vegetation in rivers add considerable resistance to flows and reduce the river's capacity to convey flood flow. They may also change the sediment transport processes. Hence the hydraulic effect of vegetation has been investigated extensively over the last decade.

Vegetation zones show two distinct features. At first, a markedly low velocity area is produced due to large form drag. Fine sediment may deposit in this region and may not be resuspended. Secondly, this slower flow may move laterally, owing to the lateral pressure difference, it then interacts with the surrounding faster flows, resulting in the expansion of low velocity region.

When computer simulation is undertaken of a heavily vegetated river this must be taken into account in the model. In this case the momentum equation contains an additional term compared to Equation (4.2).

$$\frac{\partial U_i}{\partial t} + \frac{\partial U_i U_j}{\partial x_j} = g_i - \frac{1}{\rho}\frac{\partial P}{\partial x_i} + \frac{\partial}{\partial x_i}\left(\nu\frac{\partial U_i}{\partial x_j} - \overline{u_i u_j}\right) - F_i$$

(4.9)

$$F_i = C_d b U_i \sqrt{U^2 + V^2 + W^2}$$

where F_i is the form drag in x_i direction, C_d is the drag coefficient, b is the leaf area density (leaf area in m^2 per m^3 of vegetation stand volume, characterizing the degree of filling of the volume by leaves, trunks, branches, etc.)

It should be noted that the mean quantities in Equation (4.9) represent the values averaged not only over a time interval but also over volume. Since differentiation in space and averaging are not commutative, the averaging procedure introduces several unknown terms. One of these is the spatial average of the gradient of fluctuating pressure, for which Wilson & Shaw (1977) gave the expression of F_i to show form drag. Following this approach Tsujimoto & Shimizu (1994) discuss the suspended sediment transport in the presence of vegetation over a flood plain.

Many researchers, for example, Shimizu & Tsujimoto (1995) among others, have demonstrated that two-dimensional models work well when the vegetation effects are expressed in a similar way to those in a three-dimensional model. The required functions to introduce form drag are given below.

$$F_x = C_d b h U \sqrt{U^2 + V^2}, \quad F_y = C_d b h V \sqrt{U^2 + V^2}$$

(4.10)

where h is the local water depth, U and V the depth averaged velocity components.

4.3. Turbulence Modelling

In this section, two approaches to modelling turbulent flows are discussed with the emphasis on the application to compound channels.

One method employs turbulence models for the Reynolds stresses appearing in the RANS equations to close the set of equations. Such turbulence models can be classified with varying degrees of complexity from simple eddy viscosity models to the Reynolds Stress equation models in which the turbulent stresses are determined from their transport equations. Local properties of flows in compound open channels are mainly affected by gravity, bed shear stress, and resistance due to vegetation and water depth variation. Hence the sophisticated turbulence models are only necessary when turbulent stresses play the crucial role in the momentum transport. In other words, very simple turbulent models work well when turbulent stresses are of minor importance and as long as bed shear stress is estimated correctly. Thus, the selection of a suitable turbulence model depends on the information required, the necessary accuracy and the cost. It should be noted that transport of suspended sediment, heat or water-quality variables usually requires refined turbulence models because the turbulence diffusion process is one of the dominant transport processes, which is different from momentum transport.

The second method solves all the fluid motions that are larger than the numerical grid size with a simple model for the subgrid scale motions, based on the spatially averaged Navier-Stokes equations. This approach is called Large Eddy Simulation (LES) and can offer detailed flow information at the expense of a large amount of computer resources. The advantage of LES is to resolve large scale motions that carry a large portion of energy and exert a significant influence on flow resistance and sediment transport. Large scale motions interact with complex river geometry and show strong anisotropy, which turbulence models find difficult to model. Extension of the LES concept to river flow is also described. Recalling the large width to depth ratio of river flows we may expect that large organized planform vortices that develop between the main channel and flood plain or along the interface of the vegetated zone can be captured by simplified two-dimensional LES assuming that the flow variation over the depth is small.

4.3.1. Eddy viscosity models

The first attempt at turbulence modelling was to directly correlate the Reynolds stresses with the mean strain rate through the eddy viscosity in a manner analogous to molecular diffusion. The eddy viscosity ν_t is defined as,

$$-\rho\overline{u_i u_j} = \rho\nu_t \left(\frac{\partial U_i}{\partial x_j} + \frac{\partial U_j}{\partial x_i} \right) - \frac{2}{3}\rho k\delta_{ij}, \quad k = \frac{1}{2}\overline{u_i u_i} \qquad (4.11)$$

where k is the turbulent kinetic energy, δ_{ij} the Kronecker delta.

Thus the main issue in modelling turbulent flows is how to determine the eddy viscosity. Based on dimensional analysis and physical insight, the eddy viscosity can be characterised by two parameters, usually taken as a velocity scale (U) and a length scale (ℓ). Therefore one can assume:

$$\nu_t = CU\ell \qquad (4.12)$$

where C is a constant of proportionality. The turbulence models belonging to this class are subdivided according to the additional number of transport equations used to calculate the velocity scale and length scale.

Due to its simplicity and numerical stability, many of the applications have been made with the standard two equations k–ε model. However, it is mathematically proven that no turbulence model using an isotropic eddy viscosity concept yields secondary currents of Prandtl's second kind. These are important in straight prismatic compound channel flows. Hence, isotropic eddy viscosity models have mainly been applied to the calculation of meandering compound channel flows, whereas the non-linear k–ε models and more sophisticated models have been used for studies in straight compound channel flows.

4.3.1.1. Zero-equation model

In river hydraulics, the distribution of eddy viscosity over water depth is sometimes specified based on the logarithmic velocity profile and triangular shear stress distribution as,

$$v_t = \kappa u_* z'(1 - z'/h) \tag{4.13}$$

where κ is the von Karman constant, u_* the local friction velocity, z' the distance from the river bed and h the local water depth.

The integration of the distribution provides a simple estimate of the local value of the eddy viscosity.

$$v_t = \kappa u_* h/6 \tag{4.14}$$

When a zero-equation model is employed the bed shear stresses τ_{bx}, τ_{by} are evaluated as follows.

$$\tau_{bx} = \rho C_f U \sqrt{U^2 + V^2}, \quad \tau_{by} = \rho C_f V \sqrt{U^2 + V^2} \tag{4.15}$$

Here C_f is the friction coefficient and horizontal velocities U, V are evaluated at the distance z' from the bed. The distance z' has the same magnitude as the roughness height of the bed.

Jin *et al.* (1997) carried out three-dimensional numerical simulation of flows in a meandering compound channel with a simple eddy viscosity model as expressed in Equation (4.14). Figure 4.1(a) shows the plan view of the sine-generated channel. In Fig. 4.1(b) velocity distributions at three cross-sections are compared while Fig. 4.1(c) shows the comparison of the measured and calculated water surface levels. These results clearly show that the overall characteristics of flow, over the fixed bed, are captured by their model.

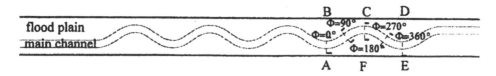

Fig. 4.1 (a) Plan-view of sine-generated compound meandering channel.

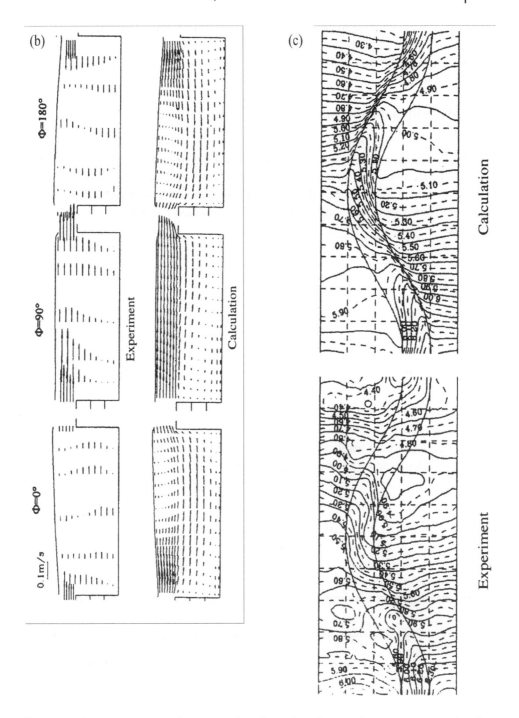

Fig. 4.1 (b) Comparison of measured and predicted secondary currents at the three sections as indicated in the previous figure and (c) Water surface levels (heights are shown in mm).

Fukuoka *et al.* (1998) employed the expression of equation (4.13) in their three-dimensional calculation of meandering channels to show good agreement with their experimental data. They also confirmed the limited contribution of turbulence to the mean flow field.

4.3.1.2. Two-equation models

This category includes two famous models, i.e., the $k-\varepsilon$ model and the $k-\omega$ model. The eddy viscosity in the $k-\varepsilon$ model is calculated using two turbulence quantities taken to characterize the turbulence field, the turbulence kinetic energy k and its dissipation rate ε. Two additional partial differential transport equations for k and ε are solved simultaneously with the continuity equation and the three momentum equations.

(1) Linear $k-\varepsilon$ models
The transport equations for k and ε are,

$$\frac{\partial k}{\partial t} + \frac{\partial U_i k}{\partial x_i} = \frac{\partial}{\partial x_i}\left[\left(\frac{\nu_t}{\sigma_k} + \nu\right)\frac{\partial k}{\partial x_i}\right] + P_{rod} - \varepsilon \tag{4.16}$$

$$\frac{\partial \varepsilon}{\partial t} + \frac{\partial U_i \varepsilon}{\partial x_i} = \frac{\partial}{\partial x_i}\left[\left(\frac{\nu_t}{\sigma_\varepsilon} + \nu\right)\frac{\partial \varepsilon}{\partial x_i}\right] + \frac{\varepsilon}{k}\left(c_{1\varepsilon}P_{rod} - c_{2\varepsilon}\varepsilon\right) \tag{4.17}$$

$$P_{rod} = -\overline{u_i u_j}\frac{\partial U_i}{\partial x_j} \tag{4.18}$$

where repeated indices imply the summation over the three coordinates. The eddy viscosity is calculated by the following relation.

$$\nu_t = c_\mu \frac{k^2}{\varepsilon} \tag{4.19}$$

The model constants are usually set as below.

$$c_\mu = 0.09, \quad c_{1\varepsilon} = 1.44, \quad c_{2\varepsilon} = 1.92, \quad \sigma_k = 1.0, \quad \sigma_\varepsilon = 1.3 \tag{4.20}$$

The $k-\varepsilon$ model calculates the wall shear stress τ_{bx}, τ_{by} using the mean velocity components and the turbulent energy at the grid point nearest to the bed.

$$\tau_{bx} = \frac{\rho\kappa c_\mu^{1/4}k^{1/2}U}{\ln Ez^+}, \quad \tau_{by} = \frac{\rho\kappa c_\mu^{1/4}k^{1/2}V}{\ln Ez^+}, \quad z^+ = \frac{u_* z}{\nu} \tag{4.21}$$

The above equations are derived from the following logarithmic velocity profile.

$$\frac{U}{u_*} = \frac{1}{\kappa}\ln(Ez') \tag{4.22}$$

Jin *et al.* (1998) applied the standard $k-\varepsilon$ model with the assumption of hydrostatic pressure distribution in the depth direction to the simulation of flow and sediment transport in a meandering channel. They found that the standard $k-\varepsilon$ model led to the bed deformation differing from the experiment. To solve this problem, they introduced the anisotropy in the eddy viscosity which comes from the algebraic stress model this resulted in better agreement with the measurement.

(2) Non-linear $k-\varepsilon$ models

The $k-\varepsilon$ model cannot reproduce turbulence-driven secondary flows in straight compound channels. Thus, a more advanced turbulence model which retains the non-linear terms is indispensable for simulating the flow field in straight prismatic channels.

Shiono & Lin (1993) used the non-linear $k-\varepsilon$ model of Speziale (1987), which takes the following form for fully developed flow.

$$-\overline{u_1 u_2} = \nu_t \frac{\partial U_1}{\partial x_2}, \quad -\overline{u_1 u_3} = \nu_t \frac{\partial U_1}{\partial x_3},$$

$$-\overline{u_2^2} = k^{1/2}\ell \frac{\partial U_2}{\partial x_2} + C_D \ell^2 \left[\frac{1}{12}\left(\frac{\partial U_1}{\partial x_2}\right)^2 - \frac{1}{6}\left(\frac{\partial U_1}{\partial x_3}\right)^2 \right]$$

$$+ C_E \ell^2 \frac{1}{3} \left[\left(\frac{\partial U_1}{\partial x_2}\right)^2 + \left(\frac{\partial U_1}{\partial x_3}\right)^2 \right],$$

$$-\overline{u_3^2} = k^{1/2}\ell \frac{\partial U_3}{\partial x_3} + C_D \ell^2 \left[\frac{1}{12}\left(\frac{\partial U_1}{\partial x_3}\right)^2 - \frac{1}{6}\left(\frac{\partial U_1}{\partial x_2}\right)^2 \right]$$

$$+ C_E \ell^2 \frac{1}{3} \left[\left(\frac{\partial U_1}{\partial x_2}\right)^2 + \left(\frac{\partial U_1}{\partial x_3}\right)^2 \right],$$

$$-\overline{u_2 u_3} = \frac{1}{2}k^{1/2}\ell \left(\frac{\partial U_2}{\partial x_3} + \frac{\partial U_3}{\partial x_2}\right) + \frac{1}{4}C_D \ell^2 \left(\frac{\partial U_1}{\partial x_2}\right)\left(\frac{\partial U_1}{\partial x_3}\right) \quad (4.23)$$

where x_1 represents the primary direction, C_D, C_E are the model constants, ℓ is the length scale calculated using k and ε. The free surface effects are taken into account by the increase in the dissipation rate at the free surface, which results in a decrease in the eddy viscosity at this location. The dissipation rate specified is expressed as,

$$\varepsilon = \frac{c_\mu^{3/4} k^{3/2}}{\kappa} \left(\frac{1}{y'} + \frac{1}{0.07h} \right) \quad (4.24)$$

where y' is the distance from the nearest side wall.

Their results show reasonable agreement with measured data about the mean velocities and the turbulent stresses near the interface between the main channel and flood plain. However, their calculation fails to reproduce the maximum velocity dip in the main channel. This weak point is common to non-linear $k-\varepsilon$ models due to their inherent difficulties in simulating the free surface.

Hosoda *et al.* (1998) used another type of non-linear $k-\varepsilon$ model originally proposed by Yoshizawa (1984), which is expressed as,

$$-\overline{u_i u_j} = v_t S_{ij} - \frac{2}{3} k \delta_{ij} - \frac{k}{\varepsilon} v_t \sum_{\beta=1}^{3} c_\beta \left(S_{\beta ij} - \frac{1}{3} S_{\beta \alpha \alpha} \delta_{ij} \right)$$

$$S_{ij} = \frac{\partial U_i}{\partial x_j} + \frac{\partial U_j}{\partial x_i}, \quad v_t = c_\mu \frac{k^2}{\varepsilon},$$

$$S_{1ij} = \frac{\partial U_i}{\partial x_r} \frac{\partial U_j}{\partial x_r}, \quad S_{2ij} = \frac{1}{2} \left(\frac{\partial U_r}{\partial x_i} \frac{\partial U_j}{\partial x_r} + \frac{\partial U_r}{\partial x_j} \frac{\partial U_i}{\partial x_r} \right), \quad S_{3ij} = \frac{\partial U_r}{\partial x_i} \frac{\partial U_r}{\partial x_j} \quad (4.25)$$

$$c_\mu = \min \left[0.09, \frac{0.3}{1 + 0.5(\min[20, S])^{1.5}} \right], \quad S = \frac{k}{\varepsilon} \sqrt{\frac{1}{2} \left(\frac{\partial U_i}{\partial x_j} + \frac{\partial U_j}{\partial x_i} \right)^2}$$

They applied the model to overbank flow with a shallow water depth over flood plain. Their simulation differs from the previous studies in that unsteady flow simulation was undertaken to investigate the relationship between the planform vortices and the secondary currents. Their model, however, produced a different mean velocity distribution from the measurement, indicating that further refinement is needed.

The above turbulence models are developed for high Reynolds number flows. Hence they are not applicable to low Reynolds number flows, which occur on the flood plain under shallow water conditions. To simulate such flows, the turbulence models suitable for application to low Reynolds simulations are necessary.

Sofialidis & Prinos (1998a) utilized a non-linear, low Reynolds number $k-\varepsilon$ model proposed by Craft *et al.* (1993) to calculate the flows at low relative water depth. Although it includes many functions to take account of the low Reynolds number effects, secondary flows are considerably underestimated. They modified the original model with reasonable agreement. Sofialidis & Prinos (1998b) adopted a non-linear, low Reynolds number type of $k-\omega$ model proposed by Patel & Yoon (1995), which is:

$$-\overline{u_i u_j} = v_t S_{ij} - \frac{2}{3} k \delta_{ij}$$

$$- c_1 v_t \frac{1}{\omega} \left(S_{ik} S_{kj} - \frac{1}{3} S_{kq} S_{kq} \delta_{ij} \right) - c_2 v_t \frac{1}{\omega} \left(S_{kj} \Omega_{ik} + S_{ki} \Omega_{jk} \right)$$

$$- c_3 v_t \frac{1}{\omega} \left(\Omega_{ik} \Omega_{kj} - \frac{1}{3} \Omega_{kq} \Omega_{kq} \delta_{ij} \right) - c_4 c_\mu^2 v_t \frac{1}{\omega^2} \left(S_{ki} \Omega_{qj} + S_{kj} \Omega_{qi} \right) S_{kq}$$

$$- c_5 c_\mu^2 v_t \frac{1}{\omega^2} \left(S_{mj} \Omega_{iq} \Omega_{qm} + S_{iq} \Omega_{qm} \Omega_{mj} - \frac{2}{3} S_{qm} \Omega_{mn} \Omega_{nq} \delta_{ij} \right)$$

$$- c_6 c_\mu^2 v_t \frac{1}{\omega^2} S_{ij} S_{kq} S_{kq} - c_7 c_\mu^2 v_t \frac{1}{\omega^2} S_{ij} \Omega_{kq} \Omega_{kq} \quad (4.26)$$

$$S_{ij} = \frac{\partial U_i}{\partial x_j} + \frac{\partial U_j}{\partial x_i}, \quad \Omega_{ij} = \frac{\partial U_i}{\partial x_j} - \frac{\partial U_j}{\partial x_i},$$

$$S = \frac{1}{\omega}\sqrt{\frac{1}{2}\Omega_{ij}\Omega_{ij}}, \quad S = \frac{1}{\omega}\sqrt{\frac{1}{2}S_{ij}S_{ij}}, \quad M = \max(S, \Omega)$$

$$\nu_t = c_\mu f_\mu \frac{k}{\omega}, \quad c_\mu = c_\mu(M),$$

$$f_\mu = \left(1 + \frac{2}{R_t}\right)\left[1 - \exp\left(1 - 0.02R_t^{3/4}\right)\right], \quad R_t = \frac{k}{\nu\omega} \tag{4.27}$$

where the model coefficients $c_1 \sim c_7$ are the functions of M except $c_4 = -1, c_5 = 0$.

The equation (4.27) includes k and $\omega(\equiv \varepsilon/k)$, ω representing the inverse turbulence time scale. These two variables are solved for using $k-\omega$ transport equations written as;

$$\frac{\partial k}{\partial t} + \frac{\partial U_i k}{\partial x_i} = \frac{\partial}{\partial x_i}\left[\left(\frac{\nu_t}{\sigma_k} + \nu\right)\frac{\partial k}{\partial x_i}\right] + P_{rod} - \omega k$$

$$\frac{\partial \omega}{\partial t} + \frac{\partial U_i \omega}{\partial x_i} = \frac{\partial}{\partial x_i}\left[\left(\frac{\nu_t}{\sigma_\omega} + \nu\right)\frac{\partial \omega}{\partial x_i}\right] + c_{\omega 1}\frac{\omega}{k}P_{rod} - c_{\omega 2}\omega^2$$

$$+ \frac{2\nu}{k}\frac{\partial k}{\partial y}\frac{\partial \omega}{\partial y} + \frac{2\nu}{k}\frac{\partial k}{\partial z}\frac{\partial \omega}{\partial z}$$

$$P_{rod} = -\overline{u_i u_j}\frac{\partial U_i}{\partial x_j} \tag{4.28}$$

Here the model coefficients $c_{\omega 1}$ and $c_{\omega 2}$ are dependent on M and $\sigma_k = 1.2, \sigma_\omega = 1.4$. More detailed forms of the coefficients are given in their original paper.

4.3.2. Reynolds stress equation models

The most complex model in engineering use today is the Reynolds Stress Equation Model (RSM) or Differential Stress Model (DSM), which solves one transport equation for each of the six Reynolds stresses and therefore accounts for anisotropic turbulence features. It is obtained after multiplication of the Navier-Stokes equations for the instantaneous velocity components by the transpose of the Reynolds stresses. These equations are then summed and time-averaged to yield:

$$\frac{D\overline{u_i u_j}}{Dt} = P_{ij} - \varepsilon_{ij} + \Phi_{ij} + \Pi_{ij} + T_{ij} + D_{ij}$$

$$P_{ij} = -\overline{u_i u_k}\frac{\partial U_j}{\partial x_k} - \overline{u_j u_k}\frac{\partial U_i}{\partial x_k}, \quad \varepsilon_{ij} = 2\nu\overline{\frac{\partial u_i}{\partial x_k}\frac{\partial u_j}{\partial x_k}}, \quad \Phi_{ij} = \frac{1}{\rho}\overline{p\left(\frac{\partial u_i}{\partial x_j} + \frac{\partial u_j}{\partial xi}\right)},$$

$$\Pi_{ij} = -\frac{1}{\rho}\left(\frac{\partial \overline{p u_i}}{\partial x_j} + \frac{\partial \overline{p u_j}}{\partial xi}\right), \quad T_{ij} = -\frac{\partial \overline{u_i u_j u_k}}{\partial x_k}, \quad D_{ij} = \nu\frac{\partial^2 \overline{u_i u_j}}{\partial x_k \partial x_k} \tag{4.29}$$

This equation shows that the Reynolds stress is determined by convection, production (P_{ij}), destruction (ε_{ij}), pressure-velocity coupling (Φ_{ij}), pressure diffusion

(Π_{ij}), turbulent diffusion (T_{ij}), and molecular diffusion (D_{ij}). The most important advantage with this type of model is that the production process does not require an assumption.

Cokljat & Younis (1995a,b) applied this type of model (shown below) to fully developed turbulent flows in compound open channels.

$$\frac{D\overline{u_i u_j}}{Dt} = -\overline{u_i u_k}\frac{\partial U_j}{\partial x_k} - \overline{u_j u_k}\frac{\partial U_i}{\partial x_k} + \frac{\partial}{\partial x_k}\left(c_s\frac{k}{\varepsilon}\overline{u_k u_l}\frac{\partial \overline{u_i u_j}}{\partial x_k} + \nu\frac{\partial \overline{u_i u_j}}{\partial x_k}\right) - \frac{2}{3}\varepsilon\delta_{ij}$$

$$- c_1\varepsilon\left(\frac{\overline{u_i u_j}}{k} - \frac{2}{3}\delta_{ij}\right) - \frac{c_2+8}{11}\left(P_{ij} - \frac{2}{3}P\delta_{ij}\right) - \frac{30c_2-2}{55}k\left(\frac{\partial U_i}{\partial x_j} + \frac{\partial U_j}{\partial x_i}\right)$$

$$- \frac{8c_2-2}{11}\left(D_{ij} - \frac{2}{3}P\delta_{ij}\right)$$

$$+ c_1'\frac{\varepsilon}{k}\left(\overline{u_k u_m}n_k n_m\delta_{ij} - \frac{3}{2}\overline{u_k u_i}n_k n_j - \frac{3}{2}\overline{u_k u_j}n_k n_i\right)f\left(\frac{\ell}{n_i r_i}\right)$$

$$+ c_2'\frac{\varepsilon}{k}\left(\Phi_{km,2}n_k n_m\delta_{ij} - \frac{3}{2}\Phi_{ki,2}n_k n_j - \frac{3}{2}\Phi_{kj,2}n_k n_i\right)f\left(\frac{\ell}{n_i r_i}\right)$$

$$P_{ij} = -\overline{u_i u_k}\frac{\partial U_j}{\partial x_k} - \overline{u_j u_k}\frac{\partial U_i}{\partial x_k}, \quad P = -\overline{u_i u_j}\frac{\partial U_i}{\partial x_j},$$

$$D_{ij} = -\overline{u_i u_k}\frac{\partial U_k}{\partial x_j} - \overline{u_j u_k}\frac{\partial U_k}{\partial x_i} \tag{4.30}$$

Figures 4.2(a) and 4.2(b) compare their calculations with data recorded in the Flood Channel Facility (see Sections 2.2.1 and 3.3.4.2) of the velocity distribution in asymmetric compound channels. From these results it can be seen that their numerical

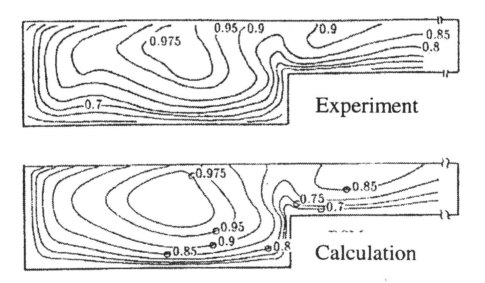

Fig. 4.2 (a) Comparison of the data obtained in the FCF and predicted primary velocity component using RSM (from Cokljat & Younis, 1995).

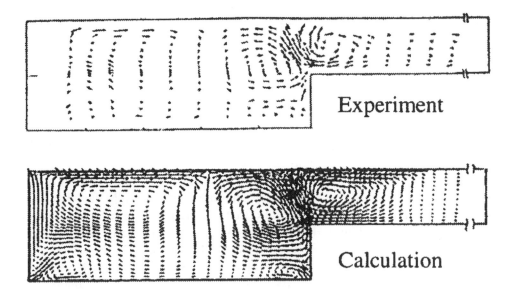

Fig. 4.2 (b) Comparison of data obtained in the FCF and predicted secondary currents using RSM (from Cokljat & Younis, 1995).

method predicts the mean velocity fields as well as the turbulence fields. They applied to their model to flows in compound trapezoidal channels with good agreement with measured mean velocity distributions. According to their comparisons it seems that a RSM may yield slightly better predictions than an algebraic stress model.

4.3.3. Algebraic stress models

The Reynolds Stress Equation Models are well suited to predicting the complex flows in natural rivers; however, they demand a lot of computer memory and CPU time. Hence engineering problems require more practical turbulence models. Rodi (1974) proposed the approximation to convection and diffusion terms in RSM.

A typical example of this type of model is the one proposed by Naot & Rodi (1982). They derived the following by applying Rodi's approximation to the classic RSM by Launder, Reece & Rodi (1974).

$$
\frac{D\overline{u_i u_j}}{Dt} = -\overline{u_i u_k}\frac{\partial U_j}{\partial x_k} - \overline{u_j u_k}\frac{\partial U_i}{\partial x_k} + \frac{\partial}{\partial x_k}\left(c_s \frac{k}{\varepsilon}\overline{u_k u_l}\frac{\partial \overline{u_i u_j}}{\partial x_k} + \nu\frac{\partial \overline{u_i u_j}}{\partial x_k} \right) - \frac{2}{3}\varepsilon\delta_{ij}
$$

$$
- c_1\varepsilon\left(\frac{\overline{u_i u_j}}{k} - \frac{2}{3}\delta_{ij} \right) - \frac{c_2 + 8}{11}\left(P_{ij} - \frac{2}{3}P\delta_{ij} \right) - \frac{30c_2 - 2}{55}k\left(\frac{\partial U_i}{\partial x_j} + \frac{\partial U_j}{\partial x_i} \right)
$$

$$
- \frac{8c_2 - 2}{11}\left(D_{ij} - \frac{2}{3}P\delta_{ij} \right)
$$

$$+ c_1' \frac{\varepsilon}{k} \left(\overline{u_k u_m} n_k n_m \delta_{ij} - \frac{3}{2} \overline{u_k u_i} n_k n_j - \frac{3}{2} \overline{u_k u_j} n_k n_i \right) f \left(\frac{\ell}{n_i r_i} \right)$$

$$+ c_2' \frac{\varepsilon}{k} \left(\Phi_{km,2} n_k n_m \delta_{ij} - \frac{3}{2} \Phi_{ki,2} n_k n_j - \frac{3}{2} \Phi_{kj,2} n_k n_i \right) f \left(\frac{\ell}{n_i r_i} \right)$$

$$P_{ij} = -\overline{u_i u_k} \frac{\partial U_j}{\partial x_k} - \overline{u_j u_k} \frac{\partial U_i}{\partial x_k}, \quad P = -\overline{u_i u_j} \frac{\partial U_i}{\partial x_j},$$

$$D_{ij} = -\overline{u_i u_k} \frac{\partial U_k}{\partial x_j} - \overline{u_j u_k} \frac{\partial U_k}{\partial x_i} \tag{4.31}$$

Krishnappan & Lau (1986) were the first to successfully apply ASM to compound channel flows. Kawahara & Tamai (1987) demonstrated that ASM can produce mean velocities and turbulent stresses and that the high level of turbulence near the edge of the flood plain is due to high energy production and not the convection of strong turbulence from near the wall. Kawahara & Tamai (1988) discussed the empirical formulae for apparent shear stress at the interface between the main channel and the flood plain based on the reduced turbulent shear stresses from three-dimensional calculations. Larsson (1988) and Naot, Nezu & Nakagawa (1993) also demonstrated the usefulness of ASM. In Chapter 2 calculated results by Nezu and co-workers using the Naot-Rodi model are compared with their experimental data. Sugiyama *et al.* (1997) applied their ASM to compound trapezoidal channels. Figure 4.3(a) shows the primary velocity and Fig. 4.3(b) the secondary

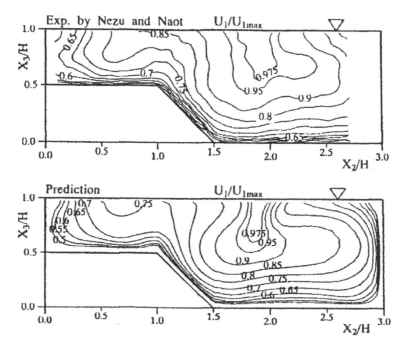

Fig. 4.3 (a) Comparison of the measured and predicted primary velocity component (from Sugiyama *et al.*, 1997).

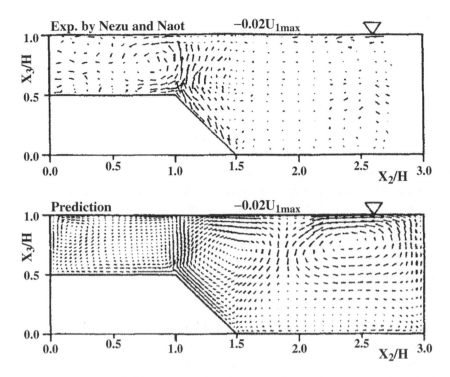

Fig. 4.3 (b) Comparison of the measured and predicted secondary currents (see Sugiyama *et al.*, 1997).

currents. These results indicate that ASM can produce good results compared to RSM at lower computational cost. This maybe due to the fact that in the ASM the production mechanism is preserved and not modelled and that the free surface effect is included in the ASM. Sugiyama, Akiyama & Tanaka (1998) employed ASM to simulate the flow in a curved compound channel. They clearly show that the development of primary and secondary flows are reasonably well reproduced using an ASM.

4.3.4. Three-dimensional Large Eddy Simulation

Three-dimensional numerical simulations of turbulent flows in straight and meandering compound channels require the solution of the RANS. Representation of turbulence in their solution requires the use of turbulence models, such as those explained in the previous section. Such models reproduce mean velocities and turbulent stresses but lack generality. They do not take account of turbulence structures directly and the model constants are usually set using a few simple flows. Hence, they may not work well for large scales of turbulent motion that are strongly affected by the boundary conditions.

The most straightforward approach to the solution of turbulent flows is direct numerical solution (DNS) of the Navier-Stokes equations, where the governing equations are discretized and solved numerically. If the calculation mesh is fine enough to resolve even the smallest scales of motion, and the numerical scheme is designed to minimize the numerical errors, one can obtain an accurate three-dimensional, time-dependent solution that is completely free from modelling assumptions. With the recent advances in computer technology, DNS has been applied to many kinds of turbulent flows, however, due to the heavy computational cost, DNS has been limited to flows at low Reynolds numbers and its application to engineering problems appears unlikely within the next decade.

Large eddy simulation (LES) is a method between DNS and RANS models. In LES, the contribution of large eddies, or grid-scales, carrying most of the turbulence energy, to momentum and mass transfer is calculated directly and only the effect of small scale turbulence is modelled using a sub-grid scale model (SGS model). Since small scales of turbulence tend to be more homogeneous than the large ones, the hope is that these models can be simpler and have wider applicability. LES is similar to DNS in that it offers a three-dimensional time-dependent solution of the Navier-Stokes equations. Its advantage is that it can provide the simulation of flows at much higher Reynolds numbers and less computational cost than DNS.

Over thirty years have passed since the first LES results by Deardorff (1970) were published. Following this the first application to compound open channel flows was undertaken by Thomas & Williams (1995a,b). They used the Smagorinsky model with a wall function to bridge the inner layer. Their results for the simulation of flows at high Reynolds numbers showed good agreement with those obtained using turbulence models. Later, Satoh & Kawahara (1998) performed another LES using the Smagorinsky model with a damping function for the eddy viscosity to study the generation mechanism of secondary currents near solid walls. Based on results obtained by LES, Satoh & Kawahara (1999) discuss the generation mechanism of turbulence-driven secondary currents.

It is well known that the Smagorinsky constant, the sole constant in the model, takes a different optimum value depending on the flow characteristics. To overcome this problem, dynamic SGS models have been proposed in which the local value of the Smagorinsky constant is updated at every time step, for example Germano (1991), Lilly (1992) and Ghosal *et al.* (1995). To date, no attempt has been made to apply a dynamic SGS model to compound channel flows.

The space filtered continuity equation and Navier-Stokes equations are written as follows:

$$\frac{\partial \bar{u}_i}{\partial x_i} = 0 \tag{4.32}$$

$$\frac{\partial \bar{u}_i}{\partial t} + \frac{\partial \bar{u}_i \bar{u}_j}{\partial x_j} = g_i - \frac{1}{\rho}\frac{\partial}{\partial x_i}\left(\bar{p} + \frac{2}{3}\rho q\right) + \frac{\partial}{\partial x_j}\left(\nu\frac{\partial \bar{u}_i}{\partial x_j} - \frac{\tau_{ij}}{\rho}\right) \tag{4.33}$$

$$\frac{\tau_{ij}}{\rho} = \left(\overline{u_i u_j} - \bar{u}_i \bar{u}_j\right) - \frac{2}{3}\delta_{ij}q, \qquad q = \frac{1}{2}\left(\overline{u_k u_k} - \bar{u}_k \bar{u}_k\right) \tag{4.34a,b}$$

where, the overbar denotes the space filtering, g_i the gravitational component, τ_{ij} the subgrid scale (SGS) stress, q the SGS turbulence energy.

The SGS stress in equation (4.34) is expressed using the original Smagorinsky model.

$$\tau_{ij} = -2\nu_t \bar{S}_{ij}, \qquad \nu_t = (C_s \Delta)^2 \left(2\bar{S}_{ij}\bar{S}_{ij}\right)^{1/2} \tag{4.35a,b}$$

$$\bar{S}_{ij} = \left(\partial \bar{u}_i/\partial x_j + \partial \bar{u}_j/\partial x_i\right)/2, \qquad \Delta = (\Delta x \Delta y \Delta z)^{1/3} \tag{4.36a,b}$$

Here the Smagorinsky constant C_s takes the value of 0.1, which is a typical value for wall turbulence. Near solid walls the filter width Δ is multiplied by the van Driest type damping function to express the correct behaviour of the eddy viscosity ν_t, whereas no correction is introduced near the free surface due to the lack of reliable information. The damping function is expressed in the following equation.

$$f = 1 - \exp\left(-y^+/25\right), \qquad y^+ = u_* y/\nu, \tag{4.37a,b}$$

where y^+ is the distance from the wall and u_* is the shear velocity.

The space filtered Navier–Stokes equations (4.33) are rearranged for the computation.

$$\frac{\partial \bar{u}_i}{\partial t} = -\frac{1}{\rho}\frac{\partial P}{\partial x_i} + H_i, \qquad H_i = -\frac{\partial \bar{u}_i \bar{u}_j}{\partial x_j} + \frac{\partial}{\partial x_j}\left(2\nu_e \bar{S}_{ij}\right) + g_i \tag{4.38a,b}$$

$$P = \bar{p} + \frac{2}{3}\rho q, \qquad \nu_e = \nu + \nu_t \tag{4.39a,b}$$

4.4. Example Applications: One Dimensional Modelling

4.4.1. Flood Channel Facility and River Dane simulations

Recently, Forbes & Pender (2000) developed subroutines that enable Ackers (1991) and James & Wark (1992) (see Section 2.7.2) to be used as the basis of conveyance calculations in one-dimensional unsteady river flow models. The subroutines were verified by undertaking steady flow modelling of experiments from the FCF experimental programme.

The geometry for experiment B26 was a 60° meandering main channel with a quasi-natural cross-section at the bend apex, shown in Fig. 4.7. This cross-section represents an average of 16 sets of field data, (Lorena, 1992). The cross-section varied linearly between bends with the cross-section being trapezoidal midway between the bends. The flume surface was a trowelled mortar finish, which had a measured 'n' value of 0.01, (Lorena, 1992). Experimental conditions were close to uniform with a tailgate being used to control the downstream water level. Experimental measurements were undertaken for a total of 16 discharges ranging from $0.04\,\text{m}^3/\text{s}$ to $1.09\,\text{m}^3/\text{s}$.

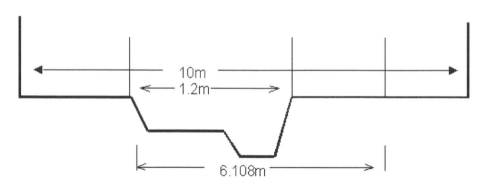

Fig. 4.7 FCF Quasi-Natural Apex Section Geometry 60° Meander.

The computer model contained 6 cross-sections located as shown in Fig. 4.8. The upstream boundary condition used was the measured discharge from the experiment and the downstream water level was set equal to that measured at the tailgate. Manning's 'n' was set to the measured value of 0.01. Two methods were used to calculate conveyance in the model; the divided channel method and a method based on James & Wark (1992). As the divided channel method does not include energy losses from secondary sources one would expect this to under-predict the observed water level at the upstream end, whereas the method based on the James & Wark (1992) technique does include energy losses from secondary sources, which should result in better water level predictions. Figure 4.9 shows this to be the case. For this experimental data, the difference is only of the order of 10 mm; however, if one considers the FCF to be a 1/10 or 1/100 scale of a real river then in the field the difference could be one to two orders of magnitude greater.

To test the significance of secondary energy losses to the modelling of real rivers Forbes (2000) applied his model to the River Dane, shown in Fig. 4.10. This model consisted of 30 surveyed cross-sections; a typical one is shown in Fig. 4.11. Manning's 'n' value was estimated by visual inspection and reference to previous computer modelling work on this reach, Ervine & McLeod (1999). Field data existed for a flood with a peak flow of $170\,\mathrm{m}^3/\mathrm{s}$ corresponding to a downstream water

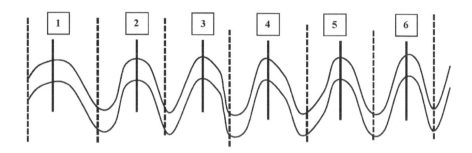

Fig. 4.8 Six Cross-Section Model with Representative Reaches.

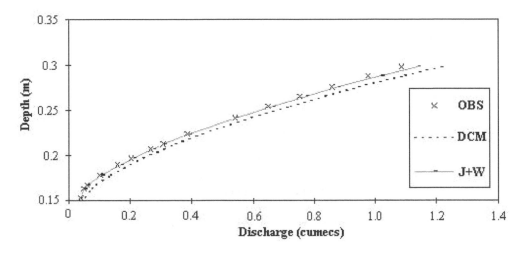

Fig. 4.9 Stage Discharge Curves For Experiment B26.

Fig. 4.10 The River Dane.

level of 13.5 m OD. Once again, a steady flow analysis was undertaken using the divided channel and James & Wark methods to calculate conveyance. On this occasion comparison of the methods showed the predicted water levels varied by up to 180 mm. It would appear therefore that significant differences in water level prediction result from the inclusion of secondary energy losses in the conveyance calculations.

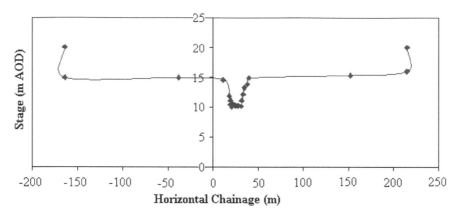

Fig. 4.11   Typical ISIS Model River Cross-Section 16.

The limitation of the modelling work reported by Forbes & Pender (2000) is that it only evaluates one of the many new methods for calculating conveyance. What is now required is consensus on which of the alternative methods discussed in Section 2.7 is the best for general application, followed by further software development and model validation.

4.5.   Example Applications: Two-Dimensional Modelling

4.5.1.   Straight channel with drop structures

Vertical drop spillways are commonly used to prevent bed erosion in steep alluvial rivers. They are installed in river reaches where the channel shape or planform has been changed to facilitate the construction of engineering works such as meander cut-offs. The Toyohira River, (see Fig. 4.12), which flows through the centre of Sapporo City, is important economically and for recreational purposes. Channel hydraulics are complicated by the fact that the cross-section is an engineered compound channel, designed to alleviate local flooding problems. It can also be seen that the river has active morphology with the low flow channel containing alternate bars that pass through the vertical drop spillways.

Between 1950 and 1973, eight vertical drop spillways were constructed to control sediment movement (Ishikari River Development and Construction Office 1993). These structures significantly improved channel stability, however, the bed slope is still very steep and during floods water velocity can reach 10 m/s, resulting in bed erosion remaining a serious problem, Yamashita *et al.* (1992). There is therefore a growing desire to re-engineer these structures to further reduce water velocity. Recent developments of numerical models enable them to be used to predict flow and bed movement in channels with vertical drop spillways, Kawashima & Fukuoka (1995) and Fukuoka *et al.* (1998). However, it is not clear if these models can be applied to the Toyohira River where the section is compound, velocities are high and the bed is very active. To evaluate the various proposals for modifying

Fig. 4.12 The Toyohira River (view from downstream, by Ishikari River Development and Construction Office).

the existing vertical drop spillways it is important to develop a computer model that can simulate this complex problem.

In this section, a numerical model is described to calculate the flow and sediment transport in compound channels with vertical drop spillways. The basic hydrodynamic equations used are the two-dimensional shallow water equations. These have been linked to sediment transport equations to successfully predict bar formation. However, previously applied numerical methods based on techniques such as upwind differencing are not effective when the channel is compound and has discontinuities such as vertical drop structures. The CIP (Cubic-Interpolated

Pseudoparticle) method proposed by Yabe (1990) is used here to calculate the flow conditions in the compound channel and over the abrupt change in level due to the vertical drop spillways. The model is verified by comparison with data from physical model experiments. Results from bank-full and over-bank flows are compared with the predictions from the computer model.

4.5.1.1. Basic equations and numerical method

The two-dimensional flow field is calculated using Equations (4.5), (4.6a) and (4.6b) Kinetic eddy viscosity is calculated simply by the following.

$$v_t = \frac{\kappa}{6} u_* h \qquad (4.55)$$

Here κ is the von Karman constant and U_* is shear velocity. The time dependent change of bed elevation is calculated from the continuity equation for bed load sediment transport.

$$\frac{\partial \eta}{\partial t} + \frac{1}{1-\lambda} \left[\frac{\partial q_{bx}}{\partial x} + \frac{\partial q_{by}}{\partial y} \right] = 0 \qquad (4.56)$$

In which λ is porosity of bed material, q_{bx} and q_{by} are bed load sediment transport rate per unit width in x and y directions, which are calculated using the formulae of Meyer-Peter Muller and Hasegawa (1980), respectively.

$$\frac{q_{bx}}{\sqrt{\left(\frac{\rho_s}{\rho} - 1\right) g d^3}} = 8 \left(\tau_* - \tau_{*c}\right)^{3/2} \qquad (4.57)$$

$$q_{by} = q_{bx} \left(\frac{v}{u} - N_* \frac{h}{r_*} - \sqrt{\frac{\tau_{*c}}{v_s v_k \tau_*} \frac{\partial n}{\partial y}} \right) \qquad (4.58)$$

Here ρ_s and d are density and diameter of bed material, τ_* is the non-dimensional bed shear stress $\left[= u_*^2/(sgd), s = \rho_s/\rho - 1\right]$, and τ_{*c} is non-dimensional critical shear stress which is calculated by Iwagaki's formula. v_s and v_k are static and kinetic friction coefficient of sand particles. The second term on the right hand side of Equation 4.58 acts as additional transverse sediment load when the secondary flow is developed. The flow field described in this paper is two-dimensional, however, because of the development of a three-dimensional bed configuration the secondary flow is taken into account when the stream line is curved. A constant value of 7 is used for N_* according to Engelund (1974), and r_* is expressed by the radius of curvature of a streamline, and it is calculated using the following equation suggested by Shimizu & Itakura (1991).

$$\frac{1}{r_*} = \frac{1}{\left(u^2 + v^2\right)^{3/2}} \left\{ u \left(u \frac{\partial u}{\partial x} - u \frac{\partial v}{\partial x} \right) + v \left(u \frac{\partial v}{\partial y} - v \frac{\partial u}{\partial y} \right) \right\} \qquad (4.59)$$

In solving momentum equations of flow, a solution technique is used, in which the equations are separated into two phases, for advection and non-advection. The

CIP method is used for advection phase, while the SOR method is used to calculate the non-advection phase coupled to the continuity Equation (1). CIP method was originally proposed by Yabe (1990) and modified for the calculation of open channel flow by Nakayama *et al.* (1998). Change in bed elevation is computed by applying central differencing to Equation (4.56).

4.5.1.2. Moveable bed experiments

Experiments using a 1/50-scale model of the Toyohira River were conducted in a flume with a compound cross-section and vertical drop spillways. Figures 4.13(a) and 4.13(b) show the outline of the experimental flume. The length and width of the flume are 20 m and 1 m, respectively. The central part of the channel is a low water channel with a width of 48 cm. The low water channel bed is covered with uniform sand of 0.2 mm diameter. The flood plains on both sides are cast in mortar. Along the low water channel, three vertical drop structures are installed at 5.3 m intervals. The vertical drop of each structure is 1.3 cm. Two experiments were bank-full and over-bank flows, with discharges of $2\ell/s$ and $7\ell/s$. These were designated as RUN-1 and RUN-2, respectively.

Sediment was supplied at a rate necessary to keep a constant bed elevation at the upper end of the flume. Experiments were continued for about an hour until the bed elevation reached its equilibrium state. At the end of RUN-1, the surface

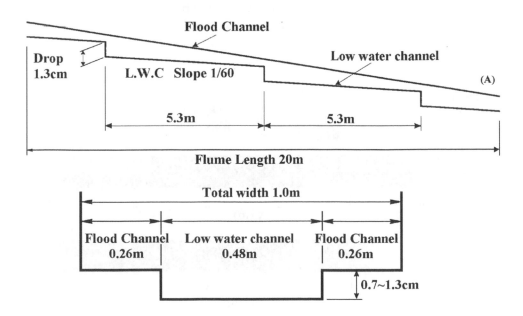

Fig. 4.13 (a) Longitudinal profile of the experimental flume and (b) Cross section of the experimental flume.

velocity was measured using a PIV technique with Styrofoam grains used as tracers. In both experiments similar bed configurations were observed between the drop structures.

Figure 4.14 shows the bed configuration (deviation from initial bed), and Fig. 4.15 shows bed elevations in the downstream direction. As can be seen from Fig. 4.16, migrating alternate bars were observed during Run-1. These are also evident in the bed elevation data presented in Figs. 4.14 and 4.15. The low flow water channel was deepest just downstream of a vertical drop, becoming shallower towards the next drop. This caused water to flow onto the floodplain about 2 m upstream of the drop. This behaviour is observed in the Toyohira River during the snow melt season. Alternate bars were not in evidence after RUN-2. The average hydraulic conditions for the two cases are plotted in Fig. 4.17, as proposed by Kuroki & Kishi (1981). It can be seen that existence or otherwise of alternate bars in each experiment agrees with the regime criteria of the diagram.

4.5.1.3. Numerical calculation

Computer simulations were conducted with conditions similar to those used in the experiments. Manning's roughness coefficient for the mobile bed of the low water channel was determined using a formula proposed by Kishi & Kuroki (1981) for flat beds, a constant value of 0.01 was used for the high water channel. Computational grids in the longitudinal and transverse directions were 122 and 28, respectively. The time step for the computation was set as 0.01 seconds. At the grid points on the vertical drop spillways, where the bed was fixed, calculation of bed elevation change was omitted from the calculation.

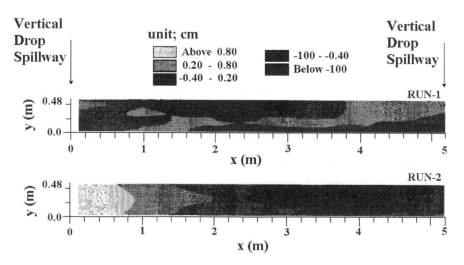

Fig. 4.14 Bed elevation shown for RUN-1 and RUN-2 shown in terms of the deviation from the initial bed.

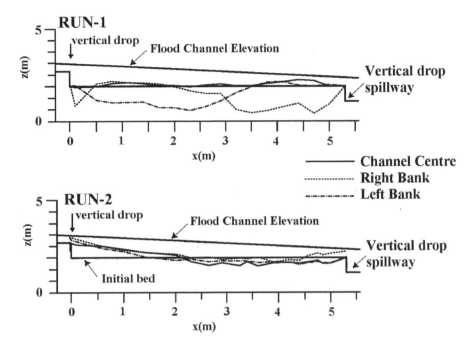

Fig. 4.15 Longitudinal bed elevation profiles for experiments RUN-1 and RUN-2.

Figure 4.18 shows the calculated water surface profile and velocity vectors under the initial conditions of RUN-1. The flow pattern of partial flooding around the vertical drop spillways observed in the experiments is well reproduced by the computer model. In order to examine the accuracy of the flow calculation, the observed bed elevation at the end of RUN-1 was given and the flow calculation of RUN-1 was conducted with fixed bed conditions. Computed results of water surface are compared with the observed data in Figs. 4.19(a) and 4.19(b). They show good agreement and thus the accuracy of the flow calculation is verified. Next, the performance of alternate bar formation was studied using the flow conditions of RUN-1. Calculations without vertical drop spillways were also conducted to compare the effect of vertical drop spillways. Calculations with and without spillways were continued for two hours, the time necessary for the bed configuration to reach equilibrium. Figures 4.20 and 4.21 show the calculated bed elevation contours with and without spillways, from 3,600 seconds to 5,400 seconds in the calculation. Figure 4.20 shows that alternate bars are migrating downstream with a constant speed. The calculated wavelength of the alternate bars was about 12 times the width of the channel. Bar height was about 1.3 cm about 1.5 times the average water depth, slightly smaller than the observed value obtained during RUN-1. For the model with vertical drop spillways the well regulated bar migration was not reproduced, Fig. 4.21. It was discovered that when the bar fronts approach the vertical drops during the computer simulation their migration speed

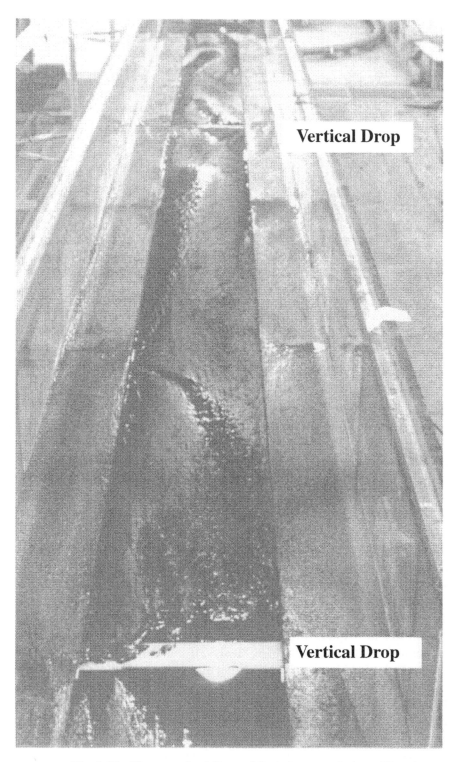

Fig. 4.16 Photograph of Channel Bed after completion of Run-1.

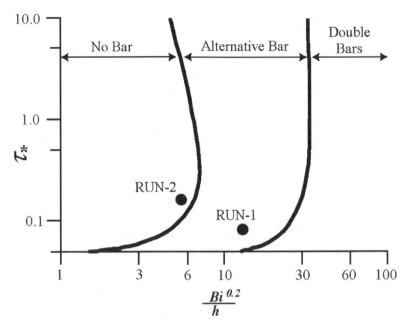

Fig. 4.17 Regime criteria of meso-scale bar.

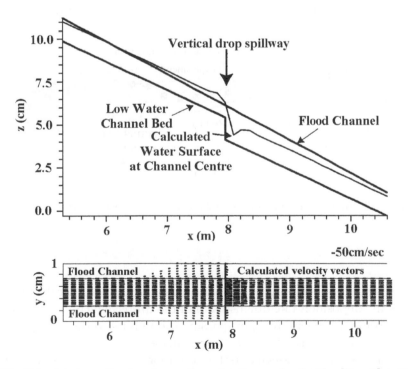

Fig. 4.18 Calculated results of water surface elevation and velocities [Run-1], initial bed condition.

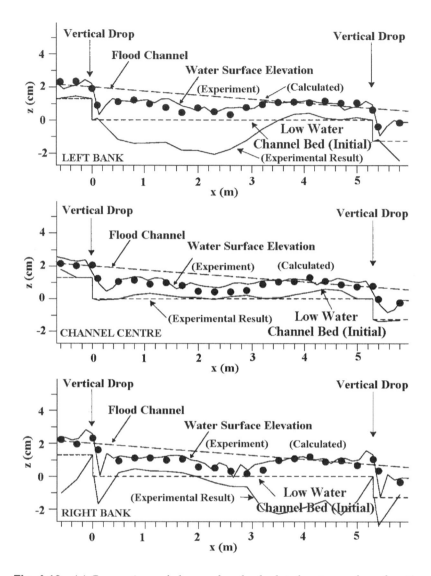

Fig. 4.19 (a) Comparison of observed and calculated water surface elevation.

decreased, and thus their regular migration was interrupted. The calculation over-estimates the deposition rate downstream of the drops, which may be caused by non-equilibrium sediment transport passing through the vertical drop spillways. This is not taken into account in the present model.

Calculation was also conducted for the conditions of RUN-2, and calculated bed contours are shown in Fig. 4.22. Alternate bars are not developed in this case, which agrees with the experimental results presented in Fig. 4.14.

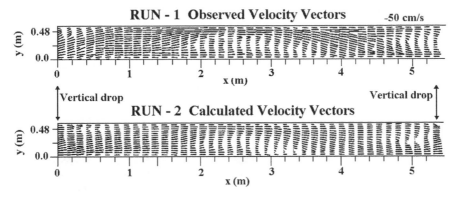

Fig. 4.19 (b) Calculated velocity vectors for RUN-1 and RUN-2.

4.5.1.4. Conclusion

Characteristics of flow and bed movement especially the migration of alternate bar formations were studied using flume experiments and computer simulation. It has been shown that the regime criterion for the formation of alternate bars is valid even where vertical drop spillways exist in the channel.

A numerical model was proposed to calculate the flow and bed deformation in a compound channel with vertical drop spillways. In the numerical model the CIP method was used to calculate flow conditions with vertical drops and a partially flooded compound channel. The applicability of the model was confirmed by comparison with the experimental results. The performance of the proposed model using the CIP method was very successful and extends the range of application of numerical models beyond that available with existing codes.

4.5.2. Straight channel with pile dikes

Pile dikes are sometimes used as training works to regulate flow in rivers, and maintain channel alignment. The dikes act as resistance to flow reducing the velocity in the dike region. This difference in flow velocity across the channel generates a lateral shear stress, the existence of which results in horizontal velocities. Ikeda & Chen (1995) have undertaken a combined physical/computing modelling exercise to investigate the behaviour of these vortices and the associated lateral transport of fluid momentum. The computer model solved the two-dimensional shallow water equations with a Large Eddy Simulation turbulence closure model.

A 12 m long by 1.2 m wide re-circulating laboratory flume was used for the experiments. Wooden cylinders with 0.5 cm diameter were attached to the bottom of the flume to simulate the dikes. These had longitudinal and lateral spacings of 5 cm (see Fig. 4.23). Fluid velocity was measured using a two-component Laser-Doppler Velocimeter, and the fluctuation of the free surface induced by the vertices was detected by two wave-gauges located at the edge of the pile dike

200 | Y. Kawahara *et al.*

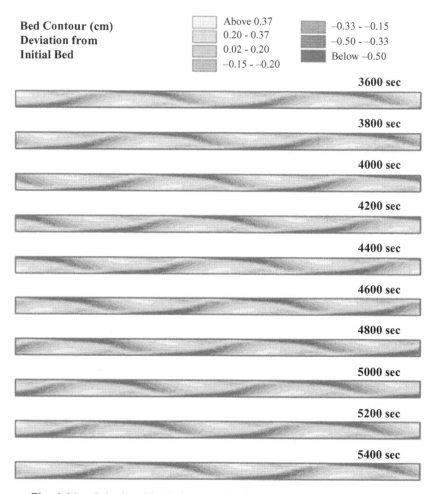

Fig. 4.20 Calculated bed elevation (without vertical drop structures).

regions. Table 4.1 provides a summary of the major hydraulic variables used in the laboratory tests.

Since the horizontal scale of the flow field considered here is much larger than the vertical scale, the turbulence can be divided into two parts: horizontal large scale vortices and local small scale turbulence typically scaled by the flow depth. The latter is termed "subdepth-scale" turbulence (SDS) because its length scale is smaller than the depth (Nadaoka & Yagi, 1993a, 1993b). In the computer model, a transport equation for the kinetic energy of SDS turbulence is employed, in which local turbulence production (induced by the bottom friction and the drag of the pile dikes) and dissipation is incorporated. The large scale horizontal vortices are computed directly by integrating the depth-averaged continuity equation and the momentum balance equations, where the lateral eddy viscosity is calculated using an algebraic equation including the kinetic energy of SDS turbulence and

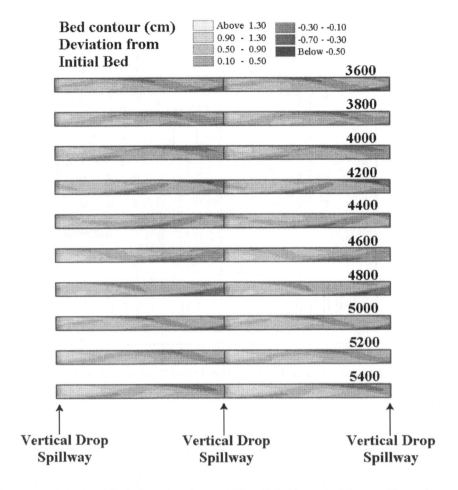

Fig. 4.21 Calculated bed elevation changes [Run 1] (with vertical drop spillways).

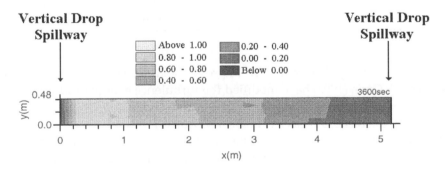

Fig. 4.22 Calculated bed elevation changes [Run 2] (with vertical drop spillways).

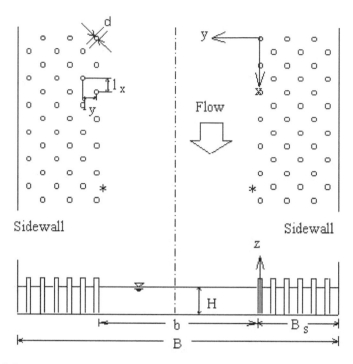

Fig. 4.23 Schematic sketch of flow field. The symbol '*' shows the location of wave gages.

Table 4.1 Major hydraulic variables of experiments.

Run	Width of b (cm)	Width of B_s (cm)	u_∞ (cm/s)	Re	Fr
1	60	30	40.5	24300	0.528
2	50	35	39.8	23900	0.528
3	40	40	35.5	21300	0.463
4	30	45	31.7	19000	0.413
5	20	50	22.8	13700	0.300
6	10	55	17.6	10600	0.230

the turbulence energy dissipation. To improve the agreement with the observations, Ikeda & Chen (1995) have modified the turbulence energy production term by considering a balance of energy production and dissipation.

   The instantaneous 2D velocity fields for Runs 1, 3 and 5 are depicted in Fig. 4.24(a), (b) and (c), respectively. The results show the flow field at about 100s from the start of numerical computation, when it has reached a statistically averaged steady state. The flow takes a stable sinuous pattern for Run 3 ($b/H = 6.67$), showing that the wavelength of each vortex is identical and the centres of vortices are staggered. For Run 1 ($b/H = 10$), i.e. the case of a larger value of b, the vortices are seen to be unstable and two vortex sheets are moving independently of

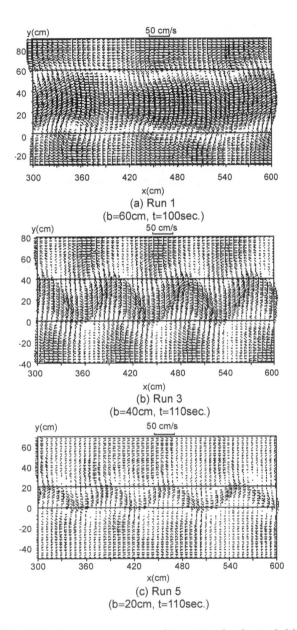

Fig. 4.24 Instantaneous two-dimensional velocity fields.

each other. For Run 5 ($b/H = 3.33$), the flow is stable. However, the scale of vortices is seen to be varying in the downstream direction. Since the turbulence model employed here assumes that horizontal scale is much larger than the vertical scale, the model is probably inapplicable to this case.

The temporally-averaged flow velocity can be compared with the laboratory tests. Figures 4.25(a), (b) and (c) show the results for Runs 1, 3 and 5 respectively.

Since the temporally-averaged velocity distribution is symmetric with respect to the channel centre at $y/b = 0.5$, the figures depicted only half the flume, in which $y/b = 0$ corresponds to the boundary of pile dike region. The numerical computation reproduces the measurement very well for Runs 1 and 3. However, the agreement becomes poor for Run 5, as previously mentioned the reason for this is that the turbulence model employed assumes that b/H is large. For Run 3, the fluid velocity normalised by u_∞ is smaller than unity, suggesting that the two vortex sheets are interacting with each other and that the lateral exchange of fluid momentum is large, where u_∞ is the depth-averaged flow velocity which is not affected by the pile dikes. For Run 1, there is a region at the channel centre where the normalised fluid velocity is unity, indicating that the two vortex sheets behave independently.

The Reynolds stress, $\overline{-u'v'}$, measured at $y/b = 0$ is shown in Fig. 4.26 against b/H. It is clear that there is a peak value of the Reynolds stress at around $b/H = 6$ both for the measurement and the numerical computation. The Reynolds stress tends to become constant for large value of b/H, which implies that two vortex sheets are independent. For b/H smaller that 5, the Reynolds stress decreases

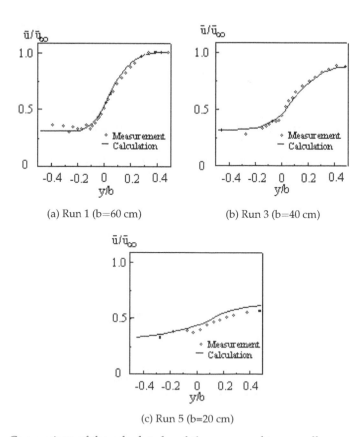

(a) Run 1 (b=60 cm)   (b) Run 3 (b=40 cm)

(c) Run 5 (b=20 cm)

Fig. 4.25 Comparison of the calculated and the measured temporally-averaged velocity.

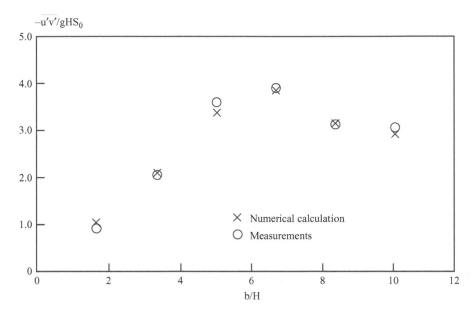

Fig. 4.26 Comparison of the measured and the calculated Reynolds stress at $y/h = 0$.

almost linearly with b/H. The agreement between the numerical computation and the measurement is very good.

The contribution of large horizontal vortices to momentum exchange is interesting. The total Reynolds stress, which can be decomposed into two parts, i.e. the contribution by SDS turbulence and that by large vortices. Figure 4.27 shows the result of decomposition for Run 3 ($b/H = 6.67$), for which the lateral exchange of fluid momentum is almost a maximum. It is clear that the large vortices occupy the major part of the total Reynolds stress, and they contribute 60% to the total Reynolds stress. The remainder is contributed by SDS turbulence.

The reason why the Reynolds stress takes a peak value at about $b/H = 6$ is clear from the above. Since the two vortex sheets are the most stable at about $b/H = 6$, (as seen in Fig. 4.24), and since the Reynolds stress is produced mainly by the large vortices, the momentum exchange at the boundary of the pile dike region becomes the largest at about $b/H = 6$.

In summary the numerical and the laboratory studies discussed here reveal the following:

- The interaction of two rows of vortex sheets increases the momentum exchange between the pile dike regions and the central region without pile dikes, accordingly the Reynolds stress at $y/b = 0$ takes a peak value at about $b/H = 6$.
- The large scale vortices contribute 60% of the total Reynolds stress.
- The turbulence model proposed originally by Nadaoka & Yagi (1993a, 1993b) and modified by Ikeda & Chen (1995) predicts 2D flow field with pile dikes well.

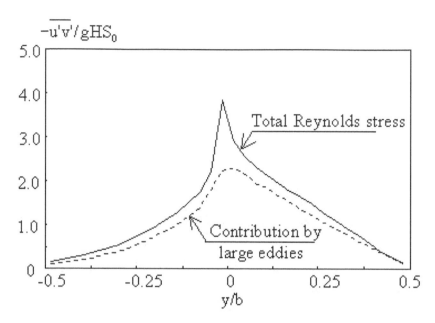

Fig. 4.27 Lateral distribution of the calculated Reynolds stress for Run 3.

4.6. Example Applications: Three Dimensional Modelling

4.6.1. Straight channel with ASM and LES

The following describes the results from a computer model of the experimental work by Tominaga & Nezu (1991), where detailed LDV measurement of velocities in a compound channel with a single flood plain is reported, (see Fig. 4.34). The experimental channel had smooth walls and a bed slope of 1/1000.

The central difference scheme is adopted for the space discretization of the above equations and the second order Adams-Bashforth method is used for the time discretization. A converged solution was obtained using the HSMAC method on a staggered grid system.

The calculation domain has a streamwise reach length of 16 cm, a water depth of 4 cm a total channel width of 20 cm, a flood plain width of 10 cm and a flood plain height of 2 cm.

Figure 4.28 shows a cross sectional view of the numerical grid. This is clustered near the bed, sidewalls and the interface of the main channel and the flood plain. The number of computational points used was $79 \times 75 \times 220$ in x, y and z directions respectively. The grid intervals non- dimensionalized by the cross-sectional averaged friction velocity and the kinematic viscosity are $\Delta x^{+} = u_{*}\Delta x/\nu \cong 30$, $\Delta y^{+} = u_{*}\Delta y/\nu \cong 22$ and $\Delta z^{+} = u_{*}\Delta z/\nu \cong 42$. These intervals allow the application of a no-slip condition at solid walls and the capturing of streaky turbulence structures near the walls. By comparison Thomas & Williams (1995a) used

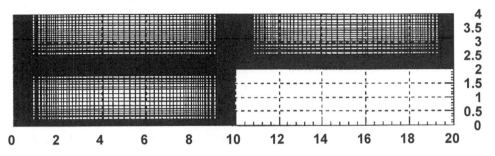

Fig. 4.28 Grid system viewed from downstream.

$127 \times 32 \times 127$ points in x, y and z directions with $\Delta x^+ \cong 47$, $\Delta y^+ \cong 31$ and $\Delta z^+ \cong 39$.

The boundary conditions specified are as follows:

- The free surface is treated as a rigid lid, assuming that the surface is flat with no shear stress and the velocity component normal to the surface is zero;
- Inflow and outflow boundaries are described using periodic conditions and no slip conditions are imposed along the solid walls.

The initial velocity field is defined as a combination of a logarithmic profile and a linear profile near the solid walls with an artificial turbulence field generated using random numbers. The secondary currents are set to zero and a hydrostatic pressure distribution is assumed. The time increment used is $\Delta t = 2.5 \times 10^{-4}$ seconds. After 220,000 time steps the calculations had developed to a steady state,

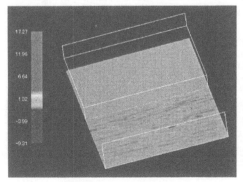
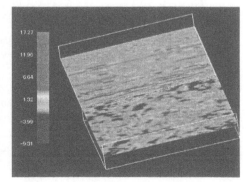

(a) Near main channel bed.　　　　　　　　(b) Near floodplain bed.

Fig. 4.35 Low and high speed streaks.

Fig. 4.29 shows the distribution of instantaneous streamwise velocities on two horizontal planes; one horizontal plane is 6.20 wall units from the channel bed and the other 5.51 wall units from the flood plain bed. The streamwise velocity is non-dimensionalised by $u'^+ = (u - \bar{u})/u_*$ where $u, \bar{u}$ are the instantaneous velocity and the local time-averaged velocity respectively.

and the statistical quantities were calculated over the following 50,000 time steps, which corresponds to 12.5 seconds.

It is found that the distance between the neighbouring low speed streaks are about 100, where the grid interval is small enough in both planes. Although the details of these structures have not been analyzed, the results indicate the validity of the grid resolution near the solid walls.

Figure 4.30 compares the distribution of the streamwise velocity. Satoh & Kawahara (SK) and Thomas & Williams (TW) results are shown along with the measured data from Tominaga & Nezu (1991). The isovels are non-dimensionalized by the maximum value. It is confirmed that the distortion of the isovels near the tip of the flood plain and the corners in the main channel are reasonably reproduced in both sets of results; however, they are underpredicted compared with the experiment. The result from TW is closer to the model than those produced by SK in that

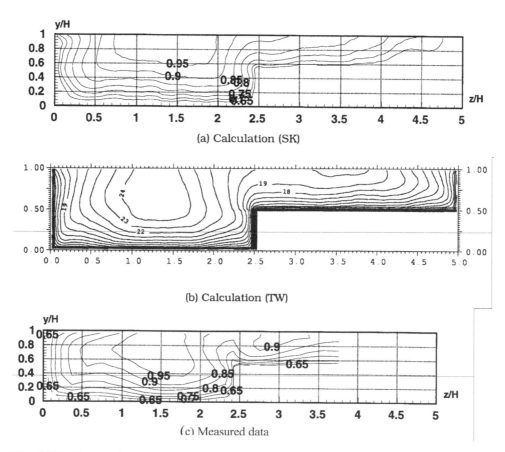

Fig. 4.30 Comparison of streamwise velocity U/U_{max} (a) Calculated by Satoh & Kawahara (b) Calculated by Thomas & Williams (1995a) and (c) Measured data by Tominaga & Nezu (1991).

the velocity has a maximum at the depth of around $z/H = 0.6$ and the high velocity region is apparent over the flood plain. The difference in the two numerical results may be a result of insufficient grid points in the region away from the walls and the shortness of the calculation domain in the streamwise direction in the SK simulation. This would be expected to lead to the underestimation of secondary currents.

It is clear, from Figure 4.31, that a pair of secondary cells near the interface is reasonably well calculated by both models. The upward motion near the tip of the flood plain shows the maximum value of 3.9 percent of the bulk primary velocity in the result of SK. This is in good agreement with the measurement; however, the size of the secondary cell in the main channel is under-estimated compared with the experiment. This indicates that the secondary currents are not as sensitive to the streaky structures near solid walls and hence the mean velocity can be predicted fairly well as long as the wall shear stress is well reproduced.

From Fig. 4.30 and 4.31, it can be seen that the SK simulation produces good agreement with the measurements. The high intensity region of turbulence energy

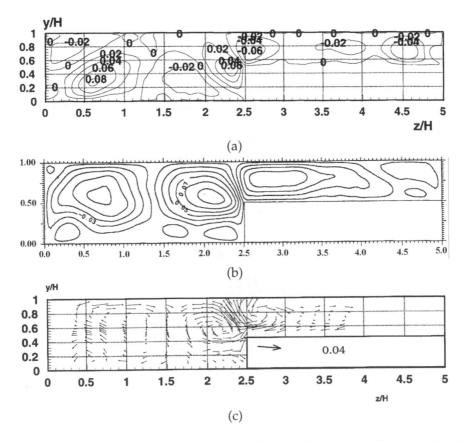

Fig. 4.31 Comparison of secondary currents (a) Calculated by Satoh & Kawahara (b) Calculated by Thomas & Williams (1995a) and (c) Measured data by Tominaga & Nezu (1991).

is well predicted near the main channel/flood plain interface. The anisotropy of the normal stresses, which controls secondary currents, is also well predicted near the interface, whereas the anisotropy near the free surface is considerably underestimated. The shear stress changes its sign near the interface in the experiment, this behaviour is reasonably well captured by the computation although the calculated level is somewhat higher than the measurement. TW also report the distributions of turbulence intensities and energy. Their simulations show good agreement.

In summary LES can reproduce the mean velocity and the turbulent stress distributions together with the instantaneous flow structure. Hence, it may provide insights that will help to better interpret experimental findings. The LES simulations are sensitive to the spatial resolution of the numerical grid and the extent of calculation domain in the streamwise direction, the differences between the two sets of simulations is larger than expected. To use LES as a research tool for turbulent flows in compound open channels, more research effort is necessary to quantify the effects of calculation conditions on the results and to develop algorithms to reduce the computational time.

4.6.2. Numerical analysis of flow and bed topography

4.6.2.1. Method of analysis

Co-ordinate transformation is used as a simple way to incorporate the effects of the complex boundary profile into the numerical analysis. Figure 4.32 shows the computation domain of the compound meandering channel to be analyzed, together with the coordinate system. The plane coordinate systems are transformed from rectangular ones (x, y) to curvilinear ones (ξ, η). As shown in Fig. 4.33, the vertical

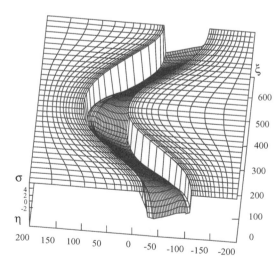

Fig. 4.32 Co-ordinate system and computational domain for compound meandering channel with moveable bed (length unit: cm).

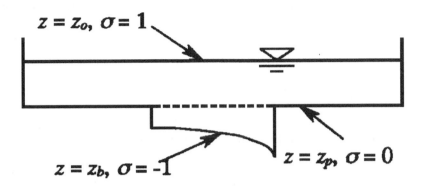

$z = z_o, \ \sigma = 1$

$z = z_b, \ \sigma = -1$

$z = z_p, \ \sigma = 0$

Fig. 4.33 Definition of a σ co-ordinate system.

coordinate system z is transformed into σ coordinate system, which is defined as

$$
\begin{cases}
z = z_0 + (z_0 - z_p)\sigma, & (\sigma \geq 0) \\
z = z_p + (z_p - z_b)\sigma, & (\sigma < 0)
\end{cases}
\tag{4.60}
$$

where z_0, z_p and z_b represent the reference plane near the water surface, flood plains and bed levels, respectively.

The values $\sigma = 1, 0, -1$ represent the reference plane for water surface, flood channel and bed heights, respectively. At this time, the scale parameters relating to the vertical axis are respectively represented by using depths above and below a flood channel. If z_0 and z_p are constant, the metric tensors can be evaluated from changes in the bed surface or the depth below the flood channel alone, without having to incorporate mesh movement. Thus, a coordinate system results in the following metric tensor matrix.

$$
\begin{bmatrix}
\xi_x & \eta_x & \sigma_x \\
\xi_y & \eta_y & \sigma_y \\
\xi_z & \eta_z & \sigma_z
\end{bmatrix}
=
\begin{bmatrix}
y_\eta/J' & -y_\xi/J' & -\left(z_\xi\xi_x + z_\eta\eta_x\right)/z_\sigma \\
-x_\eta/J' & x_\xi/J' & -\left(z_\xi\xi_y + z_\eta\eta_y\right)/z_\sigma \\
0 & 0 & 1/z_\sigma
\end{bmatrix}
\tag{4.61}
$$

$$
J = J'z_\sigma, \quad J' = x_\xi y_\eta - x_\eta y_\xi
\tag{4.62}
$$

Where subscripts in the coordinate system indicate the partial differentiation of that coordinate, t represents time, and J the Jacobian. Contravariant flow velocity is represented in the curvilinear coordinate system as follows:

$$
\begin{cases}
U = \xi_x u + \xi_y v \\
V = \eta_x u + \eta_Y v \\
W = \sigma_z W' = \sigma_z \left(\sigma'_x u + \sigma'_y v + w\right)
\end{cases}
\tag{4.63}
$$

where $u = x$-direction flow velocity; $v = y$-direction flow velocity; $U = \xi$-direction flow velocity; $U = \xi$-direction contra-variant flow velocity; $V = \eta$-direction contra-variant flow velocity and $W = \sigma$-direction contra-variant flow velocity. In Equation (4.63), σ'_x and σ'_y represent the quantities defined by $\sigma_x = \sigma'_x \sigma_z$ and $\sigma_y = \sigma'_y \sigma_z$.

4.6.2.2. Basic equations of the flow

When the coordinate system is introduced as described above, the equation of motion that expresses the non-hydrostatic pressure is as follows:

$$\frac{\partial u}{\partial t} + U_j \frac{\partial u}{\partial \xi_j} = g_x - g\xi_{j,x}\frac{\partial \zeta}{\partial \xi_j} + u_T \Delta u - \frac{1}{\rho}\xi_{j,x}\frac{\partial p}{\partial \xi_j} \tag{4.64a}$$

$$\frac{\partial v}{\partial t} + U_j \frac{\partial v}{\partial \xi_j} = g_y - g\xi_{j,y}\frac{\partial \zeta}{\partial \xi_j} + u_T \Delta u - \frac{1}{\rho}\xi_{j,y}\frac{\partial p}{\partial \xi_j} \tag{4.64b}$$

$$\frac{\partial w}{\partial t} + U_j \frac{\partial w}{\partial \xi_j} = g_z - \frac{1}{p}\sigma_z\frac{\partial p}{\partial \sigma} + u_T \Delta v \tag{4.64c}$$

$$\Delta = \frac{\partial^2}{\partial x_j^2} \quad (j = 1,2,3) \tag{4.64d}$$

where g_x, g_y, g_z = gravitational acceleration in the respective directions, ζ = water level fluctuation relative to the reference surface, p = deviation pressure from the hydrostatic pressure. The eddy viscosity v_T is expressed in the following equation using h = depth from the reference surface, z_d = height from the bottom, and $u.$ = bottom friction velocity.

$$v_T = ku_*z_d (1 - z_d/h) \tag{4.65}$$

The continuity equation is expressed as

$$\frac{\partial J U_j}{\partial \xi_j} = 0 \tag{4.66}$$

Integration of Equation (5.7) from the bed level ($\sigma = -1$) to the water surface ($z = z_o + \zeta$) yields:

$$J'\frac{\partial \zeta}{\partial t} + \frac{\partial}{\partial \xi_k}\int_{-1}^{\sigma(z_o+\zeta)} J'z_\sigma U_k d\sigma = 0 \quad (k = 1,2) \tag{4.67}$$

The kinematic boundary conditions at the water surface are given by using the contra-variant flow velocity W' in the σ-direction by Equation (5.9):

$$\frac{\partial \zeta}{\partial t} + U_k\frac{\partial \zeta}{\partial \xi_k} = W', \quad (z = z_o + \zeta) \tag{4.68}$$

In the analysis without the assumption of hydrostatic pressure, the deviation pressure p is solved quickly in the spectral space by using SMAC method for Equations

(4.64) and (4.66) (Fukuoka *et al.*, 1998). The distribution of vertical velocity w is given by the integral of Equation (4.66) from the bottom.

4.6.2.3. Basic equations of bed deformation

The bed variation is expressed by using the sediment continuity equation as follows:

$$J'\frac{\partial z_b}{\partial t} + \frac{1}{1-\lambda}\left(\frac{\partial J' q_{B\xi}}{a\xi} + \frac{\partial J' q_{B\eta}}{a\eta}\right) = 0 \qquad (4.69)$$

where $(q_{B\xi}, q_{B\eta})$ is the contra-variant sediment discharge vector as given by

$$\begin{cases} q_{B\xi} = q_B \left\{ \dfrac{U_b}{\sqrt{u_{kb}^2}} - \dfrac{1}{\sqrt{u_s u_k}}\dfrac{u_{*c}}{u_*}\left(\xi_{1,i}\xi_{k,i}\dfrac{\partial z_b}{\partial \xi_k^\tau}\right) \right\} \\[4mm] q_{B\eta} = q_B \left\{ \dfrac{V_b}{\sqrt{u_{kb}^2}} - \dfrac{1}{\sqrt{u_s u_k}}\dfrac{u_{*c}}{u_*}\left(\xi_{2,i}\xi_{k,i}\dfrac{\partial z_b}{\partial \xi_k^\tau}\right) \right\} \end{cases} \qquad (i,k=1,2) \qquad (4.70)$$

Equation (5.70) is obtained through the coordinate transformation of the longitudinal and transverse sediment discharge vector (Fukuoka *et al.*, 1993). The effects of the bed gradient are incorporated in Equation (5.70). Here, q_B = bed load, μ_s = static friction factor, μ_k = dynamic friction factor, u_{*c} = critical friction velocity, and the subscript b indicates the value at the bed. This analysis takes account of additional tractive force and changes in critical tractive force (Fukuoka *et al.* 1983) due to the longitudinal and lateral bed slope change for sediment discharge calculation.

4.6.2.4. Introduction of spectral calculation method

Because the profile of the boundary in a compound meandering channel is changed periodically in the longitudinal direction, Fourier series expanded from the 0th mode to the 7th mode are applied for the plane boundary profile, the flow field and the bed profile in this direction. The spectral collocation method is employed to solve the differential equation in the longitudinal direction. As shown in Fig. 4.32, 32 spectral collocation points were selected.

For the flow field analysis, a longitudinal differentiation was calculated by differentiating in the spectral space and then the value was inverting to the spectral collocation point through inverse Fourier transformation. The convective terms in (η, σ) plane were differentiated using an upwind difference of 1st and 3rd order in the vertical and transverse direction, respectively.

Using the procedure outlined above, velocities u, v, W, pressure p and water level ζ are calculated on the spectral points and then time-integrated in spectral space. The time integrals of velocity and water level are explicitly calculated with the Huen method having 2nd order accuracy in time. The bed deformation is very slow in comparison with the velocity change, therefore, the time variations of bed and velocity are repeatedly calculated at separate time scales in this analysis. The resistance at the walls is taken proportional to the square of the velocity near the wall, being assigned as a boundary condition for velocity. Impermeable slip-conditions are also applied at the walls. Friction velocities on the walls and the bottom are determined by dividing velocities near the walls and the bed by the resistance coefficient, respectively.

4.6.2.5. Flow and bed deformation analysis

The plane profile of the compound channel is shown in Fig. 4.32, where the total width = 4.0 m; main channel width = 0.8 m; mean main channel bank height = 0.055 m; meander length = 6.8 m; and sinuosity = 1.1. There are five and two vertical calculation points below and above the flood channel height, respectively.

Cases 4 and 5 in the large-scale laboratory experiments by Fukuoka *et al.* (1997b) were analysed. The experimental channel was 15 m long. The flood channel was covered with artificial turf to create the appropriate roughness and the main channel was filled with sand 0.8 mm in diameter. The slope of the channel bed was 1/600. The resistance coefficient at the bottom was calculated by using the equivalent roughness k_s and $z_{1/2}$ using Equation (5.12).

$$\varphi_b = 5.75 \log(z_{1/2}/k_s) + 8.5 \tag{4.71}$$

where, $z_{1/2}$ is the height in the centre, the calculation point for flow velocity, of the 1st bottom mesh. It was assumed that $k_s = 2.8$ cm in the flood channel. The value of the sand diameter was used for k_s at the bottom of the main channel but it did not include the resistance due to sand waves. Table 4.2 gives the conditions for the numerical analysis. The conditions were decided to produce a depth (i.e., relative depth) equivalent to that used in the experiment, and the resistance coefficient was fixed so that the calculated discharge would match the experimental discharge.

Figures 4.34 and 4.35 show the observed and calculated bed shapes by contour lines for Case 4, respectively. Figures 4.36 and 4.37 show similar results for Case 5. In both the cases the flow proceeds from the left to the right. The observed results show the bed shape of one wavelength in the central part of channel.

Comparing Fig. 4.36 with Fig. 4.37, the calculated scouring occurs continuously from the inner bank to the next inner bank at the maximum curvature and the calculated result is similar to the observed one. On the other hand, the bed shape in Fig. 4.35 does not agree well with the one in Fig. 4.36 In the observed results,

Table 4.2 Conditions for Analysis.

Case	Discharge (l/s)	Main Channel Depth (cm)	Flood Channel Depth (cm)	Relative Depth (Dr)
4	35.9	8.0	2.5	0.31
5	63.7	10.6	5.3	0.49

67y

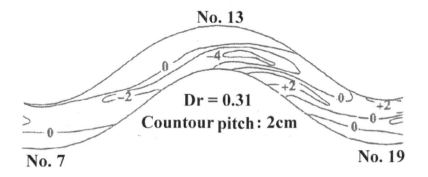

Fig. 4.34 Bed variation contour in Case 4 (Observed results).

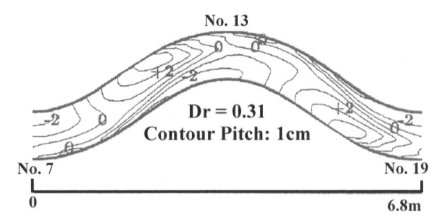

Fig. 4.35 Bed variation contour in Case 4 (Calculated results).

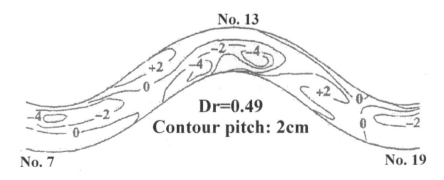

Fig. 4.36 Bed variation contour in Case 5 (Observed results).

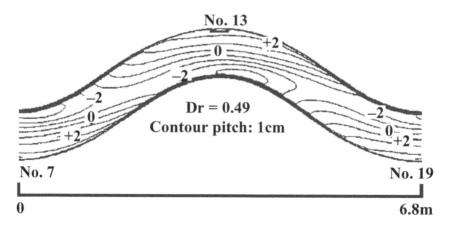

Fig. 4.37 Bed variation contour in Case 5. (Calculated results)

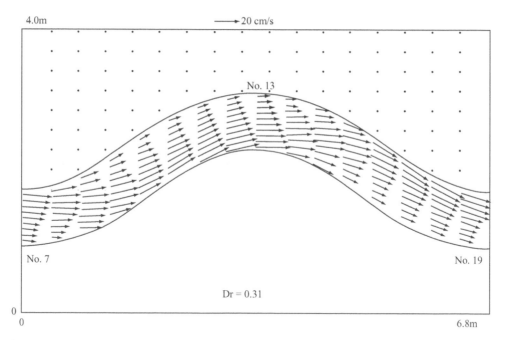

Fig. 4.38 Calculated vectors of depth-averaged velocity below flood channel height in Case 4.

bed scouring occurs at the maximum curvature. In the calculated results, a scouring region spreads out from the outer bank to the inner bank and the maximum scouring occurs near the bank at the maximum curvature. This means that the observed bed shape in $Dr = 0.49$ has the characteristics of a compound meandering flow but the calculated bed shape in $Dr = 0.31$ displays one of a single-section meandering flow.

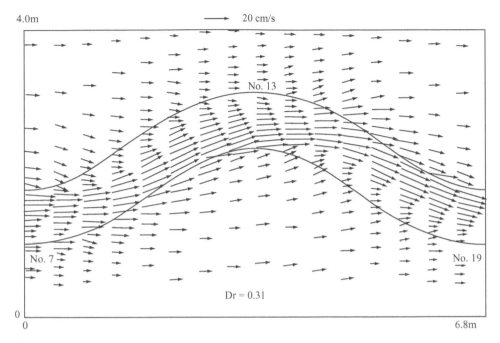

Fig. 4.39 Calculated vectors of depth-averaged velocity above flood channel height in Case 4.

Generally, the bed shape changes from one in a single section meandering channel to one in a compound meandering channel around where the relative depth becomes 0.3 (Fukuoka *et al.*, 1997c). However, this transition of bed shape is sensitive to the roughness of bed and so on. The river bed can take both shapes to this relative depth. When the effects of the secondary flow mentioned in the following section surpass the effects of the longitudinal change in the tractive force for the bed deformation, the bed shape becomes one in a compound meandering channel. The slight difference between observed and calculated results for the scouring position is due to the difference of bed roughness and the use of periodic boundary conditions for the experimental results, which were not actually periodic.

4.6.2.6. Characteristics of the flow field

Figures 4.38 and 4.39 show the calculated velocity vectors at below and above flood channel height respectively for Case 4. Figures 4.40 and 4.41 show similar results for Case 5. As shown in Fig. 4.38, flow below flood channel height concentrates near the inflection point of outer bank and scouring occurs there. On the contrary in Fig. 4.40, flow does not so concentrate and the change in velocity is small. But, the flow over bed goes towards the outer bank and bed materials move to the bank by this secondary flow. As shown in Fig. 4.39 ($Dr = 0.31$), the main flow above the height of flood channel proceeds along the main channel.

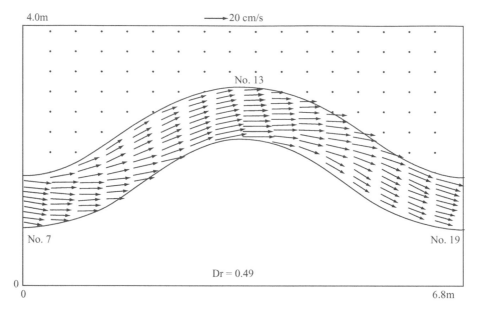

Fig. 4.40 Calculated vectors of depth-averaged velocity below flood channel height in Case 5.

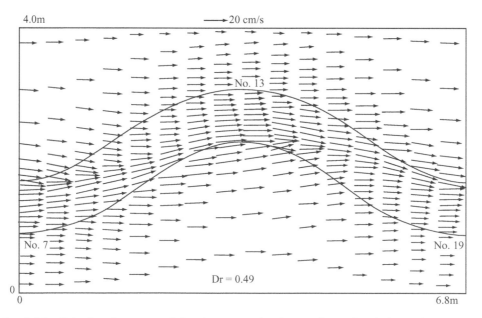

Fig. 4.41 Calculated vectors of depth-averaged velocity above flood channel height in Case 5.

The velocity over the flood channel is small and the velocity changes considerably in the transverse direction. But, the momentum exchange between flow in main channel and above flood channel is small, as the exchange in rate of flow is small in relative depth. On the contrary in Fig. 4.40, flow goes along the level of

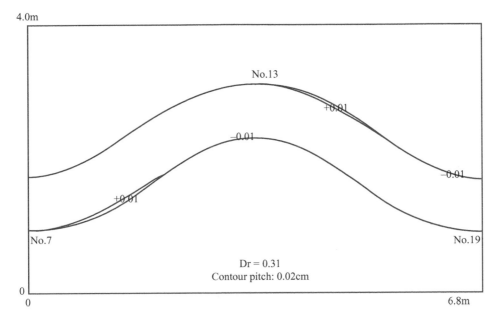

4.0m

No.13

+0.01

−0.01

−0.01

+0.01

No.7

No.19

Dr = 0.31
Contour pitch: 0.02cm

0

0

6.8m

Fig. 4.42 Calculated depth-averaged pressure deviation below flood channel height in Case 4.

flood channel and the velocity above flood channels is large. Comparing Figs. 4.39 and 4.40, it can be seen that the direction of the secondary flows is reversed from one observed in a single section meandering flow.

Figures 4.42 and 4.43 show the distribution of calculated pressure deviation for Case 4 and 5 respectively. When the relative depth is small, the pressure deviation in a main channel below the height of flood channel is small and pressure is almost equivalent to the hydrostatic condition. When the relative depth is large, however, the pressure deviation is large near the bank around the inflection point and the apex of meandering.

Figures 4.44 and 4.45 show the distribution of calculated vertical velocity at the height of flood channel for Cases 4 and 5 respectively. As shown in Fig. 4.44, flow rises up near the bank around the inflection point from lower channel and goes down to lower channel around apex. These places exist where the change in the pressure intensity is large. In Fig. 4.37, the large vertical velocity is not seen. At the same time Fig. 4.37 indicates that the bed scouring occurs continuously over the area where the faster flow goes down. The tractive force is larger over the area where the fast flow goes down and flow goes out of this area with bed materials. Therefore, in this area the bed scouring occurs.

4.6.3. Three-dimensional modelling of the flood channel facility

As discussed in Chapter 2, the velocity field during overbank flow in meandering compound channels is highly three-dimensional. At the straight section between

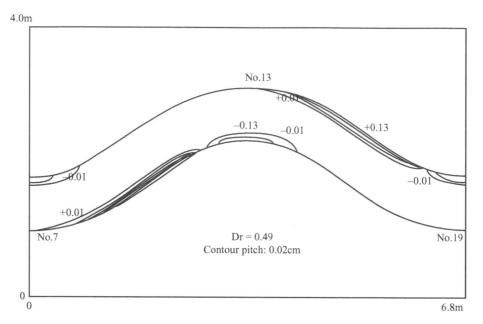

Fig. 4.43 Calculated depth-averaged pressure deviation below flood channel height in Case 5.

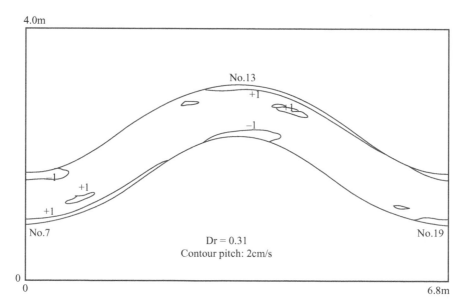

Fig. 4.44 Calculated vertical velocity at the height of flood channel in Case 4.

bends the flood plain flow roles over the main channel and initiates a rotational velocity component that grows until the next bend. As a result the secondary currents at the bend are in the opposite direction to those occurring during inbank

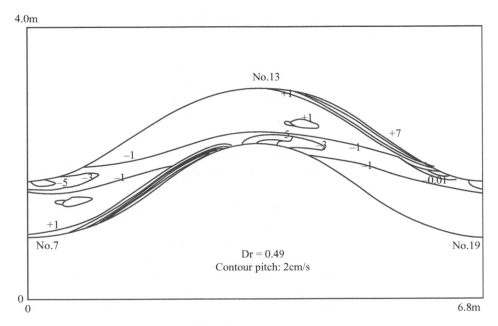

Fig. 4.45 Calculated vertical velocity at the height of flood channel in Case 5.

flow. This is because the rotation generated by the horizontal shear is stronger than those generated by the centrifugal force acting on the water as it passes round the bend. The growing helical flow structure upstream of the bend results in water coming off the flood plain and plunging vertically into the channel. As the rotation develops further in the downstream direction, the flood plain flow begins to roll over the main channel until it finally passes over it without any entrainment. Another noticeable feature of such flow is that water is vigorously expelled from the main channel onto the downstream flood plain after which it continues to travel in a skewed direction relative to the flood plain walls.

The features of such flows have been the subject of extensive physical investigation, however, few quantitative comparisons have been made between the available physical model data and computer model predictions. Recently, Morvan *et al.* (2002) presented results of one such comparison. They used the CFD code CFX to simulate experiment B23 from the FCF, series B programme. In this experiment the main channel was trapezoidal in cross-section with a depth of 150 mm, a top width of 1200 mm and banks sloping at 1:1. Four 60° meanders were contained within the 48 m long by 10 m wide flume giving a meander wavelength of 12 m. The flood plain slope was 9.96×10^{-4}. Flow conditions were uniform with a total depth of 200 mm, 50 mm on the flood plain. The measured discharge at the inlet was $0.25 \, \text{m}^3/\text{s}$.

The numerical simulation was undertaken on a non-staggered structured grid of hexahedral elements. The pressure correction term SIMPLEC, (van Doormal and Raithby, 1983) was used with the Rhie–Chow algorithm (Rhie and Chow,

1983). An algebraic multi-grid solver developed specifically for use with CFX was employed to enhance convergence (Lonsdale, 1993). In addition, a higher order numerical scheme was used to resolve the steep velocity gradients. The limiter function SMART (Gaskell & Lau, 1988) was used with QUICK (Leonard, 1979) to ensure boundedness in the solution of convection terms.

Closure of the Reynolds Averaged Navier Stokes equations was achieved using a standard $k-\varepsilon$ model (Launder & Spalding, 1974). A rigid lid approximation was used to simulate the free surface. Morvan *et al.* (2001) report that a grid independent solution was achieved using 196,800 cells over two meander wavelengths. The mesh density was increased in the locality of high velocity gradients.

Figure 4.46 shows computed and measured velocity vectors at various plan locations around a meander. In general, the comparison is good with a maximum error of around 12%. A comparison of secondary currents at the same locations is shown in Fig. 4.47. It is apparent that the computer model has captured the reversal of the secondary currents due to horizontal shear and the plunging of the flood plain flow into the main channel. It would appear therefore that provided a sufficiently refined grid is employed good quality predictions of the velocity field can be achieved using a relatively simple turbulence model.

4.6.4.   CFD modelling of the River Severn

Recently published material has reflected the increasing interest of Computational Fluid Dynamics (CFD) applied to the study of open-channel flows (Demuren, 1993; Cokljat & Younis, 1995; Sinha *et al.*,1998). Yet, few of these have attempted to apply this technology to full-scale natural channels. To date CFD application has been mostly limited to small-scale prismatic channels.

Earlier work has confirmed the potential of CFD to resolve scale problems with a high level of detail Morvan *et al.* (2002). This study reproduced a well-known laboratory experiment from the Flood-Channel Facility (FCF) programme, for an overbank flow in a 60° meandering trapezoidal channel with a depth ratio of 25%. This work has demonstrated the capacity of CFD to reproduce the detailed velocity field in overbank flows with a high degree of accuracy. It has served as a benchmark test to validate numerical and physical issues such as mesh resolution, boundary conditions and turbulence closure. The applicability of the modelling processes has therefore been established, and it is now clear that CFD has the potential to simulate the velocity field during flood flows in natural channels.

The second phase of this research programme was therefore to transfer this experience to natural rivers. Three-dimensional CFD modelling of rivers could help bridge a gap in understanding the mechanisms controlling features of the velocity field in flooded meandering channels at a large scale. This is important because the observation of such physical processes in real conditions is extremely difficult.

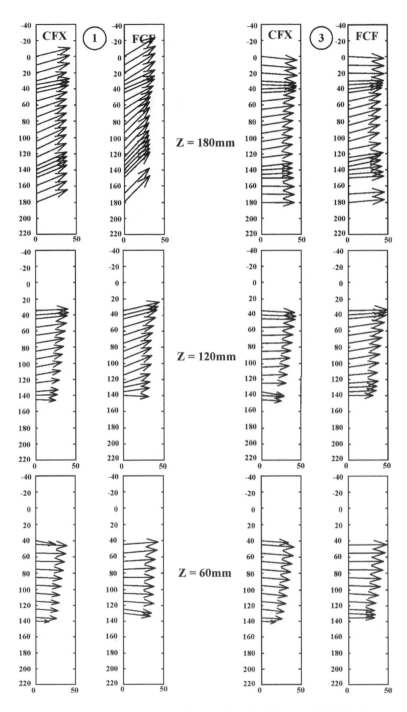

Fig. 4.46 (a) Velocity vectors at depth 60, 120 and 180 mm at FCF Sections 1 and 3.

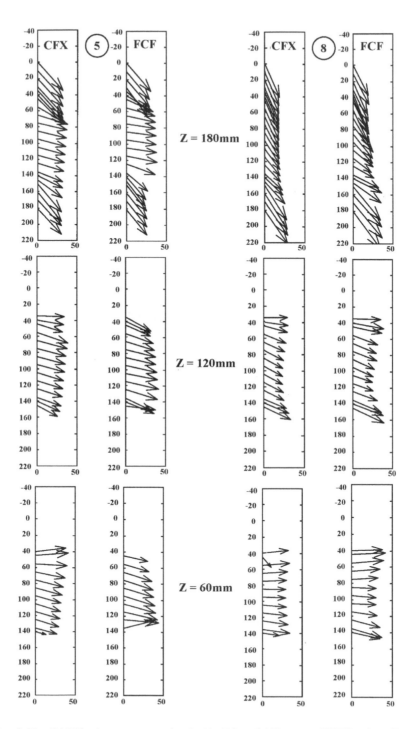

Fig. 4.46 (b) Velocity vectors at depth 60, 120 and 180 mm at FCF Sections 5 and 8.

The above numerical problem is discretized using a finite volume approach which allows the advection component to be discretized with the most appropriate numerical scheme (e.g. upwind, hybrid, QUICK) and uses a central difference scheme for the diffusion part of the equations. Pressure is calculated using a variation of the SIMPLE algorithm called SIMPLEC.

The boundaries are defined as follows: (1) inlet and outlet, and (2) solid walls. At the inlet a velocity profile is implemented and the turbulence parameters are as

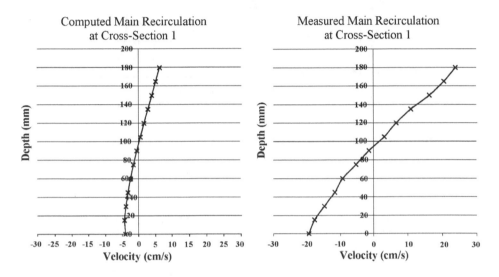

Fig. 4.47   (a) FCF B23 CFX Model: Recirculation at Cross-Section 1.

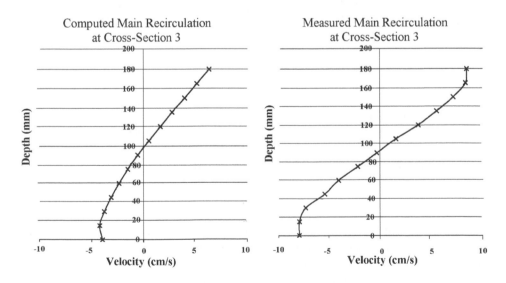

Fig. 4.47   (b) FCF B23 CFX Model: Recirculation at Cross-Section 3.

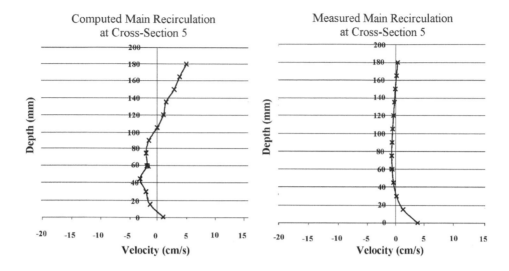

Fig. 4.47   (c) FCF B23 CFX Model: Recirculation at Cross-Section 5.

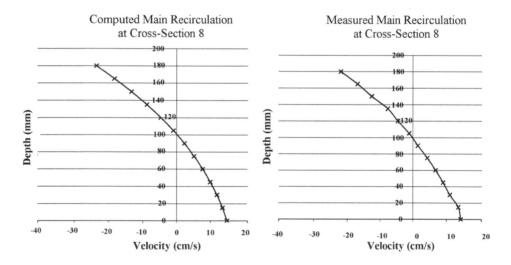

Fig. 4.47   (d) FCF B23 CFX Model: Recirculation at Cross-Section 8.

detailed in Morvan *et al.* (2002). At the outlet fully developed flow conditions are assumed and a zero normal gradient is implemented for all velocity and turbulence variables. The second point (2) implies that the free surface is implemented as a rigid lid with its initial condition specified using field data and then adjusted using the computed pressure field. At the walls the following modified law of the

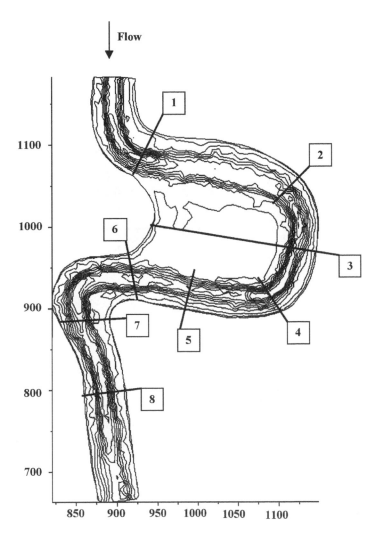

Fig. 4.48 Plan view of the River Severn and the locations of cross-sections (scale in m).

wall is introduced to account for rough hydraulic conditions:

$$\frac{u_\tau}{u_*} = y^+ \qquad \text{if } y^+ < y_0 = 11.63,$$

$$\frac{u_\tau}{u_*} = \frac{1}{\kappa} \ln(E(k_s^+)y^+) \qquad \text{otherwise,} \qquad (4.71)$$

with,

$$\forall k_s^+, E(k_s^+) = \frac{9.8}{1 + 0.3 \cdot k_s^+} \qquad (4.72)$$

where, u_τ is the tangential velocity, u_*, the shear velocity and k_s^+, the non-dimensional roughness height.

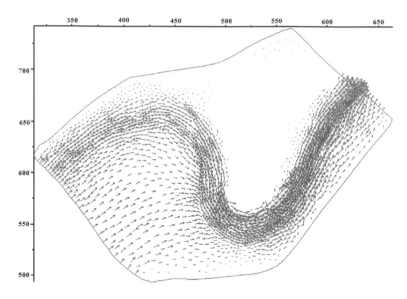

Fig. 4.49 River Severn depth-averaged velocity vectors for a flow of $100 \, m^3/s$.

The River Severn is the third longest river in Britain after the Thames and the Wye respectively. It runs along the Southeast border between England and Wales, and the Severn Estuary is a significant landmark on the map of Southern Britain. It measures 206 km, drains an area of 4330 km and has a mean annual discharge of about $62.70 \, m^3/s$ (Ward, 1981).

The section modelled in this work is located 20 km east of Shrewsbury. A single meander of about 600 m length was used for the test site. This reach is shown on Fig. 4.48. It is about 30 m wide, between 6.0 m and 7.0 m deep with respect to the right flood plain. This flood plain is bunded by an earth embankment to the South and has been lowered to provide material for embankment construction. Consequently, it is inundated reasonably frequently. Conversely, the right flood plain, which is around 2.0 m higher, is seldom flooded.

Hydraulic field data were collected for $100 \, m^3/s$ flow conditions, including water surface profile along the main channel, and velocities at cross-sections 4, 5 and 7. Bed sediment samples were also taken in the region of cross-sections 1, 2, 3, 5 and 7. Such condition gave approximately 1.5 m of water flowing over the right flood plain.

Numerical tests were implemented to determine the impact of the discretization and boundary assumptions on the solution. The velocity magnitude and direction were calculated throughout the model. However to facilitate the analysis, cross-sectional data were plotted at regular intervals (see Fig. 4.48). These are shown below for some sections, the angle being calculated from the normal to the section, positive to the left, negative to the right:

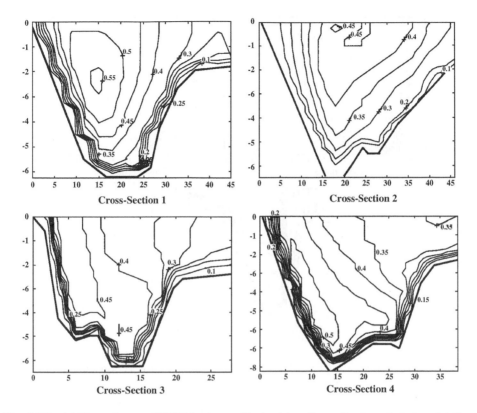

Fig. 4.50 (a) River Severn CFX Velocity profile (m/s) at Cross-Sections 1, 2, 3 and 4 ($k_s = 0.100$m, Grid CFX S-1, $k-\varepsilon$ model).

At Section 3, Figs. 4.50(a) and (c), the flow is deflected towards the outer bend, yet there is little interaction between main channel and flood plain flows, as shown on the angle plot. However, at Section 4, the flood plain flow crosses over the main channel. It shears over the inbank water, starting to generate a strong inbank.

At Section 5, Figs. 4.50(b) and (d), the shear effect is at its strongest and generates a recirculatory motion reaching an intensity of 30% of the peak local velocity. A strong helical motion exists between the walls of the main channel. This is clearly shown on the angle plot for Section 5, Fig. 4.50(d), and was confirmed by numerical tracer experiments at this location Morvan (2002). The velocity core runs along the left bank, at this location and in the upper part of the channel some water is ejected onto the left flood plain where it is accelerated (0.75 m/s) and changes direction rapidly. These observations are similar to what was observed at small scale in the FCF, Morvan *et al.* (2002) and would indicate that scale does not affect the flow mechanisms controlling this behaviour in such channels.

Validation data are difficult to come by, and when available they include a significant level of uncertainty. However some data collected in the autumn of 2000

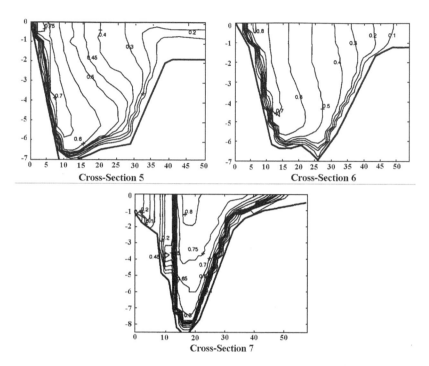

Fig. 4.50 (b) River Severn CFX Velocity profile (m/s) at Cross-Sections 5, 6, and 7 ($k_s = 0.100$ m, Grid CFX S-1, $k-\varepsilon$ model).

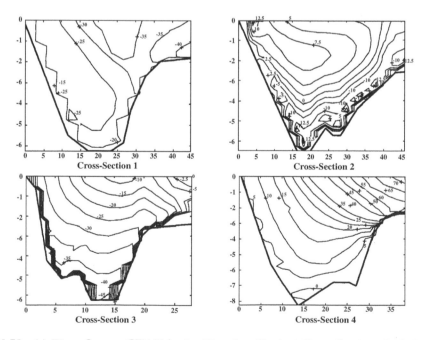

Fig. 4.50 (c) River Severn CFX Velocity Direction (deg) at Cross-Sections 1, 2, 3 and 4 ($k_s = 0.100$ m, Grid CFX S-1, $k-\varepsilon$ model).

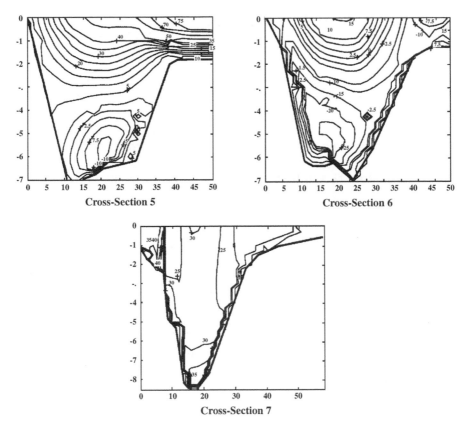

Fig. 4.50 (d) River Severn CFX Velocity Orientation (deg.) at Cross-Sections 5, 6 and 7 ($k_s = 0.100$ m, Grid CFX S-1, $k-\varepsilon$ model.)

are presented in Fig. 4.51 and used for comparison. These indicate that the model reproduces the measured velocity reasonably well. However, an in-depth analysis of the results shows significant local differences of up to 30% in velocity predictions. This is likely to be due to a combination of model and data inaccuracy, especially true at this scale (compared to flume models) and for such extreme flood events. Additionally poor resolution at the model walls greatly affects the velocity profile locally.

The model reproduces the principal features of the velocity field correctly, although full validation was not possible in the absence of a more extensive data set. This confirms the potential application of CFD as a prediction method for simulating river flood flows, although complex geometry and scale are major obstacles in the construction of suitable numerical grids.

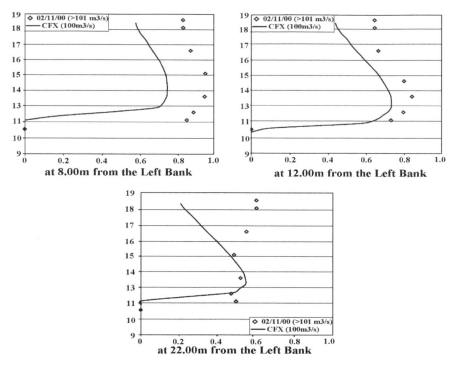

Fig. 4.51 Comparison between Field Data and River Severn CFX Model Predictions at Cross-Section 5.

4.7. Concluding Remarks

This chapter has explained the basis and the background to various techniques for the modelling of compound channel flows. The case studies examined, provide a range of applications which cover the fundamental development of the simulation methods themselves to practical investigations of engineering problems.

The collection of field data in flood conditions is extremely difficult. Because of this such data is extremely valuable where it exists, but, more often than not, it is unobtainable. Because of this, numerical simulations are set to play and increasingly dominant role in both research and in river engineering practice. It is vital then to successfully transfer the experience of modeling from the research community into the realms of applied river engineering. The scope of work described this chapter charts a progression from fundamental studies towards the application of advanced numerical methods to full-scale river conditions. All aspects of this chain are vital and each must be nurtured.

The three-dimensional CFD modelling of rivers is set to become increasingly common and is likely to become a focus of attention. At a basic level, this will occur simply because such simulations would until very recently been impossible or impractical due to the limits of computational resources. Care must now be exercised to ensure that these new tools, and the opportunities they present, are used wisely. In this regard the academic community has a vital role in disseminating examples of best practice to the wider community. There will also be benefits for the research community as the large scale application of such models can help bridge some of the gaps in understanding the mechanisms controlling features of the velocity field in flooded meandering channels at a large scale. This is especially important because the observation of such physical processes in real conditions is extremely difficult.

Finally, it is important to recognize that advanced numerical techniques may tend to lessen our reliance on physical model studies and data collection. This is a temptation that should, as far as possible, be resisted as the studies presented in this chapter all include validation of a numerical model with data obtained either in the laboratory or the field. Thus the collection of the laboratory and field data required for validation should not be neglected, as ultimately, this will undermine the application and development of advanced modelling techniques.

Chapter 5

Design Considerations

S. Fukuoka, A. Watanabe and P. R. Wormleaton

5.1. Stage-Discharge Relationship and Discharge Storage

5.1.1. Introduction

Depending on a channel's planform and cross-sectional shape, it is known that a flood flow whose discharge changes temporally is stored as it proceeds downstream, resulting in transformation of discharge and water level hydrograph such as in the peak discharge attenuation (Chow, 1956; Fukuoka, 1999; Fukuoka et al., 2000).

In a channel with a uniform cross-section, the shape of flood wave is transformed into a mild one in the process of the downstream propagation because of the time change in hydraulic quantities. On the other hand, in a flood flow the peak discharge attenuation is caused by the channel storage function inherent in an unsteady non-uniform flow. Since little is known about the relationship of flood flow characteristics and channel storage characteristics, the natural flood flow storage capacity of rivers has yet to be reflected in flood control plans. Large rivers with broad floodplain and complex planforms are assumed to have considerable storage capacity. During flooding vegetation growing in channels will raise water levels and lower a channel's discharge capacity and therefore is often cleared for that reason. Large rivers with sufficient embankment height are amenable to measures that conserve channel vegetation not just because of its environmental functions, but also to increase flood flow storage in the channel. Fukuoka et al. (2000) studied by experiment the effects on the hydraulic properties of flood flows on planform, transverse shape, and unsteadiness in the compound meandering rivers commonly seen in Japan, thus gaining a basic understanding of these effects.

Flow in a compound meandering channel was found to differ greatly from those of bankfull flow when the water depth over the floodplain is large, Ashida et al. (1989, 1990), Kinoshita (1988), Mori et al. (1989), Willetts et al. (1993) and Sellin et al. (1993). Among these differences, the maximum velocity filament in the main channel runs near the inner banks; regarding the secondary flows, the rotation is reversed from that observed in bankfull flows (Fukuoka et al., 1996; Muto et al., 1997). These changes in the flow fields are accompanied by the cessation of scouring at outer bank bed and by the scouring near the inner bank bed

(Fukuoka *et al.*, 1997a). It is generally believed that these flow characteristics are dependent on the sinuosity and the phase difference between the main channel alignment, the levee alignment, the relative depth, the ratio of roughness and the width between the flood channel and main channel. The effects of relative depth and sinuosity were shown by means of the large-scale hydraulic experiments performed by Fukuoka *et al.* (1997a, 1997b) and the flow velocity distributions calculated from actual aerial survey photographs (Fukuoka *et al.*, 1997c).

The geometrical and hydraulic conditions dramatically affect the features of the flow structure in a compound meandering channel therefore numerical analysis should be used to compute flow and bed evolution. Efforts were made to develop three-dimensional numerical models of compound meandering flows. The authors developed a numerical model that used the spectral method to express three dimensional flows in compound meandering channels with fixed bed (Fukuoka *et al.*, 1998; Watanabe *et al.*, 1999.) and it is able to express the changes in the structure of flow due to the changes in the relative depth. In this chapter, a further step has been taken on the research of Fukuoka *et al.* (1999), performing a systematic experiment on flood flows in a compound meandering channel and analyzing flood observation data for the Ota River with the objective of estimating a channel's storage capacity of a flood. Assessing both channel storage capacity and peak discharge attenuation mechanism, a new approach to flood control plans is presented.

This section also describes the numerical analysis model incorporating bed variations into a 3-D flow model (Fukuoka *et al.*, 1998; Watanabe *et al.*, 1999) and discusses its suitability for flow and bed variation against the change in the relative depth. A flow model of non-hydrostatic pressure mode is used to examine the significance of change in pressure intensity.

5.1.2. Flood flow storage and peak flow attenuation

A large open channel having a compound meandering shape (Fig. 5.1) was used to assess the mechanism of flood wave propagation, storage capacity, and peak discharge attenuation. Table 5.1 shows the specifications of the experimental channel.

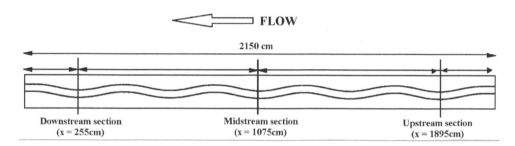

Fig. 5.1 Plan view of the experimental channel used to assess flood storage.

Table 5.1 Specifications of experimental channel.

Channel length	Total channel width	Main channel width	Channel bed slope	Sinuosity	Meander wavelngth	Flood channel height
2150 (cm)	220 (cm)	50 (cm)	1/1000	1.02	410 (cm)	4.5 (cm)

Table 5.2 Hydrograph properties.

Case	Peak discharge	Maximum relative depth	Duration of flood channel inundation	Duration of rising water period	River equivalent of peak discharge	River equivalent of duration of flood channel inundation
Case 1, 3 (Hydro A)	17 (l/sec)	0.4	1200 (sec)	600 (sec)	9622 (m^3/sec)	4.7 (hr)
Case 2 (Hydro B)	18 (l/sec)	0.41	3500 (sec)	1000 (sec)	10188 (m^3/sec)	13.6 (hr)
1983 flood in Tone River	8100 (m^3/sec)	0.56	64 (hr)	—	–	–

The floods flow having the hydrographs (Hydro A and Hydro B) indicated in Fig. 5.5 by solid lines, were released from the upstream end of the channel. Table 5.2 shows the characteristics of each hydrograph and those of a 1983 flood in the Tone River. Assuming the channel to be a 1/200-scale model, the experimental hydrograph generally satisfies Froude similarity law of flood flows in actual rivers (Fukuoka, 2000). The downstream discharge ($Qalt$) hydrograph was calculated from the temporal change in storage capacity (dS/dt), which was calculated from the temporal change in the longitudinal water level simultaneously measured between the upstream and downstream sections, and from the inflow discharge (Qin) hydrograph for each corresponding time. Water level was measured with a servo-type wave gauge meter; velocity, with a Type I electromagnetic velocimeter. The longitudinal distribution of water level was measured between the aforesaid upstream and downstream cross-sections, whereas the transverse distribution of velocity was measured at the midstream cross-section ($x = 1075$ cm) (Fig. 5.2).

The experiment was performed under three different sets of conditions. Case 1 employed an symmetrical triangular hydrograph (Hydro A), and Case 2 a hydrograph (Hydro B) with relatively less unsteadiness than in Case 1. In Case 3, the same hydrograph (Hydro A) as that in Case 1 was used, and vegetation was placed in the flood channel of a compound meandering channel to investigate the effects of riparian vegetation on storage volume.

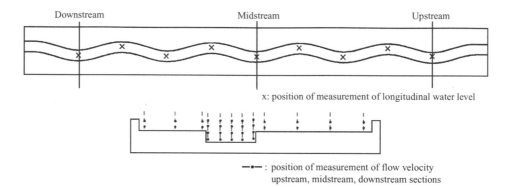

x: position of measurement of longitudinal water level

—•— : position of measurement of flow velocity
upstream, midstream, downstream sections

Fig. 5.2 Points of measurement of water level and flow velocity.

At the downstream end of the channel a porous plastic body (porosity = 91%) was installed to create sufficient resistance so that depth near the downstream end in each time would approach nearly uniform flow depth.

5.1.2.1. The effects of unsteadiness

Figures 5.3 and 5.4 show the temporal change in the longitudinal water level in Case 1 (Hydro A) and Case 2 (Hydro B). This is shown for the times, during the

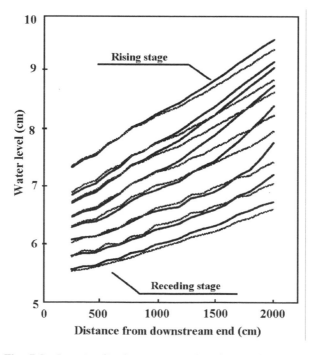

Fig. 5.3 Longitudinal water level distribution for Case 1.

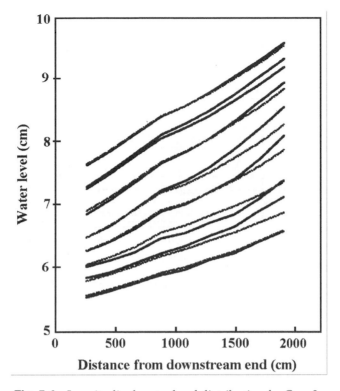

Fig. 5.4 Longitudinal water level distribution for Case 2.

rising water period (solid line) and receding water period (broken line), at which the downstream end water level was the same. These graphs demonstrate that the water surface slope is steep during the rising water period but gentle during the receding water period; the mechanism of flood wave propagation is also evident in the channel experiment, and that high unsteadiness (i.e., Hydro A) results in marked changes in water surface slope.

Figure 5.5 shows the discharge hydrographs for the upstream section time (where $x = 1895$ cm) and the downstream section (where $x = 255$ cm). These discharge hydrographs suggest that as the flow moves downstream, it is subjected to the effects of flood channel roughness, main channel alignment, and mixing of the main channel and flood channel flows, the result being that the downstream section indicates transformation in the flood hydrograph, i.e., attenuation of the peak discharge, delaying of the occurrence of peak discharge, and prolonging of the flood duration. Such hydrograph changes are prominent in high-unsteadiness (Hydrograph A). Temporal delaying of the discharge hydro graph and the attenuation of maximum discharge occur primarily because of storage of the flood flow. Storage capacity per unit time (dS/dt) is obtained by subtracting discharge Q_{out} from discharge Q_{in}. Figure 5.6 gives the ratio of temporal change in storage to inflow discharge Q_{in}. On the positive side of the graph, the value of Q_{out} is less

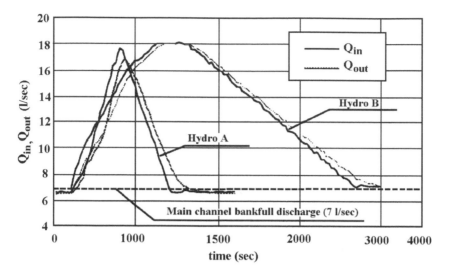

Fig. 5.5 Comparison of upstream and downstream discharges.

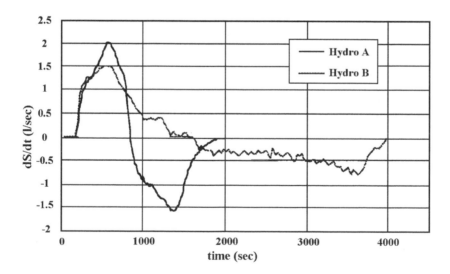

Fig. 5.6 Changes in the ratio of *dS/dt* to inflow discharge.

than that of Q_{in}, and so the flood flow is stored within the channel. Subsequently, on the negative side, Q_{in} falls below Q_{out} and so the discharge that had been stored within the channel now flows out. In Hydrograph A, dS/dt at the time of peak discharge inflow is 5% of Q_{in} whereas maximum dS/dt is 15% of Q_{in}.

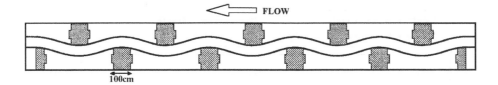

Fig. 5.7 Vegetation placed on the inner bank of the meander channel.

5.1.2.2. The effects of riparian vegetation

To investigate the storage effect of vegetation, a continuous cover of vegetation was installed on the inner bank of the meander, in the broad portion of the flood channel, where storage effect was assumed to be greatest (see Fig. 5.7). This configuration of vegetation created resistance to the flow and so raised water levels by decreasing the cross-sectional area of the flood channel.

Figure 5.8 shows the longitudinal water level distribution for Case 3, in which vegetation was placed. Figure 5.9, which is temporal change in the water surface slope in the channel's central section (i.e., the section 1,485 to 665 cm from the downstream end), shows a marked increase in water surface slope in Case 3, relative to Case 1 of Fig. 5.3, in which no vegetation was placed. In Case 1, peak surface slope occurred at the time of large mixing between the flow in the flood channel and the flow in the main channel which characterizes a compound meandering channel flow. In contrast, peak surface slope occurs in Case 3, decrease of the peak water depth, where the effect of the vegetation is greatest.

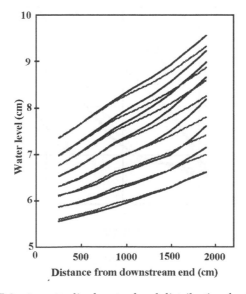

Fig. 5.8 Longitudinal water level distribution for Case 3.

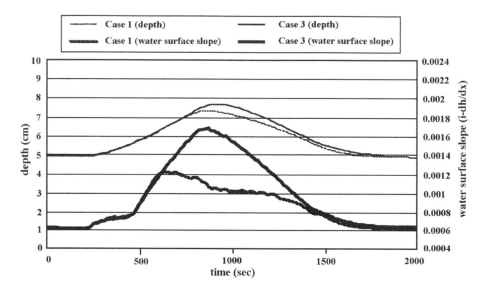

Fig. 5.9 Water depth and water surface slope hydrographs.

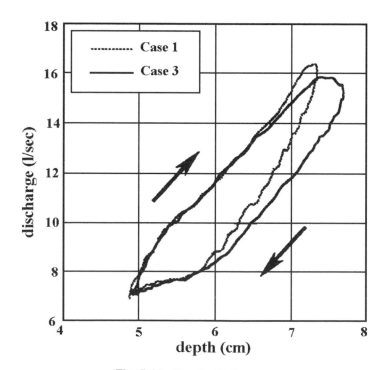

Fig. 5.10 Depth-discharge curve.

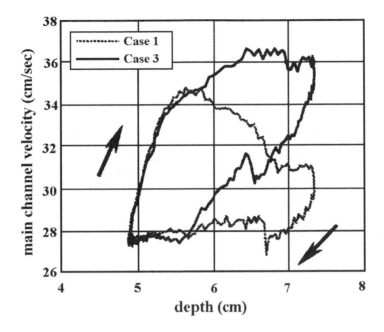

Fig. 5.11 Depth-main channel velocity curve.

Figures 5.10 and 5.11 are the depth/discharge curve and depth/main chan-nel average velocity curve for central cross-section of the channel. In Case 3, the depth/ discharge curve describes a much larger loop than in Case 1 because of a surface slope that is larger due to the effects of the vegetation. This demonstrates that, even under identical unsteady input conditions, factors that make a channel's surface slope change will result in a larger loop for the depth/discharge curve. The depth/main channel average velocity curve also shows that in Case 1, surface slope peaks at a depth of 5.5 to 6.0 cm, where the mixing effects of the compound meandering flow are great, main channel average velocity also peaks at the same time. In Case 3, in contrast, because the effects of vegetation increase along as flood channel depth increases, velocity peaks near maximum surface slope and maximum depth.

Figure 5.12, which is a hydrograph of inflow and outflow in Cases 1 and 3, shows that the attenuation of peak discharge at the downstream section is approx-imately twice as high in Case 3 than in Case 1. Figure 5.13 gives the temporal change in storage (dS/dt) in relation to Q_{in}, and shows that in Case 3, the ratio is higher than in Case 1, as much as 12% to 13% of the flood flow is stored during the rising water period. This disgram also indicated that when peak discharge occurs at the upstream section, dS/dt was 6.2% relative to Q_{in} in Case 1, but was 10.7% in Case 3. This clearly shows that riparian vegetation increases both flood flow storage and attenuation of peak discharge.

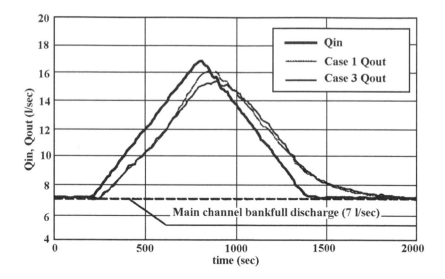

Fig. 5.12 Comparison of inflow discharge and outflow discharge for Cases 1 and 3.

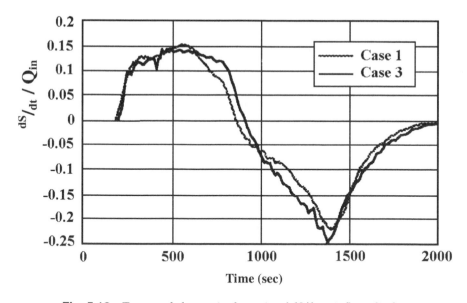

Fig. 5.13 Temporal change in the ratio of *dS/dt* to inflow discharge.

5.1.2.3. Comparison of the results of flood flow experiments and the observations from the Ota river flood

Figures 5.14 and 5.15 show the results of analysis of flood observation data obtained at the Yaguchi Observation Post No. 1 (located 11.6 km upstream from the river's

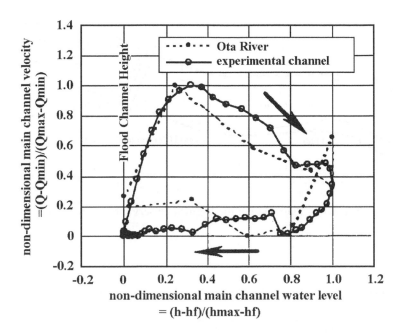

Fig. 5.14 Nondimensional depth-velocity curve for Ota River and experimental channel.

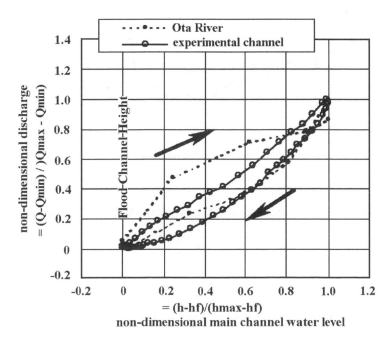

Fig. 5.15 Nondimensional depth-main channel velocity curve for Ota River and experimental channel.

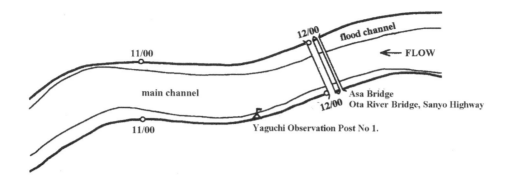

Fig. 5.16   Plan view of the Yaguchi Observation Post No. 1.

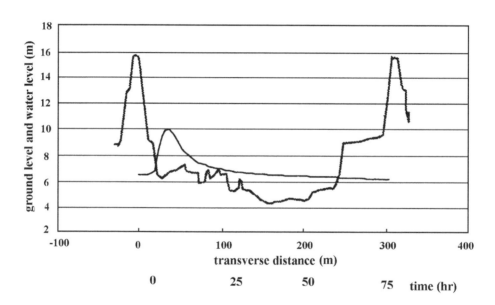

Fig. 5.17   Cross-section at the Yaguchi Observation Post No. 1 and water level hydrograph.

mouth) during July 1983 flooding and the results from the main channel Case 1. With regards to the relationship of water level to average velocity and discharge, water level is nondimensionalized with peak values. The arrows in the graphs indicate the passage of time.

Figures 5.16 and 5.17 give the plan and cross-sectional views and the flood hydrograph of the Ota River Observation station and Table 5.3 shows the

Table 5.3 Characteristics of the 1983 Ota River flood.

Flood case	Maximum relative depth	Real river equivalent peak discharge	Real river equivalent flood channel inundation time
Case 1 (Hydro A)	0.4	9622 (m³/sec)	4.7 (hr)
1983 Ota River flood (Yaguchi Observation Post No 1)	0.34	3500 (m³/sec)	6 (hr)

characteristics of the flood hydrograph. At the observation station, located in a compound, gently meandering reach of the river, a water level-main channel velocity and water level-discharge relationship indicate roughly the same characteristics as those of the experiments. This suggests that channel storage occurred during flooding in the Ota River. In order to assess the channel storage capacity of the Ota River, it is, however, necessary to measure the discharge either upstream or downstream within the reach in question and the temporal change in the longitudinal water level during a flood.

5.2. Flood Flows and Bed Variation in Meandering Channels

5.2.1. Introduction

In Japan, rivers have been planned from the viewpoint that compound profiles are preferable for river channel cross-sectional profiles with respect to such factors as embankment safety, channel stability and multiple use of rivers. Consequently, much research on compound channels has focused either on straight compound channels, in which both the embankment alignment and main channel alignment are straight, or channels in which the embankment alignment and main channel alignment meander in the same phase. The main channels and flood channels of actual rivers, however, nearly always vary in sinuosity, and there normally exists a phase difference between the embankment alignment and main channel alignment, the result being inflow and outflow between the main channel and flood channel. This has begun to be understood as a hydraulic phenomenon separate from straight compound flows and curved compound flows without phase differences (Ashida *et al.*, 1989).

To give an overview of the major research carried out thus far on compound meandering flows, with respect to experimentation on compound meandering channels with flat, fixed beds, Kiely (1990) has investigated compound flow regimes through experiment with varying relative depths Dr (= flood channel depth/total main channel depth) in compound meandering channels of constant sinuosity constructed in a straight channel. Willetts & Hardwick (1990, 1993) have demonstrated the relationship between main channel sinuosity and water level-discharge using a meandering channel comprising a combination of arcs and straight lines, while Muto *et al.* (1995 & 1996) using channels having three different sinusitis, have made detailed measurements of flow velocity distribution, the process of secondary flow development, shear layer development and other aspects of turbulence structures. Additionally, Fukuoka *et al.* (1997) have demonstrated the main flow and secondary flow structures of large-scale compound meandering flows through experiment on a scale approaching that of actual rivers, with a large flood channel height/main channel width ratio and a large flood channel roughness. The foregoing research has revealed, for instance, that in a compound meandering channel, the flows moving between the main and flood channels, in a manner determined primarily by the main channel alignment, produce a large horizontal shear force that acts roughly at the height of flood channel bed in the main channel, and the secondary flows in the main channel rotate in the opposite direction to those in a simple meandering flow; and that the maximum flow velocity filament occurs along the shortest distance between adjacent inner banks. However, these investigations addressed meandering channels with flat, fixed beds, and so do not address the issue of bed evolution.

In an experiment using mobile bed compound meandering channels, Ashida *et al.* (1989), focusing on the differences between compound and simple meandering flows, have studied the characteristics of flow and bed evolution in three scenarios: in bank flow, bankfull flow and overbank flow. However, flow and bed structures in these cases differ according to channel plan shape and hydraulic conditions, and no sufficient investigations were made of the effects on the flow and bed of differences in flood channel depth, flood channel roughness, sinuosity and phase difference, among others.

This section examines compound meandering flows in actual rivers to gain an understanding of the following, hitherto unresolved, five issues:

(1) To perform an experiment on compound meandering channel having varying embankment and main channel plan shapes and relative depths in order 10 determine the effect of these variables on flow and bed evolution.
(2) To calculate sinuosity as an indicator of channel plan shape, and relative depth and flood duration as indicators of flood characteristics, for five actual rivers each having compound section meanders, and to use these indicators to classify flows as either simple or compound meandering flows as flood flow characteristics of compound meandering flows, and to discuss bed variation of compound meandering flow and single meandering flows seen in Gonogawa river and Tonegawa river, respectively.
(3) To examine scouring depth, sedimentation depth and maximum scouring depth in the inner bank bed during flooding in the Gonogawa, Nakagawa, and Niyodogawa rivers, compound meandering rivers, to facilitate a comparison with the hydraulic phenomena observed in the experimental channel of (1).

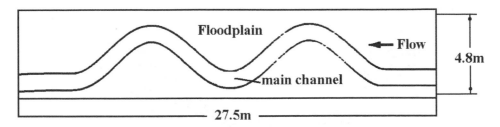

Fig. 5.18 Plan view of the experimental channel (Experiment 1).

Table 5.4 Conditions for Experiment 1.

Length of flume	27.5 m
Width of flume	4.8 m
Width of main channel	1.0 m
Channel depth	0.06 m
Thalweg wavelength	12.0 m
Sinuosity	1.17
Longitudinal slope	1/500
Discharge	68.8 l/sec
Relative depth	0.37
Median sediment size	0.8 mm

(4) To determine how unsteadiness alters flows and bed variation given a flood hydro graph in compound meandering channel having a meandering main channel.
(5) To investigate the mechanism of bed level change due to overloading sediment supply and to assess the effect of overloading sediment on the bed geometry.

5.2.2. Observations in laboratory channels and the field

Flows and bed variation in compound meandering channels are investigated experimentally using a compound meandering channel with a movable bed. A plan view of the channel is given in Fig. 5.18; the channel specifications and conditions of the experiment in Table 5.4 (this is Experiment 1). The channel consisted of a meandering main channel between straight embankments and a flood channel covered with artificial turf. Once the bed had stabilized, after approximately 8 hours of discharge, it was fixed with Comedienne diluted with toluene, after which detailed measurements of flow velocity distribution were made with an electromagnetic current meter.

5.2.2.1. Bed profiles

Bed height contour after 8 hours of discharge (Fig. 5.19) shows extensive scouring between the inner bank vicinity of the maximum curvature sections (Nos. 7

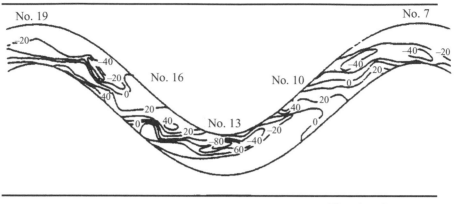

Elevations in mm

Fig. 5.19 Counter lines of bed levels with respect to initial flat bed (after 8 hours).

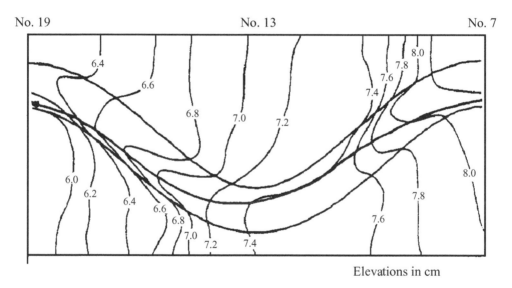

Elevations in cm

Fig. 5.20 Counter lines of water levels (after 8 hours).

and 13) and the sections directly downstream (Nos. 9 and 15), as well as sedimentation between the meander inflection point (Nos. 10 and 16) and the inner bank side of the next meander directly upstream (the vicinity of No. 18). In some cases, sediment was also washed onto the flood plain. In a channel of this sinuosity, however, sediment on the meander's outer bank experienced negligible transport, and the initial bed remained unchanged during water discharge. Bed evolution was obvious along the main channel's inner bank and along the line between inner banks.

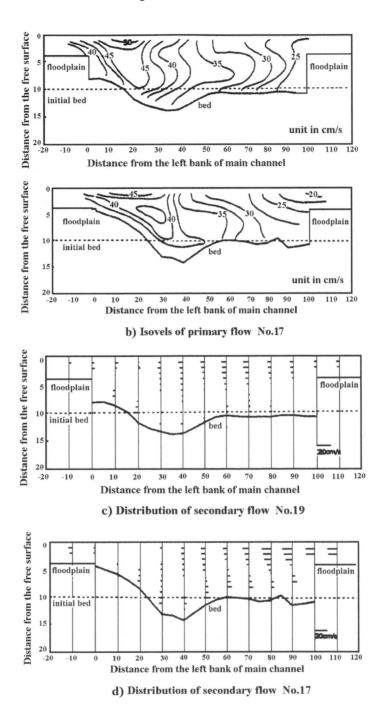

b) Isovels of primary flow No.17

c) Distribution of secondary flow No.19

d) Distribution of secondary flow No.17

Fig. 5.21 Cross sectional distribution of the velocity components in the longitudinal and transversal directions.

5.2.2.2. Water level distribution

The water level contours for the main channel (Fig. 5.21) show that water level is higher at the outer bank than at the inner bank in the sections where maximum curvature occurs (Nos. 7, 13 and 19). The solid line in the illustration represents the line of occurrence of depth-averaged maximum flow velocity in the main channel, and roughly corresponds to positions of low water level. The meander inflection section is the zone, where inflow to, and outflow from, the flood channel occurs. Considering the water surface slope at these locations of inflow and outflow it becomes, respectively, gentler and steeper than in the corresponding fixed bed experiment. This is related to differences in bed height at inflows, which correspond to sedimentation zones, flows from the flood channel spread gradually, with less loss than in the case of a fixed bed. While at outflows, which correspond to scouring zones, flows into the flood channel expand rapidly, with significant loss.

5.2.2.3. Main flow velocity distribution and secondary flow velocity distribution

In a compound meandering channel, the main channel flow is greatly affected by the flow above the flood plain bed. This is demonstrated by Fig. 5.21, the main and secondary flow velocity distributions at the maximum curvature section (No. 19) and the section near the meander inflection (No. 17). Figures 5.21(a) and 5.21(b) shows that the maximum main flow velocity occurs on the inner bank side, and that the velocity undergoes the greatest change in the vicinity of the flood channel bed (excepting the bed vicinity), where the main velocity contour line is bent. The secondary flow distribution also shows that flow direction reverses at the flood channel bed and approaches zero below it. A close look at the secondary flow distribution in Figs. 5.21(c) and 5.21(d) reveals a weak secondary flow cell directed from the outer bank toward the inner bank in the section of maximum curvature (No. 19). At No. 17, inflow from the flood channel occurs at the right bank of the main channel, but above the flood plain, a reverse-direction secondary flow cell has formed due to strong flows, which are indicative of inflow from the flood plain.

5.2.2.4. Changes in planar flow regime at different depths

Figure 5.22 shows the planar flow regimes of the upper and lower layers, the boundary between which is defined as the surface of the flood plain. A look at the upper layer at the meander inflection proximity sections (Nos. 10 and 16) reveals a flow structure that indicates inflows at the main channel's inner bank and outflows at the outer bank. Above the flood plain there are also flows heading towards the main channel. The lower layer shows a prominent flow along the main channel, despite the slight influence of the flow over the flood plain. Thus, prominent inflow and outflow between the main channel and the flood channel occurs only

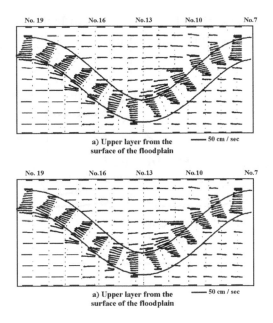

Fig. 5.22 Upper and Lower layer from the surface of the floodplain.

in the upper layer, while in the lower layer the flow travels primarily along the main channel. Figure 5.23, a comparison of longitudinal changes in the ratio of main channel lower layer discharge to total discharge for this experiment and the flat, fixed bed experiment (where $Dr = 0.31, 0.47$), shows that in contrast to the fixed bed experiment, where the ratio of main channel lower layer discharge to total discharge is roughly constant in all sections, the movable bed results in a periodicity in which the ratio was larger at the maximum curvature section and smaller at the inflection sections. In view of the roughly equivalent main channel average velocities in the sections, this periodic variation is due to differences in the sectional area of the flow.

5.2.3. Factors affecting bed variation in a compound meandering channel

5.2.3.1. Main channel alignment

An experiment was performed to determine the effect on bed evolution of a main channel's alignment and relative depth Dr (i.e. flood channel depth/total main channel depth) (this is Experiment 2). Figure 5.24 and Table 5.5 respectively give a plan view and the specifications of the experimental channel, which is a 15 m long, 4 m wide channel containing a main channel 0.8 m wide and having a sinuosity S of 1.10. The effects of main channel alignment are examined by comparing the results of Experiment 2 with those of Experiment 1 (where sinuosity = 1.17). Figure 5.25 is the bed evolution contour from Experiment 2, where relative depth

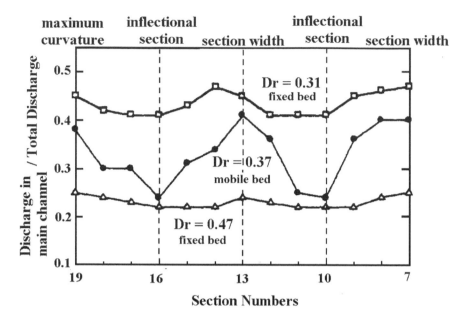

Fig. 5.23 Longitudinal variation of the ration $Q_\text{main channel}/Q_\text{total}$.

varied. Case 4 of Experiment 2, because its relative depth Dr is roughly equal to that in Experiment 1, is used to compare the two experiments with respect to the effects of sinuosity on bed evolution. As in Experiment 1, scouring occurred on the meander's inner bank and at the flow attack point near the meander inflection point, downstream from which sedimentation occurred. No large difference in scouring on the meander's inner bank was observed, but the decrease in the main channel's sinuosity caused the flow attack point in the main channel to move further downstream, which in turn also moved the locations of scouring slightly farther downstream.

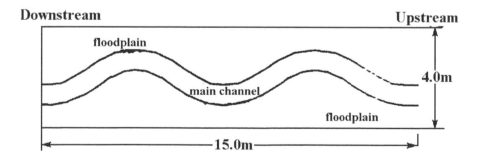

Fig. 5.24 Plan view of the experimental channel for Experiment 2.

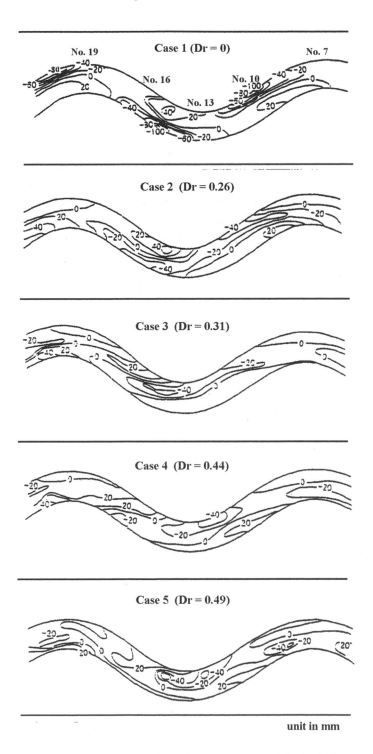

Fig. 5.25 Counter lines of bed levels with respect to initial flat bed (Experiment 2).

5.2.3.2. Relative depth

The effect of relative depth on bed evolution in a compound meandering flow is examined using Experiment 2, the conditions and results of which are respectively given in Table 5.6 and Fig. 5.25. Case 1 is a simple meandering flow in a bank-full state; Cases 2–5, compound meandering flows in the flood channel. Case 1 ($Dr = 0$) is the bed profile of a typical simple meandering flow exhibiting scouring on the meander's outer bank. Case 2 is a meandering flow of $Dr = 0.26$. Its scouring depth is less, but its bed profile is still similar to that of Case 1. This is because the flow's comparatively low relative depth allows the secondary flows caused by the centrifugal force of the main channel's flow to exceed the compound section effect of inflows from the flood channel, resulting in a simple meandering flow that creates bed scouring on the outer bank side.

In Cases 3, 4 and 5, in contrast, the structure of the compound meandering flow is such that the bed is scoured on the inner bank side. Thus, if simple and compound meandering flows were to be defined according to bed profile characteristics, with an emphasis on scouring position, then the relative depth Dr for differentiating the two would be approximately 0.30. Also, a compound meandering flow's maximum scouring depth occurs roughly at full main channel discharge. As relative depth rises, maximum scouring depth falls, reaching 40% to 60% of the scouring depth at full main channel discharge. With respect to bank erosion countermeasures, this means that in compound meandering channels, it is important to design for hydraulic quantities at full main channel discharge.

Table 5.5 Channel conditions for Experiment 2.

Length of flume	15.0 m
Width of flume	4.0 m
Width of main channel	0.8 m
Longitudinal slope	1/600
Wavelength of meandering	7.5 m
Sinuosity	1.10

Table 5.6 Hydraulic conditions for Experiment 2.

	Discharge (l/s)	Relative Depth (Dr)
Case 1	14.4	0
Case 2	24.9	0.26
Case 3	35.6	0.31
Case 4	54.1	0.44
Case 5	63.9	0.49

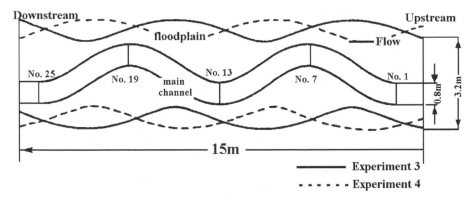

Fig. 5.26 Plan view of the experimental channel (Experiments 3 and 4).

Table 5.7 Experimental conditions (Experiments 3 and 4).

	Discharge (l/sec)	Relative Depth (*Dr*)
Experiment 3		
Case 6	19.0	0.23
Case 7	25.2	0.30
Case 8	40.3	0.40
Experiment 4		
Case 9	19.0	0.23
Case 10	25.2	0.30
Case 11	40.3	0.40

5.2.3.3. Phase differences between the embankment and main channel meanders

In the previous section the effects of main channel alignment and relative depth were examined for compound section flows in the case of a main channel meandering between straight embankments. Here, the bed evolution is investigated with the same main channel alignment as in Experiment 2 but also with embankments having a wavelength 1/4 phase ahead of the main channel (Experiment 3) and 1/4 phase behind the main channel (Experiment 4) to determine the effect of phase difference in each case. Figure 5.26 and Table 5.7 respectively give a plan view and the hydraulic conditions of the channel. Figures 5.27 and 5.28 are the bed evolution contours at relative depths of 0.23, 0.30 and 0.40 in Experiments 3 and 4, respectively. A comparison between the bed profile in the case of straight embankments (Experiment 2) and the bed profiles in those cases of experiments 3 and 4 in which relative depth was roughly equivalent shows that all three experiments display similar bed profiles. The relative depth that discriminates simple and compound meandering flows from each other, although varying slightly according to phase difference, is roughly 0.28 to 0.30.

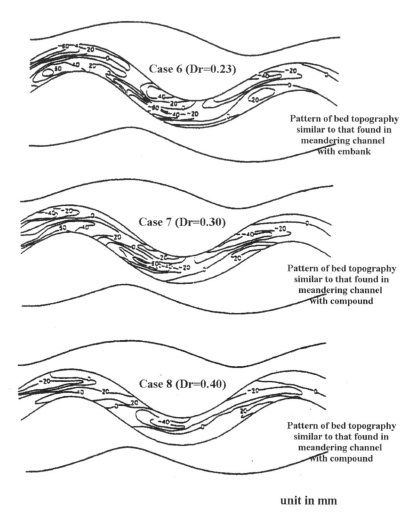

unit in mm

Fig. 5.27 Counter lines of bed levels with respect to initial flat bed (Experiment 3).

Thus, generally speaking, in a compound meandering river with a sufficiently wide flood channel, phase difference between the embankments and the main channel may be considered as having only a small effect on the dynamics of flood flow and on the bed profile of the main channel.

5.2.3.4. Rate of bedload transport

Figure 5.29 shows cumulative sediment discharge measured at 3,5,7 and 9 hours from the start of flow for different cases of relative depths (Cases 2–5) and sediment supply (Case 1 and Case 6). In every case, it shows that cumulative sediment discharge increases almost constantly after 3 hours. These gradient values define equilibrium sediment discharge for the cases of each relative depth. Each of the

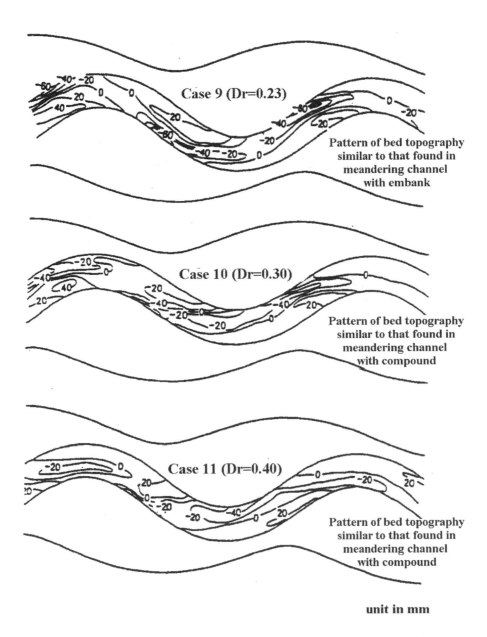

Fig. 5.28 Counter lines of bed levels with respect to initial flat bed (Experiment 4).

values is shown in Table 5.8. The largest values are found in Case 6 (bankfull flow, sediment supply), the second largest is Case 1 (bankfull flow), then Case 2, Case 3, Case 5, and Case 4 in turn.

In the case of compound channel flows with large roughness in flood channel, there is a flow interaction between main channel and flood channel varying with phase difference of the channel. The flow interaction caused the reduction of mean

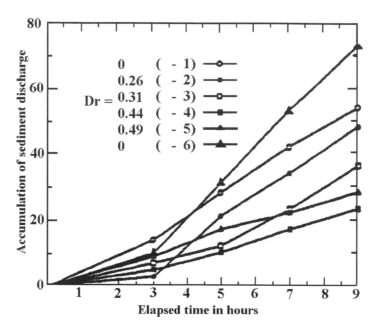

Fig. 5.29 Cumulative sediment discharge with time for Cases 1–6 (see Table 5.8).

Table 5.8 Sediment transport rates for Experiment 4.

	Sediment discharge (cc/sec)
Case 1 (without sand feed)	1.9
Case 2 (without sand feed)	1.9
Case 3 (without sand feed)	1.1
Case 4 (without sand feed)	0.8
Case 5 (without sand feed)	0.9
Case 6 (with sand feed)	2.9

velocity, and bed shear stress and resulting sediment discharge in the main channel. Thus the largest amount of sediment discharge occurred in bankfull flow condition of Case 1 and Case 6. Out of the compound channel flow cases, Case 2 ($Dr = 0.26$) has the characteristics of a single meandering flow. Because the effect of flow over the flood plain is not dominant and influence of the centrifugal force in the main channel flow is large, the maximum velocity in Case 2 occurs near the outer bank. Thus, the sediment discharge becomes larger next to Case 1.

As the relative depth is increased, sediment discharge reduces gradually. However, sediment discharge of Case 5 is a little bigger compared to Case 4. This suggests that sediment discharge was increased at the relative depth above the depth of Case 4. The increase in the relative depth causes again a single meandering flow, resulting in an increase in sediment discharge. Figure 5.30 shows the

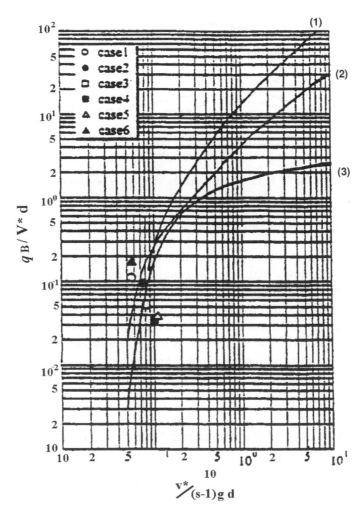

Fig. 5.30 Comparison of measured sediment transport rate with various sediment transport formulae in non-dimensional form. The equations are (1) Andru, (2) Laursen-Toch, and (3) Tarapore (see also Fig. 5.72).

sediment discharge by different sediment discharge equations and measured discharge. Dimensionless sediment discharge is plotted as a function of dimensionless shear stress. Friction velocity, u_*, is obtained from $u_*^2 = gRI$ (I is water stage gradient along meander, R is the hydraulic radius of main channel). In the case of compound channel flow flood, channel flow slows down main channel velocity so the shear stress shear velocity calculated by above relation is overestimated when compared to the actual effective values in the compound meandering channel. Figure 5.31 indicates that the dimensionless sediment discharge for single section meandering flow, Case 1 and Case 6, are larger than values calculated by discharge equations.

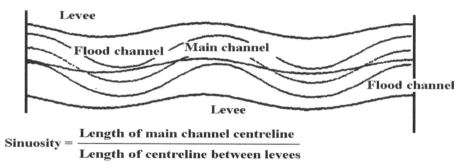

$$\text{Sinuosity} = \frac{\text{Length of main channel centreline}}{\text{Length of centreline between levees}}$$

Fig. 5.31   Sketch indicating the definition of sinuosity.

5.2.4. Characteristics of flood flows in compound meandering rivers

The previous section presented experimental examinations of the simple and compound meandering flows that occur in compound meandering channels due to differences in the relative depth, but it is necessary to determine under what conditions these two types of flows occur in actual rivers. This chapter uses survey data to determine the relationship between flood flows and bed evolution in actual rivers. The data used are surface flow velocity distribution determined from aerial photographs taken during flooding, water level time curves measured at observation stations, and the results of cross-sectional surveys taken before and after flooding.

5.2.4.1. Indicators of river plan shape and flood flow characteristics

The first step is to select numerical indicators of the characteristics of channel plan shape and flood flows, and then define numerical values. The indicator of channel plan shape used is sinuosity, while the indicators of flood characteristics are relative depth and flood duration. In the experimental channel, sinuosity was defined as the ratio of meander wavelength to straight downstream distance, but in rivers where embankments also meander, this definition of sinuosity cannot adequately express the effects of embankment meandering. For this reason, sinuosity is instead defined as the ratio of main channel centre distance to embankment centre distance (Fig. 5.31). As a plan view of the river suggests that the flood channel and main channel are roughly in phase, and that any phase differences that do exist are small, the plan shape of an actual river's compound meandering flow can be represented with sinuosity, as shown in Fig. 5.31. Also, because the left and right flood plain levels often differ in actual rivers, relative depth is calculated using the main channel's section-averaged depth and the average of both flood channel heights. Flood duration is defined as the amount of time that the flood channel is inundated, which is when the effects of the channel's compound shape manifest themselves, and is calculated from the water level time curve and the channel's lateral profile.

Table 5.9 Sinuosity Values for some Japanese Rivers.

River	Stretch	Sinuosity
Tone River	132.0k~125.0k	1.034
	124.0k~116.0k	1.010
	108.5k~100.5k	1.018
	98.0k~91.5k	1.023
	95.0k~89.5k	1.022
	92.5k~86.5k	1.021
	85.0k~75.0k	1.011
	80.0k~69.0k	1.035
	69.0k~62.0k	1.060
	66.0k~58.5k	1.023
	62.0k~53.0k	1.002
	47.5k~38.5k	1.012
	43.0k~33.5k	1.010
Ishikari River	84.0k~76.0k	1.013
	75.5k~72.0k	1.043
	70.5k~64.5k	1.075
	64.5k~56.5k	1.036
	60.5k~53.5k	1.020
	57.0k~50.0k	1.023
	50.0k~38.0k	1.003
	34.0k~29.0k	1.077
	27.0k~20.0k	1.028
Edo River	52.6k~50.0k	1.012
	42.0k~38.0k	1.002
	38.0k~34.0k	1.005
	31.6k~29.0k	1.006
	29.4k~26.2k	1.018
	25.0k~22.0k	1.027
	22.0k~18.0k	1.018
Kokai River	29.8k~27.0k	1.072
	28.7k~26.0k	1.093
	27.2k~24.0k	1.073
	26.0k~23.0k	1.030
	23.8k~20.6k	1.029
Gono River	22.2k~18.4k	1.052
	20.4k~16.2k	1.087
	18.0k~14.2k	1.025
	16.2k~11.0k	1.016

5.2.4.2. Sinuosity, relative depth and flood duration

Sinuosity was calculated for five rivers: the Tonegawa, Ishikarigawa, Tamagawa, Gounogawa and Aganogawa rivers. As Table 5.9 shows, sinuosity S was concentrated between 1.000 and 1.050 in the Tonegawa, Ishikarigawa, Tamagawa and

Table 5.10 Length of time water flowed over Flood Channel Bed and Relative Depth.

River	Observation point (Distance)	Length of time water flowed over Flood Channel Bed	Relative depth Dr
Tone River	Yatajima (181.5k)	57 Hours	0.33
August, 1981	Kawamate (150.0k)	47 Hours	0.53
	Kurihashi (130.5k)	64 Hours	0.56
	Mebuki Bridge (104.0k)	59 Hours	0.42
	Toride (85.3k)	59 Hours	0.41
	Suga (61.5 k)	61 Hours	0.42
	Sahara (40.1k)	58 Hours	0.16
Ishikari River	Hashimotocho (93.9k)	17 Hours	0.13
August, 1981	Naieohashi Bridge (76.8k)	67 Hours	0.46
	Tsukigata (58.0k)	37 Hours	0.20
	Iwamizawa Ohashi Bridge (44.5k)	75 Hours	0.39
	Ishikari Ohashi Bridge (26.6k)	88 Hours	0.58
Edo River	Nishisekiyado (58.0k)	56 Hours	0.70
August, 1981	Noda (39.0k)	Over 80 Hours	0.58
	Matsudo (19.5k)	Over 100 Hours	0.48
Kokai River	Kuroko (53.0k)	90 Hours	0.60
August, 1981			
Gono River	Kawamoto (36.3k)	Over 100 Hours	0.60
July, 1983	Yasumikyo (14.8k)	Over 120 Hours	0.76

Aganogawa rivers, and so these values can be said to represent the sinuosity of the middle and lower reaches of the rivers. The channel meanders largely in the Gonogawa river, which flows through a gorge and whose channel profile are formed by a main channel and bank terraces. Sinuosity in this section is large, ranging between 1.072 and 1.093. Table 5.10 gives the duration of inundation of the flood channel (or, in the case of the Gonogawa river, the bank terraces) and the relative depth at flood peak for the Tonegawa, Ishikarigawa, Edogawa, Kokaigawa, Gonogawa and Aganogawa rivers. The floods in question were all among the largest postwar floods in the respective rivers. In the 181.5 k to 40.1 k section of the Tonegawa river, the flood duration was 47–64 hours, meaning that the flooding lasted for 2–3 days. Relative depth at flood peak is greater upstream, and smaller at points farther and farther downstream. Downstream from Sawara (40.1 k), the river was almost entirely at bankfull discharge. In the Ishikarigawa river, flood duration tends to grow longer at points farther and farther downstream. In the Edogawa, Kokaigawa and Gonogawa rivers, the sections of the main channels are smaller relative to the Tonegawa and Ishikarigawa rivers, resulting in longer flood durations, particularly in the lower reaches of the Gonogawa river (Tanijugo,

Table 5.11 Geometrical parameters of the Tonegawa River (Japan).

Stretch	Section Studied	Wavelength (km)	Sinuosity	Phase Shift
151.3k~146.5k	149.0k	5.8	1.003	0
146.0k~139.0k	143.0k	7.0	1.012	0.064
132.0k~125.0k	129.0k	7.0	1.034	0
124.0k~116.0k	120.0k	8.0	1.010	0
108.5k~100.5k	103.5k	8.0	1.018	0.156
98.0k~91.5k	93.5k	6.5	1.023	0.267
95.0k~89.5k	91.0k	5.5	1.022	0.182
92.5k~86.5k	89.5k	6.0	1.021	0
85.0k~75.0k	79.0k	10.0	1.011	0.164
80.0k~69.0k	74.0k	11.0	1.035	0.115
69.0k~62.0k	66.0k	7.0	1.060	0.100
66.0k~58.5k	62.0k	7.5	1.023	0.035
62.0k~53.0k	58.0k	10.0	1.002	0.080
47.5k~38.5k	43.0k	9.0	1.012	0.076
43.0k~33.5k	38.0k	9.5	1.010	0.078
38.5k~25.0k	34.0k	13.5	1.008	0

14.8 k), where the riverbank terraces were continuously inundated for more than 5 days.

5.2.4.3. Meander Characteristics of the Tonegawa and Gonogawa Rivers

The stretch of the Tonegawa river investigated was between 180.0 k and 30.0 k. The cross sections studied and the wavelength and degree of sinuosity for each are indicated in Table 5.11. As a typical example, Fig. 5.32 shows the plan view of the river between 108.5 k and 100.5 k where $S = 1.018$ and $P = 0.156$. From the table it can be seen that wavelengths of around 6 km to 8 km are typical upstream of 80.0 k. Downstream of this point the wavelengths are generally larger than those upstream, the maximum being 13.5 km.

The values of sinuosity lie mainly in the range 1.000 to 1.040 where the length of the centerline of the main channel is almost the same as the length of the river course centerline, and the phase shift is less than 0.2. As the Gonogawa river is an incised river in which both riverbanks are mostly terraced, the river course meanders considerably more than the Tonegawa river. The stretch of river studied was between 29.8 k and 20.6 k and the river cross sections studied and the wavelength and degree of sinuosity for each are indicated in Table 5.12 in the same way as for the Tonegawa river. Plan views of the stretch can be seen in Fig. 5.33.

The wavelength is about 3 km. In the region of 24.0 k of this stretch, with the Kawagoe Ohashi Bridge as the dividing line, the upstream part can be classified as "meandering portion of the river course", and the downstream part as "straight portion of the river course". The sinuosity of the Gonogawa river, compared with

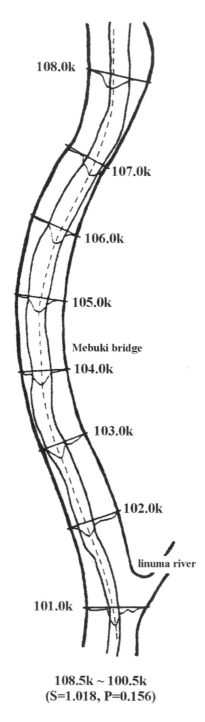

108.0k

107.0k

106.0k

105.0k

Mebuki bridge
104.0k

103.0k

102.0k

Iinuma river

101.0k

108.5k ~ 100.5k
(S=1.018, P=0.156)

Fig. 5.32 Plan view of the stretch of the Tonegawa River studied and surface velocity profile (scale: 1/40000).

Table 5.12 Geometrical parameters of the Gono River.

Stretch	Section Studied	Wavelength (km)	Sinuosity	Phase Shift
29.8k ~ 27.0k	28.6k	2.8	1.072	0.030
28.7k ~ 26.0k	27.4k	2.7	1.093	0.074
27.2k ~ 24.0k	26.0k	3.2	1.073	0.023
26.0k ~ 23.0k	23.8k	3.0	1.030	0.067
23.8k ~ 20.6k	21.8k	3.2	1.029	0.082

that of the Tonegawa river, is larger at 1.072 to 1.093 at the "meandering portion of the river course", and about the same at the "straight portion of the river course" at 1.030. The phase shift was less than 0.1, less than for the Tonegawa river. As can be clearly seen from Figs. 5.32 and 5.33, the maximum velocity filament at the surface does not occur at the outer bank of the curved portion. Particularly in the case of the Gonogawa river that has high sinuosity, it can be seen that the maximum velocity filament occurs nearer the inside levee, and for the Tonegawa river that has low sinuosity, it occurs from the centre and toward the inner bank. In the next section, these facts will be explained more quantitatively.

5.2.4.4. Categorizing floods by relative depth, sinuosity and location of maximum flow velocity

Section 5.2 explained that compound meandering rivers can be classified into simple and compound meandering ones, depending on the relative depth of the flood flows there. Here, flood flows are characterized in ways that also take into account sinuosity, which indicates the channel's plan shape. In the experimental channel, flows were categorized either as simple or compound meandering flows according to the bed scouring and sedimentation that occurred during water discharge. However, this classification method cannot be applied to actual rivers, so, simple and compound meandering flows are defined in actual rivers as follows.

As in Fig. 5.34, the indicator used is the ratio y/b, where y is the distance from the inner bank to the location of maximum flow velocity and b is the main channel width at the main channel meander apex. A y/b ratio of 0.5 or greater is interpreted as a maximum flow velocity occurring near the outer bank, indicating a simple meandering flow; a y/b ratio below 0.5, as a maximum flow velocity occurring near the inner bank, indicating a compound meandering flow. Fig. 5.35, which uses data from actual rivers showing the relationship between the ratio y/b and sinuosity S and relative depth Dr, demonstrates how a flow where $y/b \geq 0.5$, i.e., a simple meandering flow (indicated by circles in Figure 5.35), concentrates in the lower left-hand corner of the illustration. Plotting this along with the experimental results yields a solid line that roughly divides simple meandering flows from compound meandering flows. Both actual river data and experimental results indicate that when sinuosity is greater than 1.02, a simple meandering flow results at a relative depth roughly below 0.28. Let us now look at the conditions under which compound meandering flows form, namely, the group where $y/b \sim 0.5$ (indicated with triangles and squares in Fig. 5.35). The distinction between the two flows is not necessarily clear in the Tonegawa river, but in the Ishikarigawa river, as the sinuosity is higher, a maximum flow velocity occurs near the inner bank. No clear characteristics are discernible, however, for the Gonogawa river. This is believed to be because of the effect on flood flows of the many trees for flood fighting seen along the main channel banks in this section of the river.

Thus, determining sinuosity and relative depth for a compound meandering river reveals whether the flood flows in question have the characteristics of simple or compound meandering flows. It is important that the actual river data in Fig. 5.35 be further enriched, as this would yield important information for the plan shape planning of channel embankments and main channels, revetment planning, and in the predicting of whether a riverbank disaster will occur at the inner or outer bank.

5.2.4.5. Relationship between flow characteristics and bed scouring in compound meandering rivers

Discussed herein are bed variation of compound meandering flows and single meandering flows seen in compound meandering rivers.

(i) Bed scouring caused by a compound meandering flow
Survey data of the Gonogawa river is used here to ascertain what types of changes in bed height occur when a compound meandering flow forms during flooding in a compound meandering river. The Gonogawa river has a large sinuosity compared to other rivers (Table 5.9). Near the 27 km point, a compound meandering flow formed during flooding in July 1983 (peak relative depth: 0.45) (14). Figure 5.36 is a plan view of the Gonogawa river. The duration of riverbank terrace inundation was longer than in flooding in other rivers. Figures 5.37(a) and (b) give the bed profiles, measured before and after flooding, at the 27.2 k section, which is the meander apex, and the 27.0 k section, which is directly downstream. First, the pre-flooding bed profile shows that maximum scouring occurred at the outer bank at the 27.2 k point, which is roughly the vicinity of the flow attack point at normal discharge. The post-flooding bed profile reveals that at the 27.2 k point, sedimentation and scouring occurred at the outer bank and the center in the main channel, respectively, and that at the 27.0 k point, located downstream, maximum scouring occurred in the outer bank bed. This is interpreted thus: A compound meandering flow formed during flooding, shifting the location of maximum flow velocity toward the inner bank and causing scouring near the main channel center near the 27.2 k point. This repositioning of the location of maximum flow velocity caused the flow attack point on the main channel bank to move downstream, giving rise to outer bank scouring near the 27.0 k point. Such flows and the resultant bed evolution both correspond well with the experimental results, and indicate that scouring can occur in the inner bank and in the bed near the inner bank when

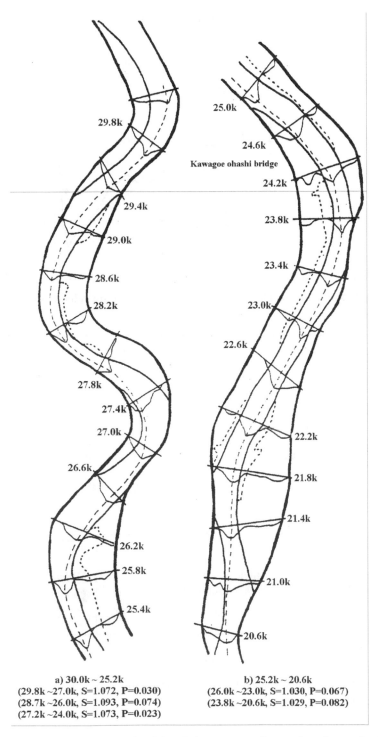

a) 30.0k ~ 25.2k
(29.8k ~27.0k, S=1.072, P=0.030)
(28.7k ~26.0k, S=1.093, P=0.074)
(27.2k ~24.0k, S=1.073, P=0.023)

b) 25.2k ~ 20.6k
(26.0k ~23.0k, S=1.030, P=0.067)
(23.8k ~20.6k, S=1.029, P=0.082)

Fig. 5.33 Plan view of the stretch of the Gobo river studies and surface velocity profile (scale: 1/20000).

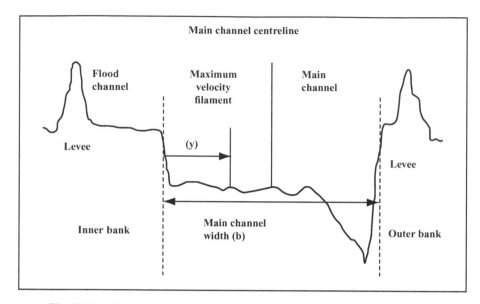

Fig. 5.34 Dimensionless lateral distance to maximum velocity filament.

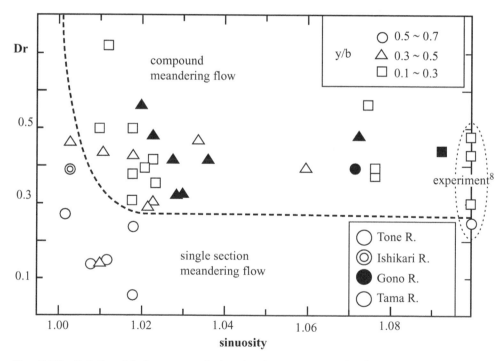

Fig. 5.35 Relationship between relative depth sinuosity and y/b in doubly compound meandering rivers.

sinuosity and relative depth are sufficiently large and when the flood duration is sufficiently long.

(ii) Bed scouring caused by a simple meandering flow

The 48.0–32.0 k section of the Tonegawa River has relative low sinuosity (1.008–1.012) and a low relative depth (Dr), which was only 0.16 at the peak of a major flood in August 1981. For this reason, a simple meandering flow is believed to have occurred in this section. Figure 5.38 contains a plan view of this section, along with the positions of extensive scouring and velocity distribution as calculated from the surface velocity vector. Extensive scouring, indicated by dark patches in the illustration, occurred at five locations, which are, from farthest upstream, Takaya (46.0 k), Ishinou (42.5 k), Sahara (40.5 k), Kousyu (39.0 k), and Tsunomiya (36.5 k). Channel alignment is characterized by differences in meander wavelengths, i.e., a 5–6 km wavelength in the 48.0–41.0 k and 36.0–32.0 k sections versus 2–3 km in the intervening 41.0–36.0 k section. Significant scouring occurs almost exclusively at the outer bank side of the main channel meander. Moreover, the surface velocity distribution during flooding indicates a general correspondence between those scouring positions and the positions of maximum velocity, which extend from outer bank to outer bank. This indicates that, in this section, simple meandering flows regularly appear during flooding and that flow is determined by main channel alignment more than by embankment alignment.

5.2.5. Unsteady flows and bed variation in a compound meandering channel

5.2.5.1. Introduction

Based on the belief that flood flows change gradually over time, water level and velocity are normally determined through non-uniform flow calculations in which maximum discharge is given. In a compound meandering channel, flow varies according to flood water level Ashida *et al.* (1989, 1990) and are far different than in a simple section meandering channel, in which water flows through the main channel only. However, the process by which, during flooding, flow and bed height

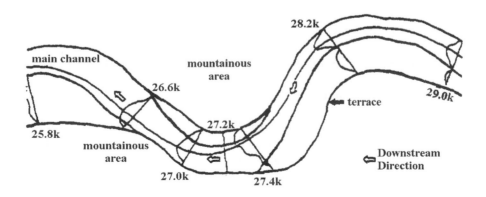

Fig. 5.36 Top view of the 29.0 ∼ 25.8 k section of the Gonogawa River.

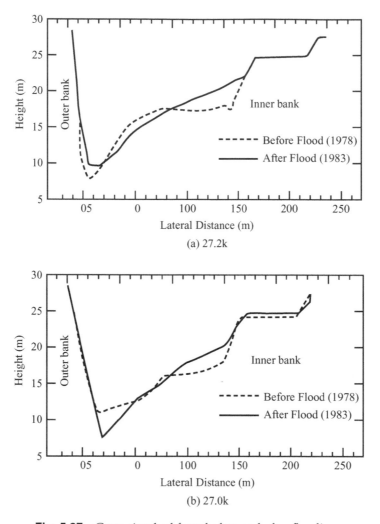

Fig. 5.37 Gono riverbed form before and after flooding.

change due to the unsteadiness of flood flows in a compound meandering channel is not fully understood. The objective of this section is therefore to determine how unsteadiness alters flows and bed variation given a flood hydrograph in a compound meandering channel having a meandering main channel.

5.2.5.2. Experimental results and considerations

In the experiment, the hydrograph in Fig. 5.40 was created in the movable bed compound meandering channel in Fig. 5.39, comprising straight embankments and a meandering main channel (sinuosity $S = 1.10$). Given the 1:100 scale of the experimental channel, this hydrograph corresponds to a large-river flood lasting for 40 hours and having a peak discharge of 6,000 m^3/s. The experiment was carried out for four different Cases in which the bed condition was varied with

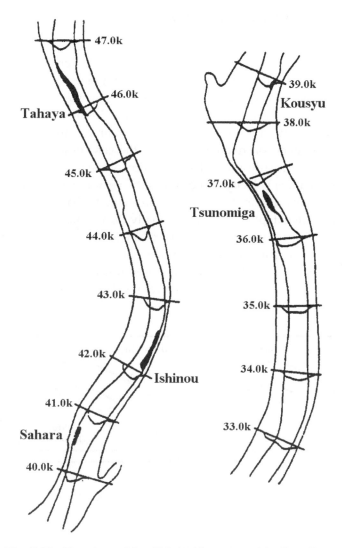

Fig. 5.38 Top view of the 47.0–33.0 k section of the Tone River.

respect to the initial bed form; Case 1 used a flat initial bed form; Case 2, the final bed form of Case 1; Case 3, a bed form equivalent to the pre-flooding bed in an actual river (i.e. bed scouring at the outer bank and bed sedimentation at the inner bank); and Case 4, an initial bed similar to that of Case 3 except that its hydro graph has a flood duration that is one-half of that of Case 3, as Fig. 5.40 shows. The measurement section was a one wavelength section in the channel's centre. With a time changing discharge, water level was measured every five minutes, and surface velocity and bed form every 40 minutes, to determine the changes over time.

5.2.5.3. Effects of unsteadiness on flow

Figure 5.41 gives temporal change in longitudinal water level. The period between 20 minutes and 180 minutes is when a compound meandering flow existed, i.e., water flowed in the flood channel. During the flood, longitudinal surface gradient remained essentially constant. Figure 5.42 shows the surface velocity distribution at peak discharge and the steady-flow surface velocity distribution for the same discharge. In a compound meandering flow, the main channel's flow regime and the bed change depend on relative depth. In the case of an unsteady flow, the main-channel flow is greatly affected by the influx of the flood channel flow, which slows the main-channel velocity and changes the vector orientation, among other changes. This velocity distribution is similar in shape to that of a steady flow with an equivalent relative depth. Thus, an unsteady flow whose water level changes gradually over time has, at a given water level, a water surface profile and velocity distribution that are essentially similar to those of a steady flow of corresponding relative depth, and so can be considered as a pseudo-steady flow. The flow and bed variation characteristics at this time are therefore also similar to those of a steady

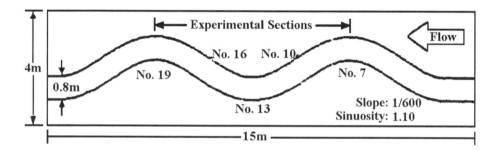

Fig. 5.39 Plan view of the experimental channel.

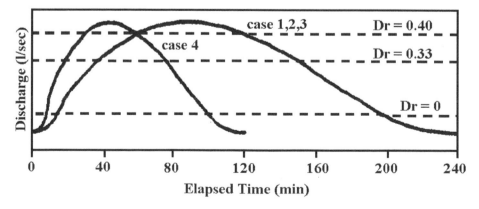

Fig. 5.40 The various hydrographs.

flow of corresponding relative depth. In Case 4, the change in discharge with time is greater than in Case 3 and occurs more rapidly although the characteristics at that time, as with Case 3, did not differ greatly from those of a steady flow of corresponding relative depth.

5.2.5.4. Effects of unsteadiness on bed topography

Initial bed form and the bed forms that resulted during flooding are shown in Fig. 5.43. The initial bed form (a), exhibiting outer-bank bed scouring and inner-bank bed sedimentation, is the type of bed form typically associated with simple meandering flows. In (b), by which time the water level has risen slightly, the flood flow exhibits the characteristics of a compound meandering flow in that velocity is greater at the inner bank. However, bed change from the initial bed form is minimal because of the short amount of time that has elapsed. After 120 minutes of flow (c), relative depth (Dr) is 0.40, and another bed characteristic associated with compound meandering flows is evident: scouring and sedimentation along a line extending from inner bank to inner bank in the vicinity of the meander's apex. Here, sediment released by scouring at the inner bank of the meander's apex has been transported downstream to the attack point (i.e., the outer-bank side of the inflection section), where it filled in the scouring depressions. A major factor in this bed variation is the length of time that the compound meandering flow existed. Once discharge and water level fell (d), the flood channel was of a sufficiently shallow depth to permit a simple meandering flow predominated by centrifugal forces to exist in the main channel; the area of scouring also moved toward the outer bank. As a result, scouring depressions formed at the attack point in the inflection section, and the inner-bank bed scouring area that had formed in stage

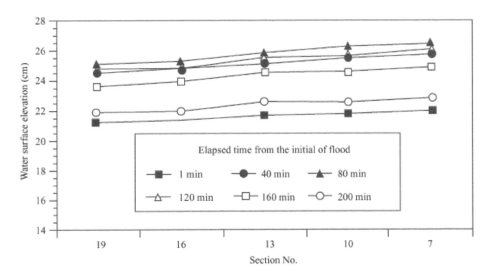

Fig. 5.41 Temporal change in longitudinal water level, elapsed time is in minutes.

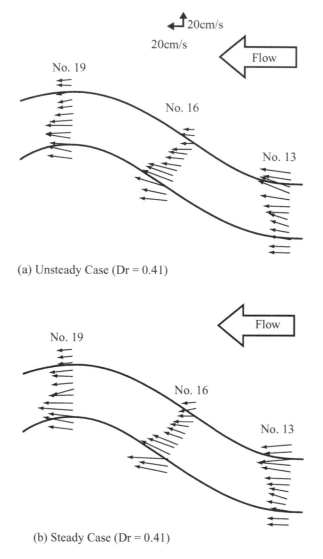

(a) Unsteady Case (Dr = 0.41)

(b) Steady Case (Dr = 0.41)

Fig. 5.42 Comparison of surface velocity distribution at 60.0 l/sec.

(c) began to be re-filled. The post-flooding bed form in Case 3 bears, at the centre of the section of maximum radius of curvature, scouring depressions left by the compound meandering flow that appeared at high water levels. Case 4, despite having a different hydrograph, yielded roughly the same results as Case 3 with respect to changes in the locations of scouring and sedimentation. However, because the compound meandering flow lasted for a relatively short period of time, Case 4 resulted in less change than Case 3: The scouring areas observed immediately after flooding at the meander's apex, in the center of the main channel, exhibited a depth of approximately 2 cm, or roughly one-half of that in Case 3.

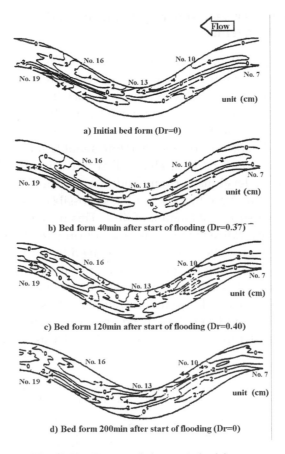

Fig. 5.43 Temporal change in bed form.

Bed change characteristics during flooding were generally similar to those of steady flows of equal relative depths. Because of the continually changing water level, however, the location and degree of change in bed form depends on relative depth at each stage and on the duration of each stage. Furthermore, particularly in the case of an unsteady compound meandering flow, vertical bed variation during flooding is extensive at the meander apex where, as Fig. 5.44 shows, the area of change undergoes considerable transverse movement. In contrast, less vertical bed variation occurs during flooding at the inflection section, where the area of change is fixed at the outer-bank side. Comparing pre- and post-flooding bed forms, the meander apex exhibits scouring areas in the main channel's centre as evidence of the flooding, whereas the bed form at the inflection section remains generally unchanged.

5.2.6. Maximum inner bank scouring depth and bed change during flood events

Although pre and post-flooding bed evolution can be ascertained from a river's lateral profile, there is no established knowledge regarding what is often most important to ascertain, namely the vertical bed variation during flooding. And while the outer bank beds of compound meandering channels have traditionally been considered susceptible to scouring, and the inner bank side safe with respect to bed scouring and bank erosion and therefore suitable for structures and close-to-nature works, Section 5.2.2.4 has shown that if, during flooding, the relative depth increases to the level that a compound meandering flow forms, maximum flow velocity will occur close to the inner bank, possibly resulting in inner bank bed scouring given a sufficient amount of time. This makes it important to more accurately determine localized inner bank bed scouring depth during flooding. To determine the extent of vertical evolution displayed by an inner bank river bed during flooding, the scouring depth was measured at various rivers by making bore holes in inner bank sandbars at meanders that are exposed at normal discharge (Fig. 5.46). The boreholes were then refilled with sand colored with a different colour for each soil layer. After flooding, those same boreholes were dug up to determine the depth to which the colored sand was washed away during

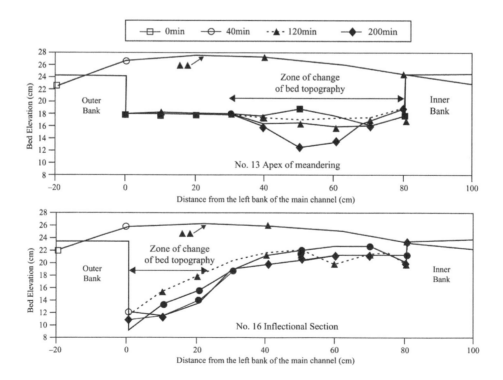

Fig. 5.44 Temporal change in channel cross-section (Case 3).

flooding. This yielded maximum scouring depth in the inner bank bed during flooding, as well as sedimentation depth during the receding of floodwater, and bed variation during flooding. Although this fieldwork was performed at rivers across Japan, the following concerns the results for the Gonogawa, Nakagawa and Niyodo rivers.

5.2.6.1. Gonogawa river investigation and results

The flooding in question occurred between July 8–13, 1997, due to a seasonal rain front. Figure 5.45 gives the lateral bed profile at the Kawamoto point, the location of a water level observation station, as well as the flood hydrograph and the precipitation at Otsu point (86.5 km). Total precipitation at this time was 301 mm at Otsu point, and peak discharge was 3,500 m^3/s at Kawamoto point. Water level fluctuation followed a pattern with two peaks. The riverbank terraces were submerged for more than 5 days, and peak relative depth was 0.40. At three points on the inner bank sandbars of the Gonogawa river, six bore holes 0.5 m in diameter and approximately 1.5 meters deep were made (see Fig. 5.46). The soil excavated from each depth was colored a different for each of six 0.2-meter layers, to a depth of 1.2 meters. The boreholes were refilled with this colored soil and initial bed height was determined by surveying. After flooding, each bore hole was examined to determine to what depth the coloured soil layers had been removed, enabling maximum scouring depth during flooding to be determined.

For this investigation we selected curved sections with a wide flood channel, a channel sinuosity greater than 1.02, a phase difference between the main channel and embankments, and inner bank sandbars exposed at normal discharge. The scouring depth survey was performed in a section (20 km to 30 km) with river terraces on both banks. Three points were surveyed: Uzumaki (near 27.0 km), Onuki (near 23.7 km), and Tazu (near 21.2 km). Figure 5.47 shows the horizontal relationship between the boreholes and the plan shape of the river containing each survey section. The sinuosity in each section was 1.093, 1.030 and 1.029, respectively. Bed materials were sand (Uzumaki point) and gravel (Onuki and Tazu points).

(i) Uzumaki point (27.0 km point) (Fig. 5.48a)
This was the point with the largest sinuosity ($S = 1.093$) in the survey section. The survey was performed at a position located on the inner bank, downstream from the proximity of the maximum curvature section. Inner bank bed scouring caused by July 1983 flooding (peak relative depth $Dr = 0.45$) occurred at the 27.2 km section, immediately upstream from this position. The survey showed a maximum scouring depth during flooding of approximately 0.2 m at all bore holes except No. 4, indicating that a compound meandering flow, the maximum flow velocity of which occurred near the meander inner bank, had at least formed at the flood peak (relative depth $Dr = 0.40$).

(ii) Onuki point (23.7 km point) (Fig. 5.48b)

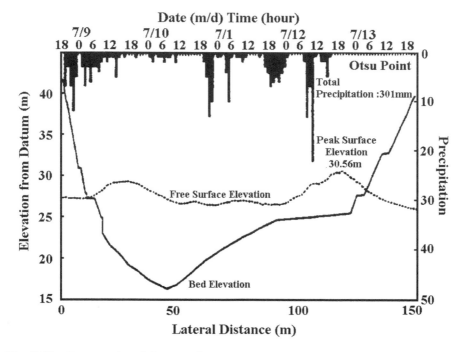

Fig. 5.45 Cross sectional shape and water stage curve at Kawamoto and rainfall in Ohtsu.

Despite the fact that bore holes 1 through 4 indicated a maximum scouring depth during flooding of 0.4 m or more, there was almost no change in inner bank bed height as measured before and after flooding. This suggests that although a compound meandering flow formed at or around the flood peak, causing inner bank scouring, the bed height recovered roughly to its pre-flooding level because of the outer bank bed scouring and inner bank bed sedimentation that occurred during the bankfull flow period, which occurs when floodwaters recede. Thus, this survey successfully demonstrated inner bank bed variation, which could not be ascertained with conventional pre and post-flooding cross-sectional surveying. Despite a sinuosity different from the Uzumaki point, the Onuki point was roughly similar positions in river course. However, there were differences in maximum scouring depth and sedimentation depth, which are believed to be due to the effect of dense trees for flood fighting on the inner bank, although this requires comparison with results obtained in other rivers.

(iii) Tazu point (21.2 km) (Fig. 5.48c)
With a different horizontal position than the other two points, Tazu point, shown in Fig. 5.48(c), is positioned at the inner bank near the inflection section. Flood scouring depth here was less than at the other two points. As the aforementioned experimental results also indicate, the inner bank bed near the inflection point is a location of almost no bed change, but rather undergoes sedimentation, regardless

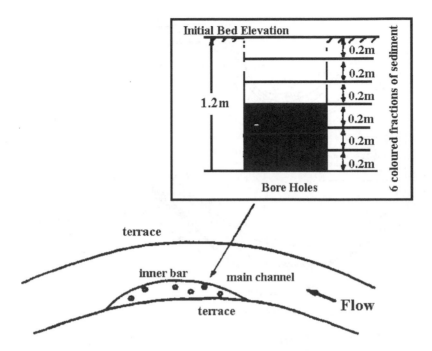

Fig. 5.46 Observation point at which holes were bored and filled with coloured layers.

of whether the flow is a compound meandering flow, a simple meandering flow, or a simple section meandering flow. The foregoing results demonstrate that in actual rivers, as well, when relative depth is large and flood duration is sufficiently long, bed scouring will occur at the inner bank side or near the maximum curvature section, with almost no bed change at or near the inflection section. These results agree well with experimental results.

5.2.6.2. Nakagawa river investigation and results

The site selected for this investigation is a sandbar formed at the inner bank of a compound meandering channel, at a point where sinuosity is great ($S = 1.02$ or higher) and the flood channel is wide. Several holes roughly 1.2 meters in depth were bored on this sandbar. The soil removed from these boring holes were colored with six different colors to indicate the layers, and then replaced back into the boring holes. After flooding, the same ground level surveying performed before flooding was carried out, and the aforesaid boring holes were re-bored to determine which colored sand layer had undergone scouring and washing away. This method made it possible to ascertain vertical bed variation and maximum scouring depth during the flood, as well as post- flood sedimentation depth. The results of this investigation were then comprehensively considered in light of channel

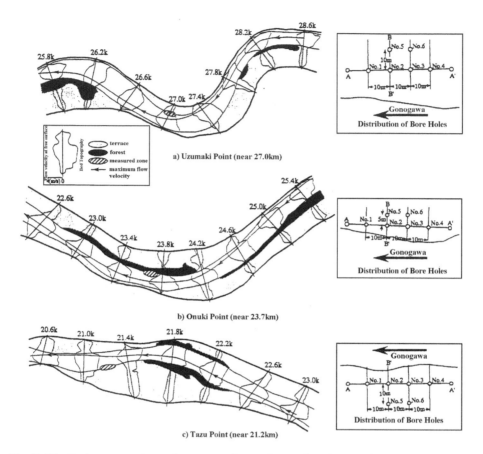

Fig. 5.47 Bed topography and water surface velocity distribution during the flood of July 1983 at Gono River and bed survey location.

alignment, flood hydrographs, vegetation distribution, bed materials, and other. factors closely related to bed variation in a compound meandering channel.

With a main watercourse length of 125 km and a drainage basin area of 874 km², the Nakagawa river is the second-largest river in Tokushima Prefecture (after the Yoshinogawa river) and one of the largest on the island of Shikoku. The upper reaches of the river run through one of the rainiest areas in Japan, and the flooding is large in scale for such a relatively small drainage basin.

The authors chose to investigate a flood caused by Typhoon No. 19 on September 16 and 17, 1997. Figure 5.49 shows bed cross-sectional form and changes over time in water level and discharge measured at the Furusho Water Level Observation Station (the 7 km point), located approximately 2 km downstream from the point investigated. Maximum water level and maximum discharge during said flood were T.P. +9.085 m and 6,100 m³/s, respectively. Relative depth determined from maximum water level was 0.53, and the duration of flood channel inundation was 19 hours. Figure 5.50 shows river planform and bed cross-sectional form in a

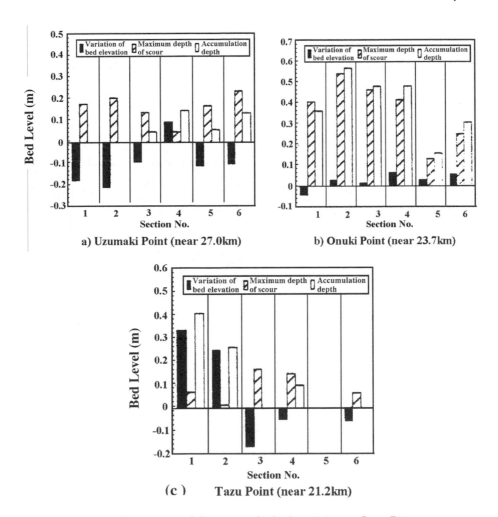

Fig. 5.48 Results of the survey for bed variation at Gono River.

section that includes the investigation point (near the 9.0 km point). The dashed line indicates the main flow at the normal time. The meander investigated has a sinuosity of 1.017. A 400 m long array of spur dikes is located on the investigation point's outer bank, which is the attack point of the flow. Bed material at the investigation point is primarily gravel.

Figure 5.51 gives the investigation sites planform and the locations of the bore holes; Fig. 5.52, the results of the investigation. These indicate that, with respect to vertical bed variation, no drop or other change in the river bed occurred in the upstream portion of the range investigated, but bed height did rise in the downstream portion. As for the maximum scouring depth, quite similar scouring occurred downstream, with extensive scouring occurring at the front face near the

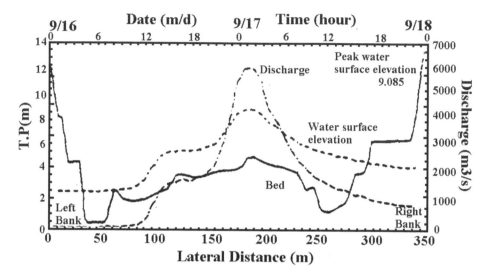

Fig. 5.49   Bed cross-sectional form, water level curve and discharge curve at the (Furusho) Point (9.0 km point).

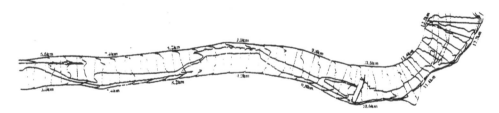

Fig. 5.50   Channel planform of the Nakagawa River near the inspection point (i.e. 9.0 km point).

main stream. The distribution of sedimentation thickness is such that it increases farther downstream. Relative depth calculated from flood marks at the investigation point was 0.29. This hydraulic quantity has indicated a boundary value separating simple and compound meandering flows. Scouring in the meander occurs more toward the center than at the inner bank. Moreover, because sinuosity is not great (1.017), maximum velocity at the flood's peak may also have occurred closer to the center than the inner bank of the main channel meander. This is inferred to explain the greater scouring depth at the channel center than at the inner bank at this specific investigation point. Sedimentation is believed to be due to the fact that the outer bank of the investigation point is an attack point (the 8.8 km point is the location of deepest bed height), and to the effects of the aforesaid spur dikes.

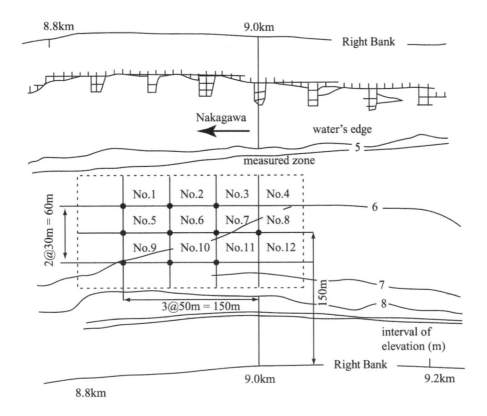

Fig. 5.51 Planform and boring hole locations at the investigation point.

5.2.6.3. Niyodogawa river investigation and results

The flood considered was one of the largest ever to occur in this river, registering a relative depth of 0.53 at the Nakashima Water Level Observation Station (located 0.5 km downstream from the investigation point) and flood channel inundation duration of 15 hours. The results of the investigation suggest a scouring depth that is greater near the main stream in the upstream portion investigated, and that decreases at points farther downstream (Fig. 5.53). In the Niyodogawa river, bed scouring at the inner bank is believed to have occurred through a mechanism different from that seen as resulting in bed variation in the Gonogawa and Nakagawa rivers. Because channel alignment is as seen in Fig. 5.54 has a sinuosity of a low 1.011, and the main channel exhibits a straight alignment for approximately 3 km upstream from the investigation point, the effects of main channel meandering are little. It is therefore believed that the flood flow was a straight or uniformly curved flow in the bankfull state and a state approaching a straight compound flow while the flood channel was inundated. Thus, the movement of alternate bars due to time change in discharge is believed to be what caused the bed variation in the Niyodogawa river.

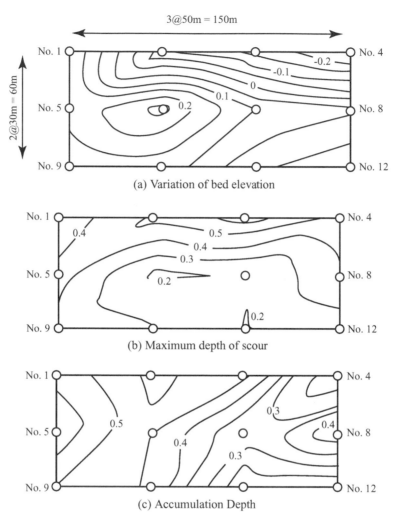

Fig. 5.52 Results of bore-hole investigation, Nakagawa River, near 9.0 km point (contour intervals in metres).

5.3. Effects of Sediment Supply on Bed topography

5.3.1. Introduction

From the velocity distribution of compound meandering channel flow of the laboratory and natural streams, the flood flow was classified into two categories, the compound meandering flow and single meandering flow. The relative depth, Dr, of about 0.3 demarcates compound and single meandering flows above and below respectively, when the sinuosity of the meandering compound channel flow is above 1.015 (Fukuoka, *et al.*, 1997, 1999). Fukuoka, *et al.* (1999) also performed numerical computations to verify the above results.

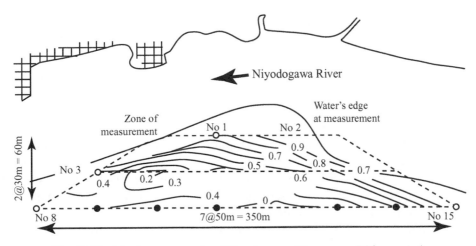

Fig. 5.53 Maximum scouring (Niyodogawa River, near 5.2 km point).

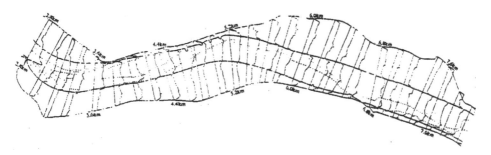

Fig. 5.54 Channel planform of the Niyodogawa River (investigation point, i.e., near 5.2 km point).

It has been investigated that the flow structure in the main channel changes significantly and sediment transport rate decreases for compound meandering flow compared to single meandering flow. (Fukuoka *et al.*, 1997). It is also important to investigate the influence of sediment overloading on flow and bed topography for the purpose of river improvement and river environment, because overloading sediment in the channel might cause bed level fluctuation.

In this section, the mechanisms of bed level change for compound meandering channels are investigated by overloading sediment supply experiments. The variables of bed level fluctuation in compound meandering channel flows include overloading sediment supply, planform of the channel and changes of relative depth.

5.3.2. Equilibrium condition of bed topography

Experiment 1 was performed to investigate reproducibility of an equilibrium bed form in a compound meandering channel. The experiments were conducted in a

S. Fukuoka *et al.*

compound meandering channel with plan geometry indicated in Table 5.13 and Fig. 5.55. The channel was 15 m long and 4 m wide, the main channel had width 0.8 m and sinuosity 1.10. Experimental conditions are indicated in Table 5.14.

The measuring reach of the channel was located downstream from section No. 7 as shown in Fig. 5.55. In this Case 1 experiment no sediment was supplied, sediment corresponding to the given flow conditions was transported in the channel.

Average bed level variation with time for Case l-2-1 and Case l-2-2 are shown in Fig. 5.57. For the above two cases, comparison of average bed level with time indicates no significant difference. Figure 5.56 displays sediment discharge measured at the downstream end of the channel. Within 3 hours from the start of the experiment, the sediment discharge had a considerable difference between Case l-2-1 and Case l-2-2, as can be seen in Fig. 5.57. The bed shape was adjusting to given flow conditions within that flow. The difference of sediment discharge for the two cases became negligible after 3 hours of flow, and no further variation of that occurred after 9 hours of flow. The flow after 9 hours implies that the

Table 5.13 Experimental channel condition.

Wavelength	7.5 m
θ_{max}	35°
Initial bed slope	1/6000
Sinuosity	1.10
Height of flood channel	5.5 cm
Diametre of bed material (uniform graded sand)	0.8 mm

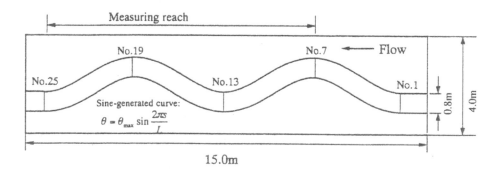

Fig. 5.55 Experimental channel plan-form.

Table 5.14 Experimental condition (Experiment 1).

Experimental case	1-1-1	1-2-1	1-3-1
	1-1-2	1-2-2	1-3-2
Discharge (l/s)	35.6	54.1	63.9
Relative depth, Dr	0.31	0.44	0.49

condition of sediment transportation was in equilibrium condition. Similar results were obtained with Case 1-1 and Case 1-3 experiments. These experimental results showed that the sediment transportation corresponding to the flow, planform and cross sectional shape of the channel made bed formation in an equilibrium condition in compound meandering channel flows.

5.3.3. Effects of overloading sediment supply

5.3.3.1. Bed deformation in the overloading sediment condition

The plan geometry of the experimental channel for Experiment 2 was the same as in Experiment 1. Conditions for the experiment and the rate of overloading sediment supply are shown in Table 5.15. Results of Experiment 1 were used as a

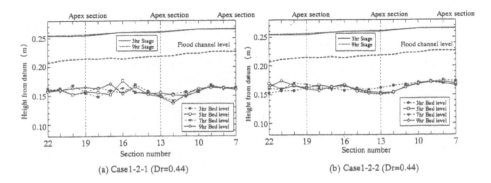

Fig. 5.56 Average bed level variation with time (Case1–2).

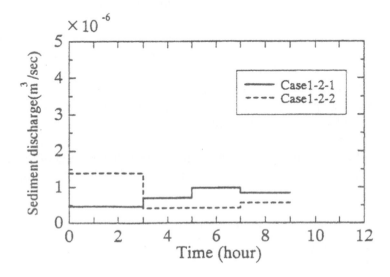

Fig. 5.57 Sediment discharge variation with time (Case 1-2-1, Case 1-2-2).

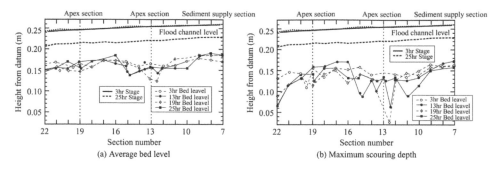

Fig. 5.58   Case 2-1 ($Dr = 0.44$, $Q_{bin} = 100\,\text{cm}^3/\text{min}$).

guide to this experiment. The results of Experiment 1 showed that the sediment discharge was $60\,\text{cm}^3/\text{min}$ for the hydraulic conditions of Case 1-2 at an equilibrium bed condition. Therefore, we decided sediment supply for Case 2-1 and Case 2-2 having rate of $100\,\text{cm}^3/\text{min}$ and $200\,\text{cm}^3/\text{min}$; about twice and three times of Experiment 1 respectively.

In order to compare the results, the experiments were performed under the conditions of the same sediment supply rate for different relative depths, $Dr = 0$ (Case 2.3) and $Dr = 0.26$ (Case 2.4). The experiments of Case 2.1, 2.2, 2.3 and Case 2.4 started with flat bed without sediment supply into the channel for first 2 hours. After that time, sediment supply was made at 0.5 m upstream of section No. 7. Water stage, bed level and sediment discharge were then measured in the various measuring reaches as indicated in Fig. 5.55. Results of bed deformation in the overloading sediment condition are show in Figs. 5.58–5.61. Figure (a) of each of these figures shows the average bed level on different sections and the associated Figure (b) shows the maximum scouring depth variation with time. It was observed from the experiments that sandbars generated by overloading advanced downstream with time. And the local bed gradient ahead of the sandbars was increased. It was also observed that maximum local scouring occurred at high velocity sections (e.g. section No. 13). These sections are the apices of the meandering channel, where maximum velocity and increment of bed gradient due to overloading sediment supply occur. At the downstream section the deposition occurs and scouring depth decreases little by little.

For Case 2.2 which is about three times the overloading sediment supply, deposition of the sediment occurred throughout the channel. Nearby the sediment supplying section, bed deposition in the inner side became larger and flow area of the

Table 5.15   Experimental conditions (Experiment 2).

Experimental case	2-1	2-2	2-3	2-4
Discharge (l/sec)		54.1	14.4	25.0
Relative depth Dr		0.44	0	0.26
Sediment supply discharge Q_{bin} (cm^3/min)	100	200	100	
Time (hour)			25	

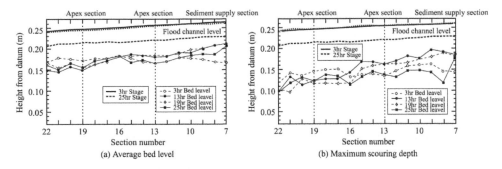

Fig. 5.59   Case 2-2 ($Dr = 0.44$, $Q_{bin} = 200\,\text{cm}^3/\text{min}$).

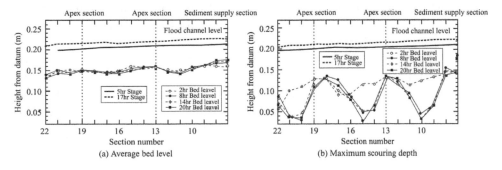

Fig. 5.60   Case 2-3 ($Dr = 0$, $Q_{bin} = 100\,\text{cm}^3/\text{min}$).

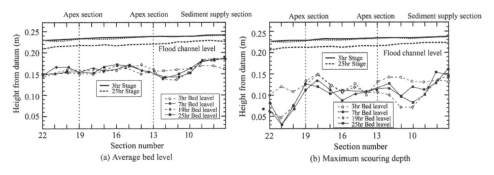

Fig. 5.61   Case 2-4 ($Dr = 0.26$, $Q_{bin} = 100\,\text{cm}^3/\text{min}$).

main channel decreased gradually with time. Even though the bed level was rising in the main channel, the water stage hardly changed. This shows that the flow was spread to the flood channel, which was wider than main channel. Case 2.3, which is a bankfull flow, maximum scouring depth of this experiment remained unchanged after 8 hours of flow.

Regarding Case 2.4 with the relative depth of $Dr = 0.26$, its maximum scouring depth became small compared to Case 2.3. The maximum scouring depth of both the cases Case 2.4 and Case 2.3 occurred at the same locations of the channel. This indicates that Case 2.4 has the flow and bed characteristics of single meandering channel flow.

For the case of overloading sediment supply, single meandering channel flow has a capability of transporting sediment to a certain extent, but in the compound meandering channel flow, sediment discharge decreases significantly and local scouring and deposition occurred. The bed elevation changes are influenced upstream and downstream of the local scouring and deposition of the channel.

5.3.3.2. Rate of sediment supply and sediment discharge

Results of measured sediment discharge are displayed in Fig. 5.62. It was deduced that sediment discharge measured at the downstream end has a close relationship with nearby bed level deformation. For Case 2.1 and Case 2.2 in Fig. 5.61(a), the sediment discharge decreased after 19 hours. This implies that sediment deposition at the upper section of downstream end occurred. Figure 5.61(b) shows the results for Case 2.3 and Case 2.4. It is seen that sediment discharge variation with time of this cases became smaller than Case 2.1 and Case 2.2, because bed topography in these cases reached equilibrium as illustrated in Figures 5.60 and 5.61. In the compound meandering channel flow, sediment discharge in time and space changed as a consequence of considerable non-equilibrium sediment transportation.

5.4. Bank Erosion in Compound Meandering Rivers

5.4.1. Flow structure and bed topography

The studies by Fukuoka *et al.* (1997, 1999) showed different flow characteristics for compound meandering channel and single meandering channel flow. The shear stresses at the imaginary horizontal interface across the main channel at the same level as the flood channel height are caused by interaction between main channel flow and flood channel flow. These horizontal shear stresses in the main channel

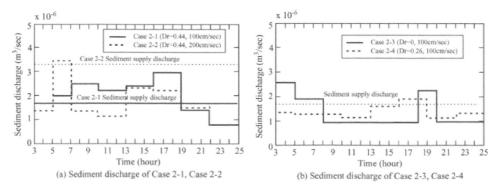

Fig. 5.62 Variation of sediment discharge with time.

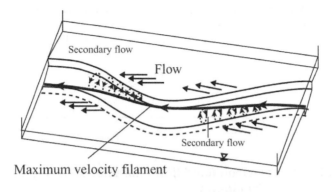

(a) Illustrations of maximum velocity
filament and secondary current

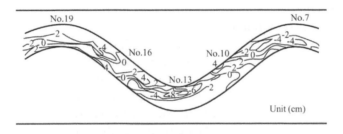

(b) Contour lines of bed level change after 9 hours
from flat bed (sinuosity = 1.17, Dr = 0.37)

Fig. 5.63 Flow structure and bed topography of a compound meandering channel flow.

produce secondary flow with the opposite rotation to the single meandering channel flow.

Flow interaction between main channel and flood channel varies in longitudinal direction, causing variation of intensity of secondary flow in that direction. As a result of flow interaction, sediment transport rate decreases with the relative depth (Fukuoka *et al.*, 1997). These flow and bed topography in a compound meandering channel are illustrated in Fig. 5.63 (after Fukuoka *et al.*, 1999). Figure 5.63(a) shows the maximum velocity filament and secondary currents. Figure 5.63(b) shows the contour lines of bed level changes from initial flatbed after 9 hours flow for the condition of sinuosity 1.17 and relative depth $Dr = 0.37$. Changes in such flow structures and bed topography occur mainly along the maximum velocity filament due to longitudinal velocity change. The apex cross section has relatively high scouring depth compared to adjacent sections as shown in Figure 5.63(b).

Table 5.16 Tama River Floods examined by the authors.

Name of Flood	Duration time of flood (hours)	Relative Depth
S57.8	30	0.50
H1.8	1	0.06
H3.8	8	0.24
H3.9	15	0.31

5.4.2. Relationship between flood flow characteristics and bank erosion in the Tamagawa river

The middle and lower reaches of the Tamagawa river contain numerous sand-bars formed of sand and gravel. Consequently, flooding of varying relative depth sometimes results in shifting of the main flow path and a change in main channel course. This section describes locations of bank damage in the Tamagawa river and considers what, if any, relationship exists with the flow categories of flood flow characteristics shown in Fig. 5.35.

The section examined stretches from 29.6–26.8 k, where damage in the form of bank scouring has been often reported. Sinuosity (S) is 1.018. Four floods are considered for this investigation (Table 5.16). On the basis of sinuosity and relative depth, the floods in August 1982 and September 1991 are classified as compound meandering flows and those in August 1989 and August 1991, as simple meandering flows. Fig. 5.64 gives the main flow paths inferred from relative depth during those floods, along with bed form and the locations of bank damage. Flooding in August 1982 (i) caused damage to the right bank in the 29.4–29.2 k section, i.e., the inner bank of the meander, downstream from which considerable scouring occurred on the right bank of the main channel. The August 1989 flood (ii) resulted in a simple meandering flow at a relative depth of 0.06, i.e., roughly the main channel bankfull state, hence a main flow line shown in the illustration is inferred.

All bank damage occurred on the outer bank of the meander, on the main flows line. When the aforesaid categories of flood flow characteristics are applied to (iii) the August and September 1991 floods are respectively inferred to be a simple meandering flow and a compound meandering flow because of their relative depths of 0.24 and 0.31, respectively; the inferred main flow line for each is given in the illustration. In each case, the site of bank damage corresponds to the main flow line. Thus corresponding to the sites of bank damage in the Tama River, the main flood flow as suggested by the flow categories used explains the locations of said damage.

The preceding suggests the following with respect to channel planning. Bank damage in a river is significantly determined by channel alignment and the river's hydraulic conditions during flooding (e.g., maximum velocity and flood duration). The location of damage can be detected to a certain extent, thereby providing data that is highly valuable to the channel alignment and revetment planning.

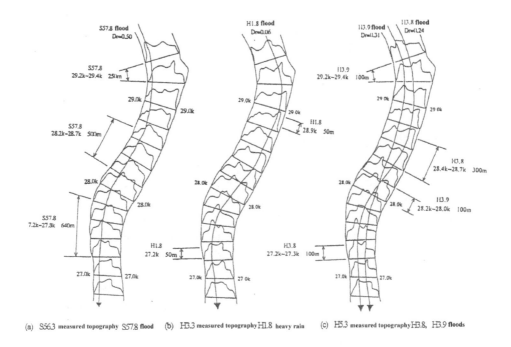

Fig. 5.64 Location of bank damage.

5.5. Effects of Main Channel Alignment on Local Scouring Depth Around Piers

5.5.1. Introduction

River plans have traditionally assumed that, during flooding, a rivers bed in a meandering channel undergoes scouring at the outer bank and sedimentation at the inner bank. Consequently, piers have often been located at the inner-bank side of a meandering channel for reasons of pier and embankment safety. However, recent research on compound meandering watercourses suggests that when, during flooding, the maximum velocity filament appears toward the inner banks and when the duration of the flood in a compound meandering channel is long, bed scouring occurs at the inner bank Fukuoka *et al.* (1997a, 1998); this has been verified by investigations of inner-bank bed variation in actual rivers Fukuoka *et al.* (1997c). It has also been shown that because conventionally-used equations for estimating scouring depth around piers apply primarily to straight channels, factors such as maximum scouring depth and optimum pier location and orientation must be re-examined in the context of flows which, like compound meandering flows, exhibit great changes in flow direction and velocity according to relative depth and sinuosity, among other variables.

In this section, model piers were placed in the main channel of a compound meandering channel, and various relative depths, pier locations, and pier shapes were used to examine the effects of these factors on local scouring depth around piers. The effects of main channel alignment are also considered by comparing the results obtained in that experiment to the results of calculations using scouring depth estimation equations for straight watercourses.

5.5.2. Experimental methodology

Figure 5.65 provides a plan view of the experimental channel, while Table 5.17 lists the specifications of that channel. In the experiments, two types of model piers were used: cylindrical piers, used in Experiment 1, and elongated piers with semicircular ends, used in Experiment 2 (Fig. 5.66). The conditions of each experiment are shown in Tables 5.18 and 5.19. In Experiment 1, local bed scouring around cylindrical model piers of two different diameters was investigated and the results compared to those of an experimental equation used for straight channel. In Experiment 2, elongated (model) piers with semicircular ends and having a ratio of width to length of 1:1.5 were used to determine bed scouring, after which the results were compared with those of Experiment 1 to determine the effect of different pier shapes and pier axis orientations. The piers were placed at the left and right banks in the section of maximum curvature and the section of meander inflection and at the location where maximum scouring depth occurs in the main channel bankfull state, after verifying that they would not affect each other. Two different pier axis orientations were examined: parallel to main channel alignment and parallel to embankment alignment. Pier diameters of 7.5% and 4.5% (i.e., ratio of pier width to river width) were used, as design standards for actual rivers prescribe an obstruction factor of no more than 7%. For a total of 10 cases of differing relative depth (Dr, defined as flood channel depth/main channel depth) and pier orientation, bed form and local scouring depth around the piers were measured in detail to determine the effect of main channel alignment on bed variation, scouring depth around piers, and water surface profile.

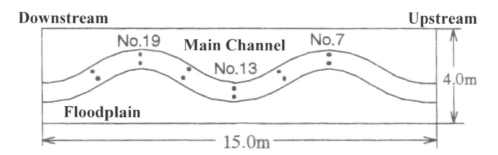

Fig. 5.65 Plan view of experimental channel.

Table 5.17 Channel specifications.

Length of flume	15.0 m
Width of flume	4.0 m
Longitudinal slope	1/600
Main channel width	0.8 m
Thalweg wavelength	7.5 m
Wavelength	6.8 m
Sinuosity	1.10
Maximum deflection angle	35°

Table 5.18 Experimental conditions (cylindrical piers).

Case	1	2	3	4	5	6
Discharge (l/s)	14.5	24.5	35.9	54.0	14.5	14.5
Relative Depth (Dr)	0	0.28	0.34	0.43	0	0
D (cm)				3.0		1.8
Location of Piers	Section of maximum curvature and inflectional section				Outer bank at inflectional section	

5.5.3. Experimental results and considerations

5.5.3.1. Longitudinal water profile and bed profile

Figure 5.67 shows the longitudinal distribution of average surface height and average bed height nine hours after the start of flooding in Case 4 ($Dr = 0.43$). To determine the piers' effect on water level and bed height, those results were

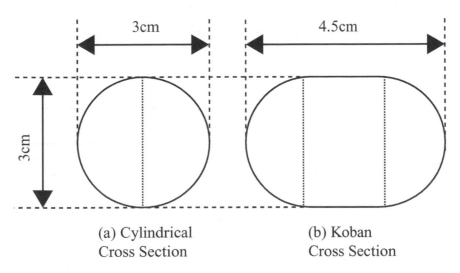

(a) Cylindrical
Cross Section

(b) Koban
Cross Section

Fig. 5.66 Plan view of model piers (a Koban is an ancient Japanese coin).

compared with the results of experiment performed with no piers in the channel. Average bed height dropped at sections where the piers were located; greater than the drop at the section of meander inflection was the drop at the section of maximum curvature. The latter section tended to undergo erosion only, with the transported sediment accumulating near the section of inflection. Longitudinally averaged water level was slightly higher at all sections due to the increase in resistance caused by the piers.

5.5.3.2. Bed variation and local scouring depth around piers

To determine what effect sediment transported due to erosion around the piers has on downstream bed variation, an experiment was also carried out with coloured sediment placed around each pier (a different colour was used for each pier). Figure 5.68 shows the dispersion of colored sediment 45 minutes and four hours after the start of a flooding experiment where $Dr = 0.43$. These hydraulic conditions produced a compound meandering flow in which the location of maximum velocity extended from inner bank to inner bank.

After 45 minutes of flooding Fig. 5.68(a), sediment created by scouring at the outer bank of the section of meander inflection and at the inner bank of the section of maximum curvature had been transported along the banks. After four hours, Fig. 5.68(b), the coloured sediment had been transported even farther downstream, depositing at the inner bank of the meander-inflection section according to the location of maximum velocity and longitudinal change in maximum velocity. At

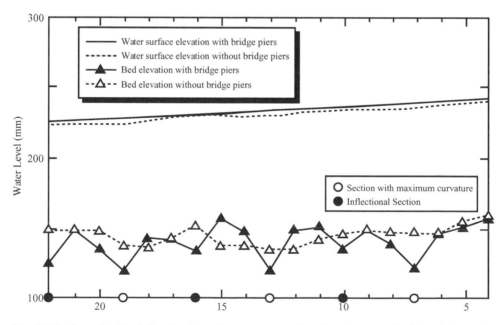

Fig. 5.67 Longitudinal distribution of average water level and average bed height in main channel sections (Case 4).

the outer bank of the section of maximum curvature, in contrast, velocity was low and almost no sediment transport was observed immediately after the start of flooding. After four hours of flooding, the colored sediment gradually spread throughout the channel, and a band of colored sediment formed along a line connecting inner-bank locations where bed variation occurred; change in bed variation thereafter was little. Thus, although sediment transport due to scouring around piers does affect bed form in the early stages of scouring, a bed form determined primarily by the compound meandering alignment becomes established with the passage of time.

Figure 5.69(a) shows the bed contour after nine hours of flooding without piers and Fig. 5.69(b) under the same conditions as Case 4. Although Case 4 shows some scouring around the piers and sedimentation downstream from the piers is apparent, no significant overall bed form difference is seen between the two cases. Figure 5.70 shows longitudinal change in maximum scouring depth around (a) cylindrical piers in flows having various relative depths, (b) elongated piers with semicircular ends when $Dr = 0$, and (c) elongated piers with semicircular ends when $Dr = 0.44$. As can be seen, scouring depth was greatest with a simple flow in the main channel bankfull state (Case 1). Specifically, local scouring depth was greatest at the outer banks of the sections of meander inflection (i.e., sections Nos. 10, 16, and 22), as these were the attack points of the flow. In contrast, in the three cases of compound meandering flow (i.e., Cases 2, 3, and 4), scouring

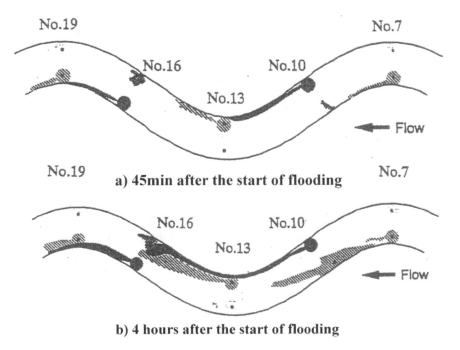

a) 45min after the start of flooding

b) 4 hours after the start of flooding

Fig. 5.68 Dispersion of coloured sediment ($Dr = 0.43$, 45 min and 4 h after start of flooding).

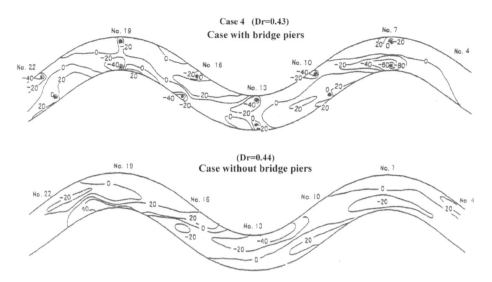

Fig. 5.69 Bed variation contour in Case 4 (with and without piers).

depth was low overall in comparison to the simple-flow cases and roughly the same as these three cases. This is because once water enters the flood channel and a compound meandering flow results, the main channel's velocity is reduced by slower-moving water entering from the flood channel.

In the case of the elongated piers with semicircular ends Fig. 5.70(b), in the main channel bankfull state, the axis of the pier located at the section of meander inflection axis is oriented at an angle to the main flow axis, and so scouring depth was greater than with cylindrical piers. However, in Cases 7 and 9, in which water level was at the main channel bankfull state, maximum scouring depths of the various sections were roughly equal, and almost no difference due to pier axis orientation was observed.

In Cases 8 and 10, in which a compound meandering flow had formed, velocity in the main channel was uniformly lower overall. Consequently, scouring depth was roughly the same at the sections of maximum curvature and meander inflection. Furthermore, because the lower velocity (relative to that of the simple flow) meant lower tactile force, the effect of pier shape and pier axis orientation was even less significant. The preceding indicates that hydraulic quantities at main channel bankfull discharge should be addressed as external forces in the design of piers-particularly setting depth- in compound meandering rivers.

5.5.3.3. Scouring depth estimation

Compound meandering channels differ importantly from straight channels in that the location of the maximum velocity filament depending on sinuosity (S) and relative depth (Dr). This longitudinal variation in maximum velocity causes bed

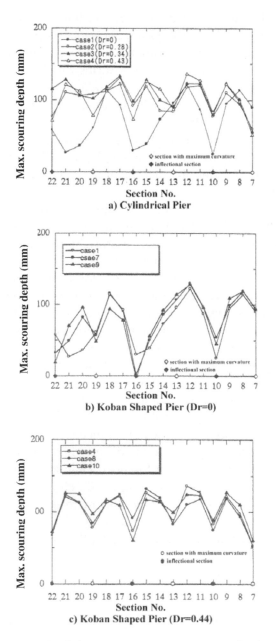

Fig. 5.70 Longitudinal change in maximum scouring depth in each case.

variation, with scouring depth and sedimentation height varying from one location to another. Therefore, the first prerequisite for assessing local scouring depth around piers in compound meandering channels is to address the bed variation that arises from the alignment of compound meandering flows. Because the effect on downstream bed form of scouring-induced sediment transport attributable to piers is limited to locations immediately around the piers, local scouring depth around piers can be viewed as arising from two separate causes: the bed variation characteristics of compound meandering flows and the piers themselves.

The authors therefore decided to assess scouring depth attributable to piers by eliminating bed variation caused by channels alignment. To do this, bed variation experiment was first carried out without the use of piers, Figure 5.71(a), and the resultant scouring and sedimentation were defined as bed variation caused by the compound meandering flow. Next, after placing the piers in the experimental channel and repeating the experiment, Figure 5.71(b), the scouring depth difference (Zs) before and after pier placement was defined as the scouring depth attributable to the piers.

Figure 5.72 compares the values of Zs obtained experimentally for cylindrical model piers of differing diameters with results obtained by scouring depth estimation equations previously proposed for straight channels. Three representative equations were used: Andru, Laursen-Toch, and Tarapore As the results in Fig. 5.72 demonstrate, once the effects of compound meandering flows on bed variation characteristics have been assessed in advance and the increase in scouring depth attributable to piers is assessed, it is possible to roughly estimate maximum scouring depth in a compounding meandering channel using the equation of Laursen- Toch (where $K = 1.35$), one of the more widely-used scouring equations for cylindrical piers in straight channels. However, when $h_0/D = 2$ ($Dr = 0$) and 3.5 ($Dr = 0.44$), Zs/D is larger than when calculated with the estimation equation. The reason is as follows: In a simple meandering flow where $Dr = 0$, the bed's secondary flows are larger, resulting in a particularly great scouring depth at the outer bank in the section of meander inflection, which is an attack point. After the flood channel becomes inundated but while the relative depth is still low, the inflow of slower water from the flood channel reduces the main channel's velocity and thereby reduces scouring depth. However, once relative depth increases further, the flood channel's flow also increases in velocity to the point where velocity in the main channel becomes uniform, resulting in bed variation throughout the channel and causing scouring depth to rise once again to the point

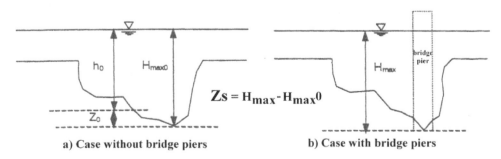

a) Case without bridge piers   b) Case with bridge piers

Fig. 5.71 Definition of scouring depth Zs attributable to the piers.

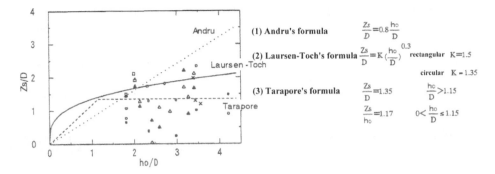

Fig. 5.72 Comparison of observed results and results from scouring depth estimation equations for straight channels (cylindrical piers).

where it is roughly the same at each pier position. This phenomenon is also evident from the experimental results of bed variation performed by Fukuoka *et al.* (1997b), in which various relative depths were used in a compound meandering channel.

With respect to scouring attributable to piers, the preceding discussion suggests that if bed variation caused by the various types of flows can be estimated in ways that account for main channel alignment, then it would be possible to roughly estimate the total scouring depth that would occur if piers were present. As the work of Fukuoka & Watanabe (1998) has reached the stage of using three-dimensional analysis to determine, theoretically, quantitative bed variation in the main channels of compound meandering channels, it may be possible to use such analysis to determine scouring depth around cylindrical piers with a considerable degree of accuracy.

5.6.   Concluding Remarks

This Chapter describes a series of state-of-the-art engineering applications in compound meandering channels. Fundamental to them is the combination of basic science leading to firm, well grounded and rationale engineering outcomes. The Chapter draws together, within a practical context, the three earlier Chapters on flow, sediment dynamics and morphology, and numerical modeling, illustrating their combined importance in understanding and predicting the behaviour of rivers in flood.

The concept of flood plain storage is well known but its effect is frequently neglected or underestimated be designers. The analysis and Case Studies presented in this chapter demonstrate that it can be a major contributor in flood wave attenuation and that it should, as a matter of course, always be considered in flood impact studies.

The simple phenomenological distinction that has been drawn between simple and compound meandering flows is a practically important and one which is very significant in understanding the behaviour of a flood channel. It is therefore necessary for engineers working in bank erosion control to appreciate and understand the different conditions which give rise to the two flow types which largely depend on relative depth and sinuosity. At a simple level, an appreciation of the differences between the two types of flow, especially the recognition that the maximum velocity filament point moves from the inner bank in a simple meandering channel flow to the outer bank in a compound meandering flow is crucial in the optimum positioning of bank protection measures. Similarly, the position of maximum scour, on the channel bed will also move depending on the nature and classification of the flow.

Looking to the future, this Chapter demonstrates how a combination of hydraulic model studies and river engineering case studies can be combined in the pursuit of practical engineering outcomes. Each approach contributes significantly and essentially to the overall effort. The net effect is one of hope because a through and systematic treatment of the complex problems has been shown to yield significant engineering dividends which form the basis for future improvements in practice. This points to a future in which further advances in understanding and practice can be made by a following this path. As engineers prepare to meet future challenges there is clearly a need for a broad-based dialogue and shared decision making with other disciplines and interests. We will serve our societies best when the contributions we make are informed by state-of-the-art technical insights which have been founded and rationale analysis and direct exposure to the challenge of engineering natural and regulated rivers.

Chapter 6

Conclusions and Recommendations

S. Ikeda and I. K. McEwan

6.1. Concluding Remarks

In the last two decades, increasingly advanced laboratory and field measurement techniques have been developed, largely based on progress in electronics. The appearance of laser-Doppler velocimetry has allowed us to measure the details of turbulent flow field. It also now possible to measure instantaneous turbulent flow field by detecting the movements of fine particles, which technique is known as particle image velocimetry (PIV). The development of remote sensing technique and the usage of satellite such as global positioning system (GPS) have made our field measurement much easier than the past, and the entire flow field can now be routinely measured grasped by using these techniques.

The development of computers has also provided many of benefits to hydraulicians and hydraulic engineers. The computers have been coupled with the development of turbulence modelling such as k-ε turbulence model, large eddy simulation (LES) and increasingly direct numerical simulation (DNS). Flow field can now be simulated even for highly unsteady and anisotropic geophysical turbulence which is encountered in rivers.

The development of sedimentary science is also important. The classical sediment transport mechanics such as lateral sediment transport has allowed to develop theoretically sound river mechanics, for example, self-formed rivers to explain the size and the shape of channel cross section, river meandering and braiding.

Changes in our societies are also important factors for river engineering; in the UK the pressures on land have led to a growth in the use of flood plains for habitation, which has tended to increase the human impacts and consequences of flood events. The predictions of flow field, flood stage and inundation are therefore important in the UK whereas in Japan there are other problems to be solved. In Japan, people are prohibited from living on floodplains and the government owns floodplains bounded by levees. The construction of dams in the upstream has changed the environment of rivers. In the past three decades, these rivers have tended to degrade, and artificial regulation of flows has decreased the flow discharge at floods, allowing vegetation to grow along river courses and providing depositions of fine sediments on the flood plains. Modern river management requires a knowledge of the characteristics of flow affected by vegetation.

In Chapter 2, both micro- and macro-structures of flow for straight, curved and meandering compound channels are treated. In straight compound channels, secondary flow of the 2nd type which is induced by the imbalance of Reynolds stress, is treated both theoretically and experimentally. The isovels of the primary flow are skewed by the secondary flow, which exchanges longitudinal fluid momentum in the cross-section. Another important phenomenon is plan-form organized vortices induced by shear instability. Since flow is retarded on the flood plains, the lateral distribution of depth-averaged flow velocity possesses inflectional points near the edges of the boundaries of the main channel and the flood plains. The plan-form vortices increase the lateral transport of fluid momentum and suspended materials. The vortices can be simulated by employing depth-averaged momentum balance equations and turbulence kinetic energy equation for SDS (sub-depth scale) turbulence. For curved and meandering compound channels, the secondary flow of the 1st kind induced by the imbalance of centrifugal force and the lateral pressure gradient adds more complexity to the flow field. Laboratory studies are abundant on the flow structures (e.g. research programme on a large flood channel facility in UK), and several interesting phenomena have been found. In addition to these detailed complex structures of flow, prediction of resistance coefficients, conveyance capacity, stage-discharge relationship are very important from practical point of view, for which several engineering approaches are also presented herein.

Chapter 3 considered sediment processes, which are important issues in considering fluid/sediment interaction and morphology of compound channels. One of the major points of interest is the effect of relative depth ratio on the sediment transport rate. Several laboratory tests in straight channels with graded sediments have shown that the mass transport rate may increase as the flow goes overbank. The reason for this phenomenon is still not clear, and further study is required. In contrast, the laboratory tests in meandering compound channels have indicated that immediately above bankfull a sudden decrease in sediment transport rate is observed, and it seems that above 0.2 of the depth ratio the transport rate begins to increase as the flow discharge increases. Research on the self-formation of free plan form channels has been developed based on theoretically-sound mechanics in the last two decades, without employing unproven extremal hypothesis such as an assumption of maximum sediment transport rate. The self-formed straight channels considered the redistribution of bed shear stress by turbulence-induced lateral diffusion of fluid momentum. This theory has been applied successfully to rivers with bank vegetation and rivers with bank groins. A mechanistic model for the development of river meanders has also been developed, and the theory has been applied to explain coupled growth of free planform and alternate bars.

The development of turbulence modelling, coupled with the rapid advance of computer technology has enhanced numerical simulations as a tool in engineering research and design. Chapter 4 discussed various techniques on the computer simulation of flow, those generally used were reviewed briefly at first, and some new techniques developed for simulating natural geophysical flows, characterized by anisotropic turbulence field, were introduced. Since natural geophysical flows usually have large width/depth ratio, depth-averaged modelling is sometimes useful. However, if flows in channels with a curvilinear plan form are treated, we have to include centrifugal force-induced secondary flow, which inherently requires at least quasi-3D analysis. When organized motion of flow turbulence is concerned, Large Eddy Simulations should be used. Thus, suitable turbulence models should be chosen according to the purpose and level of research or practical design. The new advanced methods described in Chapter 4 have been applied to various conditions and flow fields to illustrate their performance.

Practical application of research is vital. Chapter 5 illustrates the process of taking forward basic research results to the point where they have a direct bearing on practical engineering issues. The tools used to do this are traditional engineering methods drawn from appropriate combinations of laboratory modelling, field observation and numerical analysis. Findings include a demonstration of the importance of flood plain storage in attenuating flood waves and some important results, drawn for flow studies, which relate to the likely position of maximum scour during flood events.

6.2. Future Work

As described in the text and the concluding remarks, theoretical analysis, laboratory tests and numerical simulations on flow and sediment transport in compound open channels have developed remarkably in the last two decades. However, there are many significant topics yet to be tackled. Since natural geophysical flows typically seen in compound channels are characterized by unisotropic turbulence, three-dimensionality and complex boundaries, there is a desire to develop a new turbulence model, which can include these characteristics and hopefully be coupled with sediment transport for practical purposes. The effects of flood plains on the sediment transport in main channels are still not yet fully exploited, to the subject of which the model developed should be applied.

The existence of flood plains provides sink or source of materials such as fine sediments, organic materials and nutrients. The exchange of these materials between the main channel and the floodplain is one of the major agencies for maintaining eco-system in compound channels. The behaviour of these materials should therefore be studied in more detail for adequate control of habitats and to sustain the eco-system. In treating environments in rivers such as eco-system, a new synthetic approach might be useful in addition to the traditional analytical methods. River beds in main channels provide habitats for various insects, fish, birds and other animals and therefore predictions of bed scour and deposition, accretion of banks, sediment size distributions, degradation and aggradations are all important from ecological point of view. It should also be noted that these phenomena are closely correlated with vegetation, since obstacles such as trees, houses and river structures are increasingly being found on flood plains, the development of an accurate and accepted method for predicting stage-discharge relations is therefore necessary from engineering point of view.

All the authors of this book wish it to become a stimulus for young scientists and engineers who are interested in river mechanics.

References

1. Abril, B. (1995). 'Numerical modelling of turbulent flow in compound channels by the finite element method.' *Hydra 2000, Proceedings of 26th IAHR Congress,* London, UK, **5**, 1–6.
2. Abril, B. (1997). *'Numerical river modelling of turbulent flow, sediment transport and flood routing using the finite element method.'* PhD Thesis, University of Birmingham, England, UK.
3. Abril, J. B. and Knight, D. W., (2004). Stage-discharge prediction for rivers in flood applying a depth-averaged model, *Journal of Hydraulic Research, IAHR,* **42**, No. (6), 616–629.
4. Ackers, P. (1991). 'Hydraulic design of straight compound channels.' *Report SR 281,* **1**, 1–131 and **2**, 1–139, HR Wallingford, UK.
5. Ackers, P. (1992a). '1992 Gerald Lacey memorial lecture — Canal and river regime in theory and practice 1929–1992'. *Proceedings of the Institution of Civil Engineers: Water, Maritime & Energy,* **96**(3), Paper 10019, 167–178.
6. Ackers, P. (1992b). 'Hydraulic design of two stage channels.' *Proceedings of the Institution of Civil Engineers: Water, Maritime, and Energy,* **96**(4), Paper 9988, 247–257.
7. Ackers, P. (1993a). 'Stage discharge functions for two-stage channels: The impact of new research.' *Journal of the Institute of Water and Environmental Management,* **7**(1), 52–61.
8. Ackers, P. (1993b). 'Flow formulae for straight two-stage channels.' *Journal of Hydraulic Research, IAHR,* **31**(4), 509–531.
9. Ackers, P. and White, W. R. (1990). *Sediment Transport: The Ackers and White Theory Revised,* Report SR237, HR Wallingford Ltd, UK.
10. Ackers, P. and White, W. R. (1973). 'Sediment transport: new approach and analysis.' *Journal of the Hydraulic Division,* ASCE, **99**(HY11), 2041–2060.
11. Akanbi, A. A. and Katopodes, N. D. (1988). 'Model for flood propagation on initially dry land.' *Journal of Hydraulic Engineering,* ASCE, **114**(7), 689–706.
12. Alhamid, A. A. I. (1991). *Boundary shear stress and velocity distributions in differentially roughened trapezoidal channels.* PhD Thesis, The University of Birmingham, England, UK.
13. Anderson, M. G., Walling D. E. S. and Bates, P. D. (1996). *Floodplain Processes,* Wiley.
14. Ashida, K. (1969). 'Study on bed deformation — bed deformation due to drop of water stage at the downstream end.' Annuals, Disas. Prev. Res. Inst., Kyoto Univ., No.12B, pp. 437–447, 1969 (in Japanese).
15. Ashworth, P. J., Bennett, S. J., Best, J. L. and McLelland, S. J. (1996). *Coherent flow structures in open channels.* Wiley.
16. Asselman, N. and Middelkoop, H. (1995). 'Floodplain sedimention – quantities, patterns and processes.' *Earth Surface Processes* **20**(6), 481–499.
17. Atabay, S. (2001). *Stage-discharge, resistance and sediment transport relationships for flow in straight compound channels.* PhD Thesis, The University of Birmingham, England, UK.

18. Atabay, S. and Knight, D. W. (1999) 'Stage discharge and resistance relationships for laboratory alluvial channels with overbank flow.' In Jayawardena, A.W., Lee, J. H. W. and Wang., Z. Y. (eds.), *Proceedings of the 7th International Symposium on River Sedimentation*, Hong Kong, 1998, 223–229.

19. Atabay, S., Knight, D. W. and Seckin, G., (2004), Influence of a mobile bed on the boundary shear in a compound channel, *River Flow (2004), Proc. 2nd Int. Conf. on Fluvial Hydraulics, 23–25 June, Napoli, Italy.* (eds.), Greco, M. Carravetta, A., and Morte, R. D., Vol. 1, 337–345.

20. Atabay, S., Knight, D. W. and Seckin, G., (2005), Effects of overbank flow on fluvial sediment transport rates, *Water Management, Proceedings of the Instn. of Civil Engineers*, London, 158, WM1, March, 25–34.

21. Ayyoubzadeh, S. A. (1997). *Hydraulic aspects of straight compound channel flow and bed sediment transport*, PhD thesis, The University of Birmingham, England, UK

22. Babaeyan-Koopaei, K., Ervine, D. A., Carling, P. A. and Cao, Z. (2002) 'Velocity and turbulence measurements for two overbank flow events in River Severn'., *Journal of Hydraulic Engineering*, **128**(10), 891–900.

23. Babaeyan-Koopaei, K. and Valentine, E. M. (1995). 'Experimental assessment of a rational regime theory, *Proceedings of the XXVIth I.A.H.R. Congress, Hydra 2000*, London, pp. 360–365.

24. Bates, P. D., Anderson, M. G., Price, D., Hardy, R. J. and Smith. C. N. (1996). 'Analysis and development of hydraulic models for floodplain flows'. In Anderson, M. G., Walling, D. E. and Bates, P. D. (eds.), *Floodplain Processes*, John Wiley and Sons, Chichester, pp. 215–254.

25. Bathurst, J. C. (1979). Distribution of boundary shear stress in rivers. In Rhodes, D. D. and Williams, G. P. (eds.), *Adjustments of the fluvial system*. Harper Collins, 95–116.

26. Bathurst, J. C., Thorne, C. R. and Hey, R. D. (1977) 'Direct measurement of secondary currents in river bends.' *Nature*, **269**, 5628, 504–506.

27. Bennett, S. J., Alonso, C. V., Prasad, S. N. and Romkens, M. J. M. (2000). 'Experiments on headcut growth and migration in concentrated flows typical of upland area.' *Water Resources Research*, **36**(7), 1911–1922.

28. Bennett, S., Best, J. and McLelland, S. (eds.), *Coherent Flow Structures in Open Channels*, J Wiley, Chapter **28**, 581–608.

29. Bettess, R. and Fisher, K. R. (1999). 'Lessons to learn from the UK river restoration projects – RIBAMOD River basin modelling management and flood mitigation Concerted Action.' In Casale, R., Samuels, P. and Bronstert, A. (eds.), *Proceedings of the 2nd Workshop on Impact of Climate Change on flooding and Sustainable River Management.*

30. Bettess, R. and White, W. R. (1987). 'Extremal hypotheses applied to river regime'. In Thorne, C. R., Bathurst, J. C. and Hey, R. D. (eds.), *Sediment Transport in Gravel-Bed Rivers*. Wiley.

31. Bhowmik, N. G. (1982). 'Shear stress distribution and secondary currents in straight open channels.' In Hey, R. D., Bathurst, J. C. and Thorne, C. R. (eds.), *Gravel bed rivers*. Wiley.

32. Bhowmik, N. G. and Demissie, M. (1982). 'Carrying capacity of floodplains.' *Journal of Hydraulic Division*, ASCE, **108**(HY3), 443–452.

33. Blalock, M. E. and Sturm, T. W. (1981). 'Minimum specific energy in compound open channel.' *Journal of Hydraulics Division*, ASCE, **107**(HY6), 699–717.

34. Bousmar, D. and Zech, Y. (1999). 'Momentum transfer for practical flow computation in compound channels.' *Journal of Hydraulic Engineering*, ASCE, **125**(7), 696–706.

35. Bousmar, D., (2002), 'Flow modelling in compound channels: momentum transfer between main channel and prismatic or non-prismatic floodplains', *PhD Thesis*, Universitie catholique de Louvain.

36. Bousmar, D. and Zech, Y., (2002), Periodical turbulent structures in compound channels, *Proc. RiverFlow 2002, Louvain*, (eds.), Bousmar, D. and Zech, Y., Vol. 1, Balkema, pp. 177–185.

37. Bousmar, D. and Zech, Y., (2004), Velocity distribution in non-prismatic compound channels, *Proc. Instn of Civil Engineers*, Water Management, Vol. 157, 99–108.

38. Bradshaw, P. (1987). 'Turbulent secondary flows.' *Annual Review of Fluid Mechanics*, **19**, 53–74.

39. Bousmar, D., Atabay, S., Knight, D. W. and Zech, Y., (2006), Stage-discharge modelling in mobile bed compound channels, *RiverFlow 2006*, September, Lisbon, Portugal.

40. Brooks, A. N. and Hughes, T. J. R. (1982). 'Streamline upwind/Petrov Galerkin formulations for convection dominated flows with particular emphasis on the incompressible Navier-Stokes equations.' *Computer Methods in Applied Mechanics and Engineering*, **32**(1–3), 199–259.

41. Brown, F. A. (1997). *Sediment transport in river channels at high stage*. PhD Thesis, University of Birmingham, UK.

42. Cassells, J. B. C. (1998). *Hydraulic characteristics of straight mobile bed compound channels with uniform sediment*, PhD Thesis, University of Ulster, UK.

43. Cassells, J. B. C., Lambert, M. F. and Myers, R. W. C. (2001) "Discharge prediction in straight mobile bed compound channels", *Proc. ICE, Water and Maritime Engineering*, **148**(3), 177–188.

44. Cassells, J. B. C., Lambert, M. F. and Myers, W. R. C., (2001), Discharge prediction in straight mobile bed compound channels, *Proc. Instn. of Civil Engineers, Water and Maritime Engineering*, London, 158, WM1, March, pp. 25–34.

45. Chang, H. H. (1983). 'Energy expenditure in curved open channels.' *Journal of Hydraulics Engineering*, ASCE, **109**(7), 1012–1022.

46. Chang, H. H. (1984). 'Variation of flow resistance through curved channels.' *Journal of Hydraulics Engineering*, ASCE, **110**(12), 1772–1782.

47. Chang, H. H. (1988). *Fluvial Processes in River Engineering*. Wiley.

48. Chow, V. T. (1959). *Open Channel Hydraulics*. McGraw-Hill.

49. Chu, V. H., Wu, J. H. and Khayat, R. E. (1991). 'Stability of transverse shear flows in shallow open channels.' *Journal of Hydraulic Engineering, ASCE*, **117**(10), 1370–1388.

50. Cokljat, D. and Younis, B. A. (1982). 'Resistance to flow in channels with overbank flood plain flow.' *Proceedings of the 1st International Conference on Channels and Channel control structures*, Southampton, pp. 4-137–4-150.

51. Cokljat, D. and Younis, B. A. (1994). 'On modeling turbulent flows in non-circular ducts.' *American Society of Mechanical Engineers* 118, 77–82.

52. Cokljat, D. and Younis, B. A. (1995). 'Second-order closure study of open channel flows.' *Journal of Hydraulic Engineering*, ASCE. **121**, 94–107.

53. Colebrook, C. F. (1939). 'Turbulent flow in pipes, with particular reference to the transition region between the smooth and rough pipe laws.' *Journal of the Institution of Civil Engineers*, **11**,133–156.

54. Colebrook, C. F. and White, C. M. (1937). 'Experiments with fluid friction in roughened pipes.' *Proceedings of the Royal Society A*, **161**, 367–381. *Conference on Hydro-Science and Engineering*, Washington DC, USA, 7–11 June, pp. 1309–1316.

55. Cunge, J. A., Holly, F. M. and Verwey, A. (1980). *Practical aspects of computational river hydraulics*. Iowa Institute of Hydraulic Research, USA, Pitman Publishing. .

56. Craft, T. J., Suga, K. and Launder, B. E. (1993). 'Extending the applicability of eddy viscosity models through the use of deformation in-variants and non-linear elements.' *Proceedings of the 5th International Symposium on Refined Flow Modeling and Turbulence Measurements*. pp. 125–132.

57. Day, T. J. (1980). *A Study of the Transport of Graded Sediments*. Hydraulic Research Wallingford, Report IT, 190.

58. Deardorff, J. W. (1970). 'A numerical study of three dimensional turbulent channel flow at large Reynolds numbers.' *Journal of Fluid Mechanics*, **41**(2), 435–452.

59. Dermirdzic, I. A., Gosman, A. D., Issa, R. J. and Peric, M. (1987). 'A calculation procedure for turbulent flow in complex geometries.' *Computers and Fluids*, **15**(3), 251–273.

60. Dey, A. K., Kitamura, T. and Tsujimoto, T. (2000). 'Gully development by two-dimensional head-cut'. *Journal of Hydraulic, Coastal & Environmental Engineering*, JSCE. (in review).

61. Dietrich, W. E. and Smith J. D. (1984). 'Bedload transport in a river meander.' *Water Resources Research*, **20**(10), 1355–1380.

62. Deitrich, W. E. and Whiting, P. (1989). 'Boundary shear stress and sediment transport in river meanders of sand and gravel.' In Ikeda, S. and Parker, G. (eds.), *American Geophysical Union*, Water Resources Monograph **12**, 1–50.

63. Einstein, H. A. (1934). 'The hydraulic or cross section radius.' *Schweizerische Bauzeitung*, **33**(8), 89–91 (in German).

64. Elliot, S. C. A. and Sellin, R. H. J. (1990). 'SERC Flood Channel facility: skewed flow experiments.' *Journal of Hydraulic Research*, IAHR, **28**(2), 197–214.

65. Engelund, F. (1964). 'Flow resistance and hydraulic radius.' *ACTA*, Ci **24**, Copenhagen

66. Engelund, F. (1974). 'Flow and bed topography in channel bends', *Journal of Hydraulic Division*, ASCE, **100**(HY11), 1631–1648.

67. Engelund, F. and Hansen, E. (1967). *A Monograph on Sediment Transport in Alluvial Streams*, Copenhagen: Teknisk Forlag.

68. Ervine, D. A., Babaeyan, K. and Sellin, R. H. J. (2000). 'Two-dimensional solution for straight and meandering overbank flows.' *Journal of Hydraulic Engineering*, **126**(9), 653–669.

69. Ervine, D. A. and Ellis, J. (1987). 'Experimental and computational aspects of overbank flood-plain flow.' *Transactions of the Royal Society of Edinburgh: Earth Sciences* Series A, **78**, 315–325.

70. Ervine, D. A. and Macleod, A. B. (1999). 'Modelling a river channel with distant flood-banks.' *Proceedings of the Institution of Civil Engineers: Water, Maritime and Energy*, Paper 11608, **136**(1), 21–33.

71. Ervine, D. A., Willetts, B. B., Sellin, R. H. J. and Lorena, M. (1993). 'Factors affecting conveyance in meandering compound flows.' *Journal of Hydraulic Engineering*, ASCE, **119**(12), 1383–1399.

72. Ferziger, J. H. and Peric, M. (1999). *Computational Methods for Fluid Dynamics*. Berlin Springer.

73. Fisher, K. R. (1992). 'Assessing the hydraulic performance of environmentally acceptable channels.' *HR Wallingford*, Report EX 1799, 1–132.

74. Fisher, K. and Dawson, H., (2003), Roughness review, *Defra and Environment Agency*, Reducing uncertainty in River flood conveyance, Report W5A-057, Defra & EA, July, 1–213.

75. Forbes, M. (2000). *Practical Application of enhanced conveyance calculations in a one dimensional river model*, PhD Thesis, University of Glasgow, UK.

76. Forbes, M. and Pender, G. (2000). 'The application of enhanced conveyance calculatiuons in flood prediction.' *Proceedings of the International Symposium of Flood Defence*, Kassel, Germany.

77. Fletcher, C. A. J. (1988). *Computational Techniques for Fluid Dynamics, Vol. 2:* Specific techniques for different flow categories. Springer-Verlag, Berlin, 1991.

78. Fukuoka, S. and Fujita, A. (1989). 'Prediction of flow resistance in compound channels and its application to design of river courses.' *Proceedings of JSCE*, **411**, 63–72 (in Japanese).

79. Fukuoka, S., Watanabe, A. and Tsumori, T. (1994). 'Structure of plane shear flow in river with vegetations.' *Proceedings of JSCE 1994–1995*, **491**, 51–60 (in Japanese), 41–50.

80. Fukuoka, S., Miyazaki, H., Ohgushi, H. and Kamura, D. (1996). 'Flow and bed topography in a meandering compound channel with phase difference between the alignment of the main channel and the levee', *Annual Journal of Hydraulic Engineering*, JSCE, **40**, 941–946.

81. Fukuoka, S., Watanabe, A. and Tsomori, T. (1993). *The Structure of Plane Flow on a Compound Channel with Trees Along the Sides of a Main Channel*. Technical Report **48**, Department of Civil Engineering, Tokyo Institute of Technology, Japan.

82. Fukuoka, S., Watanabe, A., Uesaka, T. and Tsumori, T. (1995). 'Structure of flood flow in Tone river with vegetation clusters along low-water channel bank.' *Proceedings of JSCE*, **509**/IIU-30 (in Japanese), 79–88.

83. Fukuoka, S. and Watanbe, A. (2000). 'Numerical Analysis on Three Dimensional Flow and bed Topography in a Compound Meandering Channel,' *Proceedings of the 4th International Conference on Hydroinformatics*, Iowa City, USA.

84. Gaskell, P. H. and Lau, K. C. (1988). 'Curvature-compensated convective transport: SMART, a new boundedness-preserving transport algorithm,' *International Journal for Numerical Methods in Fluids*, **8**(6), 617–641.

85. Gay, G. R., Gay, H. H., Gay, W. H., Martinson, H. A., Maeda, R. H. and Moody, J. A. (1998). 'Evolution of cutoffs across meander necks in Powder River, Montana, USA.' *Earth Surface Processes and Landforms*, **23**(7), 651–662.

86. Gessler, J. (1971). 'Aggradation and degradation.' In Shen, H. W. (ed.), *River Mechanics*, Colorado: Water Resources.

87. Gessler, D., Gessler, J. and Watson, C. C. (1998). 'Prediction of discontinuity in stage-discharge rating curves.' *Journal of Hydraulic Engineering*, ASCE, **124**(3), 243–252.

88. Germano, M., Piomelli, U., Moin, P. and Cabot, W. H. (1991). 'A dynamic subgrid scale eddy viscosity model.' *Physics of Fluids* A (Fluid Dynamics) **3**(7), 1760–1765.

89. Ghosal, S., Lund, T. S., Moin, P. and Akselvoll, K. (1995). 'A dynamic localization model for large eddy simulation of turbulent flows,' *Journal of Fluid Mechanics*, **286**, 229–255.

90. Gomez, B., Mertes, L. A. K., Phillips, J. D., Magilligan, F. J. and James, L. A. (1995). 'Sediment characteristics of an extreme flood: 1993 upper Mississippi River valley', *Geology*, **23**, 963–966.

91. Goto, T., Kitamura, T. and Tsujimoto, T. (2000). 'Study on bed-degradation and bank-erosion in straight gravel channel due to changes in boundary conditions.' *Journal of Hydraulics, Coastal and Environmental Engineering*, JSCE (in review, in Japanese).

92. Haidera, M. A. and Valentine, E. M. (1999). 'Behaviour of Alluvial Channels with Overbank Flow.' In *Proceedings of the IAHR Symposium on River, Coastal and Estuarine Morphodynamics*, Genova, Italy, pp. 153–162.

93. Haidera, M. A. and Valentine, E. M. (1999). 'A Laboratory Study of Alluvial Channels with Overbank Flow.' In *Proceedings of XXVII IAHR Congress*, Austria.

94. Haque, S. M. A. (1959). 'The effect of eddy viscosity on the velocity profile of steady flow in a uniform rough channel.' *Journal of Fluid Mechanics*, **5**, 310–316.

95. Harlow, F. H. and Welch, J. E. (1965). 'Numerical calculation of time-dependent viscous incompressible flows of fluid with free surface.' *Physics of Fluids*, **8**, 2182–2187.

96. Hasegawa, K., Asai, S., Kanetaka, S. and Baba, H. (1999). 'Flow properties of a deep open experimental channel with dense vegetation bank.' *Journal of Hydroscience and Hydraulic Engineering*, Japan Society of Civil Engineers, **17**(2), 59–70.

97. Hervouet, J. M. (1999). TELEMAC, a hydroinformatic system, Houille Blanche, **54**, (3–4), 21–28.

98. Hey, R. D. (1978). 'Determinate hydraulic geometry of river channels.' *Journal of Hydraulic Division*, ASCE, **104** (HY6), 869–885.

99. Hirt, C. W. and Nichols, B. N. (1981). 'A computational method for free surface hydro-dynamics.' *Journal of Pressure Vessel Technology-Transactions of the ASME*, **103**(2), 136–141.

100. Holden, A. P. and James, C. S. (1989). 'Boundary shear distribution on floodplains.' *Journal of Hydraulic Research*, IAHR, **27**(1), 75–89.

101. Holley, E. R. and Abraham, G. (1973). 'Laboratory studies on transverse mixing in rivers.' *Journal of Hydraulic Research*, IAHR, **11**(3), 219–253.

102. Horton, R. E. (1933). 'Separate roughness coefficients for channel bottom and sides.' *Engineering News Record*, **111**(22) 652–653.

103. Hosoda, T., Sakurai, T. and Muramoto, Y. (1998). '3-D computation of unsteady flows in compound open channels with horizontal vortices and secondary currents.' *Annual Journal of Hydraulic Engineering*, JSCE **42**, 631–636.

104. Ikeda, S. (1981). 'Self-formed straight channels in sandy beds.' *Journal of Hydraulic Division*, ASCE, **107**(HY4), 389–406.

105. Ikeda, S. (1999). 'Role of lateral eddies in sediment transport and channel formation.' In Jayawardena, A. W., Lee, J. H. W. and Wang, Z. Y. (eds.), *Proceedings of the 7th International Symposium on River Sedimentation*, Hong Kong, 1998,195–203.

106. Ikeda, S. and Kuga, K. (1997). 'Laboratory study on large horizontal vortices in compound open channel flow.' *Proceedings of JSCE*, **558**, 91–102 (in Japanese).

107. Ikeda, S., Kuga, K. and Toda, Y. (1995). 'Measurement of the instantaneous structures of periodical vortices in a compound open channel flow by Particle Image Velocimetry.' *Proceedings of the Symposium on River Hydraulics and Environments*, JSCE, pp. 33–38 (in Japanese).

108. Ikeda, S., Izumi, N. and Ito, R. (1991) 'Effects of pile dikes on flow retardation and sediment transport.' *Journal of Hydraulic Engineering*, ACSE, **117**(11), 1459–1478.

109. Ikeda, S., Murayama, N. and Kuga, T. (1995). 'Stability of horizontal vortices in compound open channel flow and their 3-D structure, Periodic vortices at the boundary of vegetated area along river bank.' *Journal of Hydraulic, Coastal and Environmental Engineering JSCE* No. 509/II-30,131–142 (in Japanese), 87–97.

110. Ikeda, S., Ohta, K. and Hasegawa, H. (1992). 'Periodic vortices at the boundary of vegetated area along river bank.' *Journal of Hydraulic, Coastal and Environmental Engineering, JSCE* 443/II-18, 47–54 (in Japanese).

111. Ikeda, S., Ohta, K. and Hasegawa, H. (1994). 'Instability-induced horizontal vortices in shallow open channel flows with an inflexion point in skewed velocity profile.' *Journal of Hydroscience and Hydraulic Engineering*, **12**(2), 69–84.

112. Ikeda, S. and Parker, G. (1989) *River Meandering*, Water Resources Monograph, American Geophysical Union **12**.

113. Ikeda, S., Sano, T., Fukumoto, M. and Kawamura, K. (2000). 'Organized horizontal vortices and lateral sediment transport in compound open channel flows.' *Proceedings of JSCE* (accepted).

114. Ikeda, S., Tanaka, M. and Chiyoda, M. (1984). *Turbulent flow in a sinuous air duct*, Departmental Report No. **14**, Foundation Engineering and Construction, Saitama University, Japan, 1–24.

115. Imamoto, H., Ishigaki, T. and Fujisawa, H. (1982). 'On the characteristics of open channel flow in bend with floodplains.' *Annuals, DPRI*, **1**, 25B-2, 529–543, Kyoto University (in Japanese).

116. Imamoto, H., Ishigaki, T. and Inada, S. (1982). *On the hydraulics of an open channel flow in complex cross section 1*, Disaster Prevention Research Institute, Kyoto University, Annual Report, **25**(57), 4, 509–527 (in Japanese).

117. Imamoto, H., Ishigaki, T. and Kinoshita, S. (1984). *On the hydraulics of an open channel flow in complex cross section 2*, Disaster Prevention Research Institute, Kyoto University, Annual Report, **27**(59), 4, 433–444 (in Japanese).

118. Imamoto, H., Ishigaki, T. and Mutoh, H. (1992). Experimental study on the turbulent mixing between main channel and flood plain flows, *Proceedings of the Annual Journal of Hydraulic Engineering*, **36**, 139–150 (in Japanese).

119. Ishigaki, T., Shiono, K., Rameshwaran, P., Scott, C. F. and Muto, Y. (2000). 'Impact of secondary flow on bed form and sediment transport in a meandering channel for overbank flow.' *Annual Journal of Hydraulic Engineering*, JSCE, **44**, 1–6.

120. Ishigaki, T., Shiono, K., Rameshwaran, P., Scott, C. F. and Muto, Y. (2000). 'Sediment transport rates for overbank flow in meandering channels.' *Annual Journal of Hydraulic Engineering*, JSCE, **44**, 849–854.

121. Ishikan River Development and Construction Office, (1993). 'Forests that foster water.' *Hokkaido Development Bureau*.

122. Izumi, N. (1993). *Channelization and drainage basin formation in cohesive soils*. PhD Thesis, University of Minnesota, USA.

123. Izumi, N. and Parker, G. (1997). 'Equilibrium cross-sectional geometry of gravel rivers transporting suspended sediment.' *Journal of Hydraulic, Coastal and Environmental Engineering*, JSCE, **565**(II-39), 43–55 (in Japanese).

124. James, C. S. (1987). 'The distribution of fine sediment deposits in compound channel systems.' *Water South Africa*, **13**(1), 7–14.

125. James, C. S. and Myers, W. R. C. (2002) "Conveyance of one- and two-stage meandering channels", *Proc. ICE, Water and Maritime Engineering*, **154**(4), 265–274.

126. James, C. S. and Wark, J. B. (1992). *Conveyance estimation for meandering channels*, Report **SR329**, HR Wallingford, UK.

127. Jin, H. S., Egasgira, S. and Liu, B. Y. (1997). '3-D numerical solution of flow in sine generated meandering compound channel.' *Energy and Water: Sustainable Development, Proceedings of the 27th IAHR Congress*, **D** 246–251.

128. Jin, H. S., Egashira, S. and Liu, B. Y. (1998). 'Modification of k-e turbulence closure and its application to meandering compound channel flow.' *Annual Journal of Hydraulic Engineering*, JSCE, **42**, 949–954.

129. Karamisheva, R. D., Lyness, J. F., Myers, W. R. C. and Cassells, J. B. C., (2005), Improving sediment discharge prediction for overbank flow, *Water Management, Proceedings of the Instn. of Civil Engineers*, London, 158, WM1, March, 17–24.

130. Karim, F. (1993). 'Bed material discharge prediction for non-uniform bed sediments.' *Journal of Hydraulic Engineering*, ASCE, **124**(6), 597–604.

131. Karim, M. F. and Kennedy, J. F. (1990). 'Menu of coupled velocity and sediment-discharge relations for rivers.' *Journal of Hydraulic Engineering*, ASCE, **116**(8), 978–996.

132. Karki, K. C. and Pantakar, S. V. (1988). 'Calculation procedure for viscous incompressible flows in complex geometries. *Numerical Heat Transfer*, **14**(3), 295–307.

133. Kawahara, Y. and Tamai, N. (1988). 'Numerical calculation of turbulent flows in compound channel flows.' *Proceedings of the 23rd IAHR*, **B**, 463–470.

134. Kezemipour, A. K. and Apelt, C. J. (1979). 'Shape effects on resistance to uniform flow in open channels.' *Journal of Hydraulic Research*, IAHR, **17**(2), 129–147.

135. Keulegan, G. H. (1938). 'Law of turbulent flow in open channels' *Journal of the National Bureau of Standards*, Research Paper 1151, **21**(6), 707–741.

136. Kiely, G. (1990). 'Overbank flow in meandering channels-the important mechanisms.' In White, W. R. (ed.), *Proceedings of the International conference on River Flood Hydraulics*, Paper F3, Wiley, 207–217.

137. Kikkawa, H., Ikeda, S. and Kitagawa, A. (1976). 'Bed deformation at river bend.' *Proceedings of JSCE*, **251**, 65–75 (in Japanese).

138. King, I. P. and Norton, W. P. (1978). 'Recent application of RMA's finite element models for two dimensional hydrodynamics and water quality.' *Finite Elements in Water Resources*, 2.81–2.99.

139. King, I. P. and Roig, L. C (1991). 'Finite element modelling of flow in wetlands.' *Proceedings — National Conference on Hydraulic Engineering*, pp. 286–291.

140. Kinoshita, R. (1984). 'Present status and future prospects of river flow analysis by aerial photograph.' *Proceedings of JSCE*, **345**, 1–20 (in Japanese).

141. Klaassen, G. J. and Van der Zwaard, J. J. (1974). 'Roughness coefficients of vegetated flood plains.' *Journal of Hydraulic Research*, IAHR, **12**(1) 43–63.

142. Knight, D. W. (1981). 'Some field measurements concerned with the behaviour of resistance coefficients in a tidal channel.' *Estuarine, Coastal and Shelf Science*, Academic Press, London, **12**, 303–322.

143. Knight, D. W. (1989). 'Hydraulics of flood channels.' In Beven, K. (ed.), *Floods: Hydrological, sedimentological and geomorphological implications*. Chichester, Wiley, Chapter 6, pp. 83–105.

144. Knight, D. W. (ed.,) (1992). *SERC Flood Channel Facility experimental data - Phase A*(1–15) Report SR314, **1–15**, HR Wallingford. (available from HR Wallingford, Oxon, OX10 8BA, UK).

145. Knight, D. W. (1999). Flow mechanisms and sediment transport in compound channels. In Wang, Z. Y., Soong, T. W. and Yen. B. C. (eds.), *Proceedings of the 1st Sino-US Workshop on Sediment Transport and Disasters, Special Issue of International Journal of Sediment Research* **14**(2), Beijing, China, pp. 217–236.

146. Knight, D. W. (2005a). River flood hydraulics: theoretical issues and stage-discharge relationships, Chapter 17, in *River Basin Modelling for Flood Risk Mitigation* [(eds.), Knight D. W., & ShamseldinA. Y.,], Balkema, pp. 301–334. .

147. Knight, D. W. (2005b). River flood hydraulics: calibration issues in one-dimensional flood routing models, Chapter 18, in *River Basin Modelling for Flood Risk Mitigation* [(eds.), Knight D. W., & Shamseldin A. Y.], Balkema, pp. 335–385.

148. Knight, D. W. and Abril, B. (1996). 'Refined calibration of a depth averaged model for turbulent flow in a compound channel.' *Proceedings of the Institution of Civil Engineers: Water, Maritime and Energy*, Paper 11017, **118**, 151–159.

149. Knight, D. W., Alhamid, A. A. I. and Yuen, K. W. H. (1992). 'Boundary shear in differentially roughened trapezoidal channels.' In Falconer, R. A *et al.,* (eds.), *Proceedings of the 2nd International Conference on Hydraulic and Environmental. Modelling of Coastal, Estuarine and River Waters*, University of Bradford, Gower Technical Press.

150. Knight, D. W. and Brown, F. A. (2001). 'Resistance studies of overbank flow in rivers with sediment using the Flood Channel Facility.' *Journal of Hydraulic Research*, IAHR, **39**(3), 283–301.

151. Knight, D. W., Brown, F. A., Ayyoubzadeh, S. A. and Atabay, S. (1999). 'Sediment transport in river models with overbank flow.' In Jayawardena, A. W., Lee, J. H. W. and Wang, Z. Y. (eds.), *River Sedimentation: Proc. Seventh International Symposium on River Sedimentation*, Hong Kong, 16–18 December, 1998 pp. 19–25.

152. Knight, D. W., Brown, F. A., Valentine, E. M., Nalluri, C., Bathurst, J. C., Benson, I., Myers, W. R. C., Lyness, J. F. and Cassells, J. (1999). 'The response of straight mobile bed channels to inbank and overbank flows.' *Proceedings of the Institution of Civil Engineers: Water, Maritime and Energy Division*, **136**(4) 211–224.

153. Knight, D. W. and Cao, S. (1994). 'Boundary shear in the vicinity of river banks.' *Proceedings of ASCE National Conference on Hydraulic Engineering*, New York, **2**, 954–958.

154. Knight, D. W. and Chlebek, J., (2006), A new perspective on sidewall correction procedures, based on SKM, *RiverFlow 2006*, September, Lisbon, Portugal.

155. Knight, D. W. and Demetriou, J. D. (1983). 'Flood plain and main channel flow interaction.' *Journal of Hydraulic Engineering*, ASCE, **109**(8), 1073–1092.

156. Knight, D. W., Demetriou, J. D. and Hamed, M. E. (1983). Hydraulic analysis of channels with flood plains. *Proceedings of the International Conference on Hydraulic Aspects*

of Floods and Flood Control, British Hydromechanics Research Association, Cranfield, Paper **E1**, 129–144.

157. Knight, D. W., Demetriou, J. D. and Hamed, M. E. (1984a). 'Stage discharge relationships for compound channels.' In Smith, K. V. H. (ed.), *Proceedings of the International Conference on Hydraulic Design of Channel Control Structures in Water Resources Engineering : Channels and Channel Control Structures*, Springer-Verlag, Heidelberg, **4**, 21–4.25.

158. Knight, D. W., Demetriou, J. D. and Hamed, M. E. (1984b). 'Boundary shear in smooth rectangular channels.' *Journal of Hydraulic Engineering*, ASCE, **110**(4), 405–422.

159. Knight, D. W. and Lai, C. J. (1985). 'Turbulent flow in compound channels and ducts.' *Proceedings of the 2nd International Symposium on Refined Flow Modelling and Turbulence Measurements*, Iowa, USA, Paper **I21**, 1–10, Hemisphere Publishing Co.

160. Knight, D. W. and Macdonald, J. A. (1979). 'Open channel flow with varying bed roughness.' *Journal of the Hydraulics Division*, ASCE, **105**(HY9), Paper 14839, 1167–1183.

161. Knight, D. W. and Patel, H. S. (1985). 'Boundary shear stress distributions in rectangular duct flow.' Journal of Hydraulics Division, ASCE, **111** Paper 19408, Hemisphere Publishing Co., pp. 29–47.

162. Knight, D. W., Samuels, P. G. and Shiono, K. (1990). 'River flow simulation: research and developments.' *Journal of the Institution of Water and Environmental Management*, **4**(2), 163–175.

163. Knight, D. W. and Sellin R. H. J. (1987). 'The SERC flood channel facility.' *Journal of the Institution of Water and Environmental Management*, **1**(2), 198–204.

164. Knight, D. W. and Shiono, K. (1990). 'Turbulence measurements in a shear layer region of a compound channel.' *Journal of Hydraulic Research*, IAHR, **28**(2), 175–196, Discussion in IAHR Journal, 1991, **29**(2), 259–276.

165. Knight, D. W. and Shiono, K. (1996). 'River channel and floodplain hydraulics.' In Anderson, M. G., Walling, D. E. and Bates. P. D. (eds.), *Floodplain Processes*, Chapter 5, Chichester, Wiley, pp. 139–181.

166. Knight, D. W., Shiono, K. and Pirt, J. (1989). 'Prediction of depth mean velocity and discharge in natural rivers with overbank flow.' In Falconer, R. A *et al.*, (eds.), *Proceedings of the International Conference on Hydraulic and Environmental Modelling of Coastal, Estuarine and River Waters*, Paper **38**, 419–428, University of Bradford, Gower Technical Press.

167. Knight, D. W., Yuan, Y. M. and Fares, Y. R. (1992) 'Boundary shear in meandering river channels.' *Proceedings of the International Symposium on Hydraulic Research in Nature and Laboratory*, Yangtze River Scientific Research Institute, China, **2**, 102–6.

168. Knight, D. W. and Yu, G. (1995). 'A geometric model for self formed channels in uniform sand.' *26th IAHR Congress, International Association for Hydraulic Research*, HYDRA 2000 **1**, 354–359 London, Thomas Telford.

169. Knight, D. W., Yuen, K. W. H. and Alhamid, A. A. I. (1994). 'Boundary shear stress distributions in open channel flow.' In Beven, K., Chatin, P. and Millbank, J. (eds.), *Physical Mechanisms of Mixing and Transport in the Environment*, Chapter 4, Wiley, 51–87.

170. Kouwen, N. and Li, R. M. (1980). 'Biomechanics of vegetative channel linings.' *Journal of Hydraulics Division*, ASCE, **106**(HY6), 1085–1103.

171. Kouwen, N., Li, R. M. and Simons, D. B. (1980) 'Velocity measurements in a channel lined with flexible plastic roughness elements.' *Technical Report CER79-80NK-RML-DBS11*, Department of Civil Engineering, Colorado State University, Fort Collins, Colorado.

172. Kouwen, N. and Unny, T. E. (1973). 'Flexible roughness in open channels.' *Journal of Hydraulics Division*, ASCE, **99**(HY5), 713–728.

173. Krishnappan, B. G. and Lau, Y. L. (1986). 'Turbulence modelling of flood plain flows.' *Journal of Hydraulic Engineering*, **112**(4), 251–266.

174. Kuroki, M. and Kishi, T. (1981). 'Structures of longitudinal vortices in wide open channels.' *Proceedings of the Japanese Society of Civil Engineers*, **312**, 83–92.

175. Lai, C. J. (1986). *Flow resistance, discharge capacity and momentum transfer in smooth compound closed ducts*, PhD Thesis, The University of Birmingham, UK.

176. Lai, C. J. and Knight, D. W. (1988). 'Distributions of streamwise velocity and boundary shear stress in compound ducts.' *Proceedings of the 3rd International Symposium on Refined Flow Modelling and Turbulence Measurements*, Tokyo, Japan, pp. 527–536.

177. Lai, C. J., Liu, C. L. and Lin, Y. Z. (2000). 'Experiments on flood-wave propagation in compound channels.' *Journal of Hydraulic Engineering*, ASCE, **126**(7), 492–501.

178. Lambert, M. F. and Myers, W. R. C. (1998). 'Estimating the discharge capacity in straight compound channels.' *Proceedings of the Institution of Civil Engineers: Water, Maritime and Energy Division*, **130**(2), Paper 10655, 84–94.

179. Lambert, M. and Sellin, R. H. J. (1996). 'Velocity distribution in a large-scale model of a doubly meandering compound river channel.' *Proceedings of the Institution of Civil Engineers: Water, Maritime and Energy Division*, **118**, Paper 11530, 10–20.

180. Lane, E. W. (1955). 'Design of stable channels.' *Transactions of the* ASCE, **120**, 1234–1260.

181. Lauder, B. E., Reece, G. J. and Rodi, W. (1975). 'Progress in the development of a Reynolds Stress turbulence closure.' *Journal of Fluid Mechanics*, **68**(3), 537–566.

182. Larsson, R. (1988). 'Numerical simulation of flow in compound channels.' *Proceedings of the 3rd International Symposium on Refined Flow Modelling and Turbulence Measurements*, pp. 537–544.

183. Launder, B. E. and Spalding, D. B. (1974). 'The numerical computation of turbulent flow.' *Computer Methods in Applied Mechanics and Engineering*, **3**, 269–289.

184. Laverdrine, I. (1996). *Evaluation of 3D models for river flood applications*, HR, Wallingford Report **TR 6**.

185. Lavedrine, I., (1997). *Application of 3D models to river flood problems* HR, Wallingford Report **TR 26**.

186. Lean, G. H. and Weare, T. J. (1979). 'Modelling two-dimensional circulating flow.' *Journal of Hydraulics Division, Proceedings* ASCE, **105**(HY1), 17–26.

187. Lee, P. J., Lambert, M. F. and Simpson, A. R., (2002), Critical depth prediction in straight compound channels, *Proc. Instn. of Civil Engineers, Water and Maritime Engineering*, London, Vol. 154, No. 4, Dec., pp. 317–332.

188. Leendertsee, J. J. (1967). *Aspects of a computational model for long period water wave propagation*. Memo **RM 5294** PR Rand Corporation, Santa Monica, USA.

189. Leonard, B. P. (1988). Three order multi-dimensional Euler/Navier-Stokes Solver. AIAA/ASME/SLAM/APS 1st National Fluid Dynamics Congress, pp. 226–231.

190. Leopold, L. B. and Wolman, M. G. (1960). 'River meanders.' *Geological Society of American Bulletin*, **71**, 769–794.

191. Leschziner, M. A. and Rodi, W. (1979). 'Calculation of strongly curved open channel flow.' *Journal of the Hydraulics Division*, ASCE, **105**(HY10). Paper 14927, 1297–1314.

192. Liao, H. and Knight, D. W. (2006) Analytic stage-discharge formulae for flow in straight prismatic open compound channels, *Journal of Hydraulic Engineering*, ASCE (submitted).

193. Lilly, D. K. (1992). 'A proposed modification of the Germano subgrid scale closure method.' *Physics of Fluids* **4**(3), 633–635.

194. Lin, B. and Shiono, K. (1992). 'Prediction of Pollutant Transport in Compound Channel Flows.' *Proceedings of Second International Conference on Hydraulic and Environmental Modelling of Coastal, Estuarine and River Waters*, Bradford, Ashgate Publishing Ltd. Aldershot, pp. 373–384.

195. Liu, X., Knight, D. W., Gong, T. and Jiu, B. (1999). 'Lateral turbulent dispersion in compound channels: some experimental and field results.' In Jayawardena, A. W.,

Lee, H. W. and Wang, Z. Y. (eds.), 'Environmental Hydraulics'. *Proceedings of the 2nd International Symposium on Environmental Hydraulics*, Hong Kong, December 1998, pp. 369–374.

196. Lonsdale, R. D. (1993). 'Algebraic multigrid solver for the Navier-Stokes equations on unstructured meshes.' *International Journal of Numerical Methods for Heat and Fluid Flow*, 3(1), 3–14.

197. Lorena, M. (1992). *Meandering compound flow*, PhD Thesis, The University of Glasgow, UK.

198. Lotter, G. K. (1933). 'Considerations on hydraulic design of channels with different roughness of walls.' *Transactions of the All-Union Scientific Research Institute of Hydraulic Engineering*, Leningrad, 9, 238–241 (in Russian).

199. Lynch, D. R. and Gray, W. G. (1980). 'Finite element simulation of flow in deforming regions.' *Journal of Computational Physics*, 36(2), 135–153.

200. Lyness, J. F., Myers, W. R. C. and Wark, J. B., (1997), The use of different conveyance calculations for modelling flows in a compact compound channel, *Journal of the Institution of Water and Environmental Management*, 11(5), 335–340.

201. Lyness, J. F., Myers, W. R. C. and O'Sullivan, J. (1998). 'Hydraulic characteristics of meandering mobile bed compound channels.' *Proceedings of the Institution of Civil Engineers: Water, Maritime & Energy*, 130(4), Paper 11545, 179–188.

202. Macklin, M. G. and Dowsett, R. B. (1989). 'The chemical and physical speciation of trace metals in fine-grained overbank flood sediments in the Tyne basin, Northeast England.' *Catena* 16(2), 135–151.

203. Macklin, M. G. and Klimek, K. (1992). 'Dispersal, storage and transformation of metal contaminated alluvium in the upper Vistula basin, South-West Poland.' *Applied Geography*, 12, 7–30.

204. Manning, R. (1891). 'On the flow of water in open channels and pipes.' *Transactions of the Institution of Civil Engineers of Ireland*, Dublin, 20, 161–207.

205. McGahey, C. and Samuels, P. G., (2003), Methodology for conveyance estimation in two-stage straight, skewed and meandering channels, *XXX IAHR Congress, Thessaloniki, Volume C1 of Proceedings published by IAHR*.

206. McGahey, C., (2006), A practical approach to estimating the flow capacity of rivers, *PhD thesis*, The Open University, UK. .

207. McGahey, C. and Samuels, P. G., (2003), Methodology for conveyance estimation in two-stage straight, skewed and meandering channels, XXX IAHR Congress, Thessaloniki, Volume C1 of Proceedings published by IAHR.

208. Ackers, P. (1990). "SERC-FCF Phase C experiments: feasibility study of tests with sediments", Report to SERC-FCF Management Committee, June.

209. McGuirk, J. and Rodi, W. (1977). 'A mathematical model for a vertical jet discharging into a shallow lake.' In *Proceedings of the 17th Congress of the International Association for Hydraulic Research, Hydraulic Engineering for Improved Water Management*, Germany August 1977, 1, 579–586.

210. McKeogh, E. J. and Kiely, G. K. (1989). 'Experimental comparison of velocity and turbulence in compound channnels of varying sinuosity.' *Proceedings of the International Conference on Channel Flow and Catchment Runoff*, In Yen, B. C. (ed.), *Channel Flow Resistance: Centennial of Manning's formula*, Water Resources Publications, Colorado, USA, pp. 393–408.

211. McKeogh, E. J. and Kiely, G. K. (1989). 'A comparison of velocity measurements in straight, single meander and multiple meander compound channels.' *Proceedings of the International Conference on Channel Flow and Catchment Runoff, Centennial of Manning's Formula and Kuichling's Rational Formula*, University of Virginia, USA.

212. Melling, A. and Whitelaw, J. H. (1976). 'Turbulent flow in a rectangular duct.' *Journal of Fluid Mechanics*, 78(2), 289–315.

213. Meyer, L. and Rehme, K. (1994). 'Large scale turbulence phenomena in compound rectangular channels.' *Experimental Thermal and Fluid Science*, **8**(4), 286–304.

214. Michioku, K., Takemoto, O. and Hirota, M. (1999). 'Flow resistance in an open channel with an alternative riffle-pool arrangement.' *Journal of Hydroscience and Hydraulic Engineering*, **17**(1), 87–101.

215. Milhous, R. T. (1973). *Sediment transport in gravel-bottomed stream*. PhD Thesis, Oregon State University, U.S.A.

216. Molinaro, P. and Natale, L. (1994). Proceedings of the Specialty Conference on Modelling of Flood Propagation Over Initially Dry Areas, Speciality Conferences ENEL-DSR-CRIA, Milan.

217. Morvan, R., Tanguy, N., Vilbe, P. and Calvez, L. C. (2000). 'Pertinent parameters for Kautz approximation.' *Electronics Letters*, **36**(8), 769–771.

218. Morvan, H., Pender, G., Wright, N. G. and Ervine, D. A. (2002). "Three-dimensional hydrodynamics of meandering compound channels", *Journal of Hydraulic Engineering*, **128**(7), 674–682.

219. Myers, W. R. C. (1978). 'Momentum transfer in a compound channel.' *Journal of Hydraulic Research*, IAHR, **16**(2), 139–150.

220. Myers, W. R. C. and Brennan, E. K. (1990). 'Flow resistance in compound channels.' *Journal of Hydraulic Research*, IAHR, **28**(2), 141–155.

221. Myers, W. R. C., Knight, D. W., Lyness, J. F., Cassells, J. and Brown, F. A. (1999). 'Resistance coefficients for inbank and overbank flows.' *Proceedings of the Institution of Civil Engineers: Water, Maritime and Energy Division*, **136**(2), 105–115.

222. Myers, W. R. C. and Lyness, J. F. (1997). 'Discharge ratios in smooth and rough compound channels.' *Journal of the Hydraulics Division*, ASCE, **123**(HY3), 3, 182–188.

223. Nadaoka, K. and Yagi, H. (1993). 'Horizontal large-eddy simulation of river flow with transverse shear by SDS and 2DH model.' *Proceedings of the JSCE*, **473**, 35–44, (in Japanese).

224. Nakagawa, H., Tsujimoto, T. and Murakam, S. (1986). 'Non-equilibrium bed load transport along side bank.' *Proceedings of the 3rd International Symposium on River Sedimentation*, USA, pp. 1059–1065.

225. Nansen, G. C. and Hicken E. J. (1986). 'A statistical analysis of bank erosion and channel migration in western Canada.' *Geological Society of America Bulletin*, **97**, 497–504.

226. Naot, D. and Rodi, W. (1982). 'Numerical simulations of secondary currents in channel flow.' *Journal of Hydraulic Engineering*, **119**(3), 390–408.

227. Naot, D., Nezu, I. and Nakagawa, H. (1993). 'Hydrodynamic behaviour of compound rectangular open channels.' *Journal Hydraulic Engineering*, ASCE, **119**(3), 390–408.

228. Naot, D., Nezu, I. and Nakagawa, H. (1993). 'Calculation of compound open channels.' *Journal Hydraulic Engineering*, ASCE, **119**(12), 1418–1426.

229. Naot, D. and Rodi, W. (1982). 'Calculation of secondary currents in channel flow.' *Journal of the Hydraulics Division*, ASCE, **108**(HY8), 948–968.

230. Neary, V. S., Sotiropoulos, F. and Odgaard, A. J. (1999). 'Three-dimensional numerical model of lateral intake inflows.' *Journal of Hydraulic Engineering*, **125**(2), 126–140.

231. Nelson, J. M. and Smith, J. D. (1989). 'Flow in meandering channels with natural topography.' In Ikeda, S. and Parker, G. (eds.), *River Meandering*, American Geophysical Union, Water Resources Monograph 12, Balkema, pp. 69–102.

232. Nezu, I. (1996). 'Experimental and numerical study on 3-D turbulent structures in compound open channel flows.' In Chen, C. J. (ed.), *Flow Modelling and Turbulence Measurements*, Hemisphere Pub Co-op, pp. 65–74.

233. Nezu, I., Kadota, A. and Nakagawa, H. (1997). 'Turbulent structure in unsteady depth-varying open-channel flows.' *Journal of Hydraulic Engineering*, ASCE, **123**(9), 752–763.

234. Nezu, I. and Nakagawa, H. (1984). 'Cellular secondary currents in straight conduit.' *Journal Hydraulic Engineering*, ASCE, **110**(2), 173–193.

235. Nezu, I. and Nakagawa, H. (1989). 'Self forming mechanism of longitudinal sand ridges and troughs.' *Proceedings of the 23rd IAHR Congress*, Ottawa, Vol. B, pp. 65–72.

236. Nezu, I. and Nakagawa, H. (1993). *Turbulence in Open-Channel Flows*. IAHR Monograph Series, A. A. Balkema, Rotterdam, 1–281.

237. Nezu, I., Nakagawa, H. and Abe, T. (1995). 'Secondary currents and exchange processes in compound open-channel flows.' *Proceedings of the 26th IAHR Congress*, London, 1, 45–50.

238. Nezu, I., Nakagawa, H. and Saeki, K. (1994a). 'Coherent structures in compound open-channel flows by making use of particle-tracking visualization technique.' *Proceedings of the Symposium of Fundamentals and Advancements in Hydraulic Measurements and Experimentation*, ASCE, Buffalo, USA, pp. 406–415.

239. Nezu, I. and Nakayama, T. (1997). 'Space-time correlation structures of horizontal coherent vortices in compound open-channel flows by using particle-tracking velocimetry.' *Journal of Hydraulic Research*, IAHR, 35(2), 191–208.

240. Nezu, I. and Naot, D. (1995). 'Turbulent structures and secondary currents in compound open channel flow with variable depth floodplain.' *Proceedings of the 10th Symposium On Turbulent Shear Flows*, Pennsylvania, pp. 7.7–7.12.

241. Nezu, I. and Onitsuka, K. (1998b). '3-D turbulent structures in partly vegetated open channel flows.' In Lee, J. H. W., Jayawardena, A. W. and Wang, Z. Y. (eds.), *Environmental Hydraulics, Proceedings of the 2nd International Symposium on Environmental Hydraulics*, Hong Kong, December 1998, Balkema, pp. 305–310.

242. Nezu, I. and Onitsuka, K. (2001). 'Turbulent structures in partly vegetated open-channel flows: LDA and PIV measurements.' *Journal of Hydraulic Research*, IAHR, 39(6), 629–643.

243. Nezu, I., Onitsuka, K. and Iketani, K. (1999). 'Coherent horizontal vortices in compound open channel flows.' In Singh, V. P., Seo, I. W and Sonu, J. H. (eds.), *Hydraulic Modelling*, Water Resources Publications, Colorado, pp. 17–32.

244. Nezu, I., Onitsuka, K. and Sagara, Y. (1999). 'Turbulent structures in compound open channel flows with various relative depth between main channel and flood plain.' *Journal of Japan Society of Fluid Mechanics*, 18(4), 228–237, (in Japanese).

245. Nezu, I., Onitsuka, K., Sagara, Y. and Iketani, K. (2000). 'Effects of relative depth between main-channel and floodplain on turbulent structure in compound open-channel flows.' *Journal of Hydraulic, Coastal and Environmental Engineering*, 649/II51, 1–15 (in Japanese).

246. Nezu, I. and Rodi, W. (1985). 'Experimental study on secondary currents in open channel flow.' *Fundamentals and computation of 2D and 3D flows: Transport and mixing in rivers and reservoirs. Proceedings of the 21st IAHR Congress*, Melbourne, 2B, 115–119.

247. Nezu, I., Saeki, K. and Nakagawa, H. (1994). 'Turbulence measurements in compound open-channel flows by using particle-image velocimeter.' *Proceedings of the 9th APD-IAHR Congress*, Singapore, 1, 421–428.

248. Nichols, B. D., Hirt, C. W. and Hotchkiss, R. S. (1980). '*SOLA-VOF: A solution algorithm for transient fluid flow with multiple free boundaries*', Los Alamos Scientific Laboratory Report, LA-8355.

249. Nokes, R. I. and Wood, I. R. (1988). 'Vertical and lateral turbulent dispersion: some experimental results.' *Journal of Fluid Mechanics*, 187, 373–394.

250. Nordin, C. F. (1971). *Statistical properties of dune profile*. USGS, Professional paper 562-F, United States Government Printing Office, USA.

251. Odgaard, J. A. (1986). 'Meander flow model: Development and Applications, Parts I and II.' *Journal of Hydraulic Engineering*, ASCE, 112(12), 1117–1136.

252. Omran, M. and Knight, D. W., (2006), Modelling the distribution of boundary shear stress in open channel flow, *RiverFlow 2006*, September, Lisbon, Portugal.

253. Pantakar, S. V. (1980). *Numerical Heat Transfer and Fluid Flow*, McGraw Hill.

254. Pantakar, S. V. and Spalding, D. B. (1972). 'A calculation procedure for heat, mass and momentum transfer in three-dimensional parabolic flows.' *International Journal of Heat and Mass Transfer*, **15**(10), 1787–1806.

255. Parker, G. (1978) 'Self-formed straight rivers with equilibrium banks and mobile bed. Part 1. The sand-silt river.' *Journal of Fluid Mechanics*, **89**(1), 109–125.

256. Patel, V. C. and Yoon, Y. J. (1995). 'Application of turbulence models to separated flow over rough surfaces.' *Journal of Fluids Engineering, Transactions of the ASME*, **117**(2), 234–241.

257. Pavlovskii, N. N. (1931). 'On a design formula for uniform movement in channels with nonhomogeneous walls.' *Transactions of the All-Union Scientific Research Institute of Hydraulic Engineering*, Leningrad, **3**, 157–164 (in Russian).

258. Perkins, H. J. (1970). 'The formation of streamwise vorticity in turbulent flow.' *Journal of Fluid Mechanics*, **44**(4), 721–740.

259. Rameshwaran, P., Spooner, J., Shiono, K. and Chandler, J. H. (1999). 'Flow Mechanisms in two-stage meandering channel with mobile bed.' *Proceedings of IAHR Congress* Graz, Austria, August, D6. *Scientific Research Institute*, **2**, 102–107, Wuhan China.

260. Rezaei, B. (2006). Overbank flow in compound channels with prismatic and non-prismatic floodplains, *PhD thesis*, The University of Birmingham, UK.

261. Rhie, C. M. and Chow, W. L. (1983). Numerical study of the turbulent flow past an airfoil with trailing edge separation. *AIAA Journal*, **21**(11), 1525–1532.

262. Rhodes, D. G. (1991). *An experimental investigation of the mean flow structure in wide ducts of simple rectangular and compound trapezoidal cross-section in particular zones of high lateral shear, 4 volumes*, PhD Thesis, The University of Birmingham, UK.

263. Rhodes, D. G. and Knight, D. W. (1994a). 'Distribution of shear force on boundary of smooth rectangular duct.' *Journal of Hydraulic Engineering*, ASCE, **120**(7), 787–807.

264. Rhodes, D. G. and Knight, D. W. (1994b). 'Velocity and boundary shear in a wide compound duct.' *Journal of Hydraulic Research*, IAHR, **32**(5), 743–764.

265. Rhodes, D. G. and Knight, D. W. (1995a). 'Lateral shear in a compound duct.' *26th IAHR Congress, HYDRA 2000*, **1**, Thomas Telford, 51–56.

266. Rhodes, D. G. and Knight, D. W. (1995b). 'Lateral shear in a wide compound duct.' *Journal of Hydraulic Engineering*, ASCE, **121**(11), Technical Note, 829–832.

267. Rhodes, D. G. and Knight, D. W. (1996). 'Distribution of local friction factor in a compound duct.' *Proceedings of the 10th IAHR-APD Congress, National Hydraulic Research Institute*, Malaysia, **2**, 220–227.

268. Rodi, W. (1984). *Turbulence models and their application in hydraulics — A state of the art review*. International Association of Hydraulic Research, Delft, The Netherlands, 1984.

269. Samuels, P. G. (1989). 'Backwater length in rivers.' *Proceedings of the Institution of Civil Engineers*, London, Part 2: Research and Theory, **87**, 571–582.

270. Satoh, H. and Jawahara, Y. (1998). 'Large eddy simulation of turbulent flow in a prismatic compound open channel.' *Journal of applied mechanics JSCE*, **1**, 673–682 (in Japanese).

271. Samuels, P. G. (1990). 'Cross-section location in 1-D models.' In White, W. R. (ed.), *Proceedings of the International Conference on River Flood Hydraulics*, Wallingford, Wiley Paper **K1**, 339–350.

272. Sellin, R. H. J. (1964). 'A laboratory investigation into the interaction between the flow in the channel of a river and that over its floodplain.' *La Houille Blanche*, **7**, 793–802.

273. Sellin, R. H. J., Ervine, D. A. and Willetts, B. B. (1993). 'Behaviour of meandering two-stage channels.' *Proceedings of the Institution of Civil Engineers: Water, Maritime and Energy*, **101**(2), 99–111.

274. Sellin, R. H. J. and Willetts, B. B. (1996). 'Three-dimensional structures, memory and energy dissipation in meandering compound channel flow.' In Anderson, Walling and Bates (eds.), *Floodplain Processes* Chichester Wiley, Chapter 8, 255–297.

275. Sellin, R. H. J. and van Beesten, D. P. (2004). "Conveyance of a managed vegetated two-stage river channel", *Proc. ICE, Water Management*, **157**(1), 21–33.

276. Shakir, A. (1992). *An experimental investigation of channel planforms*, PhD Thesis, University of Newcastle upon Tyne, England, UK.

277. Shiono, K., Al-Romaih, J. S. and Knight, D. W. (1999). 'Stage-discharge assessment in compound meandering channels.' *Journal of Hydraulic Engineering*, ASCE, **125**(1), 66–70.

278. Shiono K. and Knight, D. W. (1988). 'Two dimensional analytical solution for a compound channel.' *Proceedings of the 3rd International Symposium on Refined Flow Modelling and Turbulence Measurements*, Tokyo, Japan, pp. 591–599.

279. Shiono, K. and Knight, D. W. (1989). 'Transverse and vertical Reynolds stress measurements in a shear layer region of a compound channel.' *Proceedings of the 7th International Symposium on Turbulent Shear Flows*, Stanford, USA, pp. 28.1-1–28.1.6.

280. Shiono K. and Knight, D. W. (1990). 'Mathematical models of flow in two or multi stage straight channels.' In White, W. R. (ed.), *Proceedings of the International Conference on River Flood Hydraulics*, Wallingford, Wiley Paper **G1**, 229–238.

281. Shiono, K. and Knight, D. W. (1991). 'Turbulent open channel flows with variable depth across the channel.' *Journal of Fluid Mechanics*, **222**, 617–646.

282. Shiono, K. and Muto, Y. (1998). 'Complex flow mechanisms in compound meandering channels with overbank flow.' *Journal of Fluid Mechanics*, **376**, 221–261.

283. Shiono, K., Muto, Y, Imamoto, H. and Ishigaki, T. (1994). 'Flow structure in meandering compound channel for overbank.' *Proceedings of the 7th International Symposium on Application of Laser Techniques to Fluid Mechanics*, Lisbon Portugal, 28.2.1–28.2.8.

284. Shiono, K., Muto, Y., Knight, D. W. and Hyde, A. F. L. (1999). 'Energy losses due to secondary flow and turbulence in meandering channels with overbank flow.' *Journal of Hydraulic Research*, IAHR, **37**(5), 641–664.

285. Simons, D. B. and Richardson, E. V. (1961). 'Forms of bed roughness in alluvial channels.' *Journal of Hydraulic Engineering*, ASCE, **87**(HY3).

286. Simm, D. J. (1995). 'The rates and patterns of overbank deposition on a lowland floodplain', in Foster, I. D. L., Gurnell, A. M. and Webb, B. W. (eds.), *Sediment and Water Quality in River Catchments*, Wiley, Chichester, pp. 247–264.

287. Smith, C. R. (1996). 'Coherent flow structures in smooth wall turbulent boundary layers; facts, mechanisms and speculation.' In Ashworth, P. J., Bennett, S. J., Best, J. L. and McLelland, S. J. (eds.), *Coherent Flow Structures in Open Channels*, ppWiley, 1–39.

288. Sinha, S. K., Sotiropoulos, F. and Odgaard, A. J. (1998). 'Three-Dimensional Numerical Model for Flow through Natural Rivers.' *Journal of Hydraulic Engineering*, **124**(1), 13–24.

289. Sleigh, J. (1988). 'An Unstructured Finite-Volume Algorithm for Predicting Flow in Rivers and Estuaries.' *Computers and Fluids*, **27**(4).

290. Sofialidis, D. and Prinos, P. (1998b). 'Compound open channel flow modelling with nonlinear low Reynolds $k - E$ models.' *Journal of Hydraulic Engineering*, **124**(3), 253–2621.

291. Sofialidis, D. and Prinos, P. (1999). 'Numerical study of momentum exchange in compound open channel flow.' *Journal of Hydraulic Engineering*, **125**(2), 152–165.

292. Soil Conservation Services. (1963). *Guide for selecting roughness coefficient n values for channels*. Soil Conservation Service, US Department of Agriculture, Washington DC.

293. Speziale, C. G. (1987). 'On non-linear k-l and k-e model of turbulence.' *Journal of Fluid Mechanics*, **178**, 459–475.

294. Stein, C. J. and Rouve, G. (1988). '2D LDV technique for measuring flow in a meandering channel with wetted floodplains - A new application and first results.' *Proceedings of the International Conference on Fluvial Hydraulics*, pp. 5–10.

295. Sugiyama, H., Akiyama, M., Kamezawa, M. and Noguchi, D. (1997). 'The numerical study of turbulent structure in compound open-channel flow with variable-depth flood plain.' *Journal of Hydraulic Coastal and Environmental Engineering* JSCE, **565** II-39, 73–83 (in Japanese).

296. Sugiyama, H., Akiyama, M. and Tanaka, M. (1998). 'Numerical study of turbulent structure in curved open-channel flow of flood plains.' *Journal of Applied Mechanics JSCE,* **1,** 683–692.

297. Sutherland, A. J. (1967). 'Proposed mechanism for sediment entrainment.' *Journal of Geophysical Research,* **72,** 6183–6194.

298. Svidchenko, A. B. and Pender, G. (2000).'Flume study of the effect of relative depth on the incipient motion of coarse uniform sediments.' *Water Resources Research,* **36,** 619–628.

299. Swindale, N. R. (1999). *Numerical modelling of river rehabilitation schemes.* PhD Thesis, Nottingham.

300. Tamai, N., Asaeda, T. and Ikeda, Y. (1986). 'Study on generation of periodical large surface eddies in a composite channel flow.' *Water Resources Research,* **22**(7), 1129–1138.

301. Tamai, N., Asaeda, T. and Ikeda, H. (1986). 'Generation mechanism and periodicity of large surface-eddies in compound channel flow.' *Proceedings of the 5th Congress, APD-IAHR,* Seoul, Republic of Korea, pp. 61–74.

302. Tang, X., Knight, D. W. and Liao, H. (2001). 'Analysis of velocity and stage-discharge relationships for overbank flow in preparation.

303. Tang, X. and Knight, D. W., (2001), Analysis of bed form dimensions in a compound channel, *Proc. 2nd IAHR Symposium on River, Coastal and Estuarine Morphodynamics,* September, Obihiro, Japan, pp. 555–563.

304. Tang, X. and Knight, D. W., (2006), Sediment transport in river models with overbank flows, *Journal of Hydraulic Engineering,* ASCE, Vol. 132, No. 1, January, pp. 77–86.

305. Task Force, ASCE (1963). 'Friction factors in open channels, Task Force Report of the Hydromechanics Committee.' *Journal of the Hydraulics Division,* ASCE, **89**(HY2).

306. Tchamen, G. W. and Kawahita, R. A. (1988). 'Modelling wetting and drying effects over complex topography.' *Hydrological Processes,* **12,** 1151–1182.

307. Technology Research Centre (1994). *Proposed guidelines on the clearing and planting of trees in rivers.* Technology Research Centre for Riverfront Development, River Bureau, Ministry of Construction, Japan, pp. 1–144.

308. Thomas, T. G. and Williams, J. J. R. (1995a). 'Large eddy simulation of turbulent flow in an asymmetric compound open channel.' *Journal of Hydraulic Research,* **33**(1), 27–41.

309. Thomas, T. G. and Williams, J. J. R. (1995b). 'Large eddy simulation of symmetric trapezoidal channel at a Reynolds number of 430,000.' *Journal of Hydraulic Research* **33**(6), 825–842.

310. Thorne, C. R., Hey, R. D. and Newson, M. D. (1997). *Applied fluvial geomorphology for river engineering and management.* Wiley.

311. Toebes, G. H. and Sooky, A. A. (1967). 'Hydraulics of meandering rivers with flood-plains.' *Journal of Waterways and Harbours Division,* ASCE, **93,** WW2, 213–236.

312. Tominaga, A., Liu, J., Nagao, M. and Nezu, I. (1995), Hydraulic characteristics of unsteady flow in open channels with floodplains, *Proc 26th Congress,* IAHR, London, Vol. 1, pp. 373–378.

313. Tominaga, A. and Nezu, I. (1991). 'Turbulent structure in compound open channel flows.' *Journal Hydraulic Engineering,* ASCE, **117**(1), 21–41.

314. Tominaga, A., Nezu, I. and Kobatake, S. (1989). 'Flow measurements in compound channels with fibre-optic Laser Doppler Anemometer.' *Proceedings of the Workshop on Instrumentation for Hydraulic Laboratories,* Canada Centre for Inland Waters, Burlington, Canada, pp. 45–50.

315. Tominaga, A. and Knight, D. W., (2004), Numerical evaluation of secondary flow effects on lateral momentum transfer in overbank flows, *River Flow 2004, Proc. 2nd Int. Conf. on Fluvial Hydraulics, 23–25 June, Napoli, Italy* (eds.), Greco, M., Carravetta, A. and Morte, R. D., Vol. 1, pp. 353–361.

316. Townsend, D. R. (1968). 'An investigation of turbulence characteristics in a river model of complex cross-section.' *Proceedings of the Institute of Civil Engineers*, **40**, 155–175.

317. Townson, J. M. (1974). 'An application of the method of characteristics to tidal calculations in (x-y-t) space.' *Journal of Hydraulic Research*, **12**(4), 499–530.

318. Tracy, H. J. (1965). 'Turbulent flow in a three-dimensional channel.' *Journal of the Hydraulics Division*, ASCE, **91**(HY6), 9–35.

319. Tsujimoto, T. (1996). 'Coherent fluctuations in a vegetated zone of open channel flow : causes of bedload lateral transport and sorting.' In Ashworth, P. J., Bennett, S. J., Best, J. L. and McLelland, S. J. (eds.), *Coherent flow structures in open channels*, Wiley, 375–396.

320. Tsujimoto, T. (1999). 'Fluvial processes in a stream with vegetation.' *Journal of Hydraulic Research*, IAHR, **37**(6), 789–803.

321. Tsujimoto, T. and Kitamura, H. (1994). 'Experimental study on mechanism of transverse mixing in open channel flow with longitudinal zone of vegetation along side wall.' *Proceedings of the JSCE*, **491**, 61–70 (in Japanese).

322. Tsujimoto, T. and Shimizu, Y. (1994). 'Flow and suspended sediment in a compound channel with vegetation.' *Proceedings of the 1st International Symposium on habitat hydraulics*, 357–370.

323. Tsujimoto, T. and Tsujikura, H. (1998). 'Bed Load Transport around Vegetated Area and Development process of Island.' *Annual Journal of Hydraulic Engineering*, JSCE, pp. 457–462 (in Japanese).

324. US Corps of Engineers (1956). *Hydraulic capacity of meandering channels in straight floodways; hydraulic model investigation*, Technical Memorandum No. 2–429, US Army Engineer Waterways Experiment Station, Vicksburg, Mississippi, USA.

325. Valentine, E. M., Benson, I. A., Nalluri, C. and Bathurst, J. C. (2001). 'Regime theory and stability of straight channels with bankfull and overbank flow.' *Journal of Hydraulic Research*, **39**(3), 259–268.

326. Valentine, M. E. and Haidera, M. (2001). 'A proposed modification to the White, Bettess and Paris rational approach.' *Proceedings of the Symposium of River Coastal and Estuarine Morphodynamics*, 2nd IAHR- Japan, pp. 275–284.

327. Valentine, E. M. and Shakir, A. S. (1992). River Channel Planform: An Appraisal of A Rational Approach. *8th Congress of the Asia and Pacific Division of the International Association of Hydraulic Research*, Pune, India.

328. Valentine, E. M. and Shakir, A. S. (1992). 'An Experimental Investigation of the Effects of Sediment Load on the Geometry of Meandering Laboratory Channels.' *8th Congress of the Asia and Pacific Division of the International Association of Hydraulic Research*, Pune, India.

329. Van Doormal, J. P. and Raithby, G. D. (1984). 'Enhancements of the SIMPLE method for predicting incompressible fluid flow.' *Numerical Heat Transfer*, **7**, 147–163.

330. van Prooijen, B. C. (2004). *Shallow mixing layers, PhD thesis*, Delft University of Technology.

331. Van Rijn, L. C. (1984a). 'Sediment transport. Part I: Bed load transport', *Journal of Hydraulics Division*, ASCE, **110**(10), 1431–1456.

332. Van Rijn, L. C. (1984b). 'Sediment transport. Part II: Suspended load transport', *Journal of Hydraulic Division*, ASCE, **110**(11), 1613–1641.

333. Van Rijn, L. C. (1984c). 'Sediment transport. Part III: Bed forms and alluvial roughness.' *Journal of Hydraulic Division*, ASCE, **110**(12), 1733–1754.

334. Vanoni, V. and Brooks, N. H. (1957). 'Laboratory studies of the roughness and sus-pended load of alluvial streams.' Report No. 11, California Institute of Technology, Pasadena.

335. Versteeg, H. K. and Malalasekara, W. (1995). *An introduction to computational Fluid Dynamics – the finite volume method.* Longman Harlow, London.

336. Vreugdenhil, C. B. and Wijbenga, J. H. A. (1982). 'Computation of flow patterns in river,. *Journal of the Hydraulics Division,* ASCE, **108**(HY11), 1296–1310.

337. Wallis, S. G. and Knight, D. W. (1984). 'Calibration studies concerning a one-dimensional numerical tidal model with particular reference to resistance coefficients.' *Estuarine, Coastal and Shelf Science,* Academic Press, London, **19**(5), 541–562.

338. Walling, D. E., He, Q. and Nicholas, A. P. (1996). 'Floodplains as suspended sediment sinks'. In Anderson, M. G., Walling, D. E. and Bates, P. D. (eds.), *Floodplain Processes,* Wiley.

339. Wang, S. Y., Alonso, V. V., Brebbia, C. A., Gray, W. G. and Pinder, G. F. (1989). 'Finite Elements in Water Resource.' *Proceedings of the 3rd International Conference,* Mississippi, USA.

340. Wark, J. B., James, C. S. and Ackers, P. (1994). *Design of straight and meandering channels.* National Rivers Authority R & D Report 13, pp. 1–86.

341. Wark, J. B., Samuels, P. G. and Ervine, D. A. (1990). 'A practical method of estimat-ing velocity and discharge in compound channels.' In White, W. R. (ed.), *International Conference on River Flood Hydraulics,* Wallingford, Wiley, pp. 163–172.

342. White, R. W. and Day, T. J. (1982). 'Transport of graded gravel bed material.' in Hey, R. D. *et al.,* (eds.), *Gravel Bed Rivers,* Wiley, pp. 181–213.

343. White, W. R., Bettess, R. and Paris, E. (1982). 'Analytical approach to river regime.' *Journal of the Hydraulics Division,* ASCE, **108**(HY10), 1179–1193.

344. White, W. R., Paris, E. and Bettess, R. (1980). 'The frictional characteristics of alluvial streams: a new approach.' *Proceedings of the Institution of Civil Engineers, Part 1: Design and Construction* **69**(2), 737–750.

345. Wijbenga, J. H. A. (1985). 'Determination of flow patterns in rivers with curvi-linear coordinates.' In *IAHR Preprint. Proceedings of the 21st Congress, 2, Theme B (Part 1), Fundamentals and Computation of 2-D and 3-D flows; Transport and Mixing in Rivers and Reservoirs, Australia,* Institution of Engineers Australia, 1985, Session 3(B), 131–138.

346. Willetts, B. B. and Hardwick, R. I. (1990). 'Model studies of overbank flow from a meandering channel.' *International Conference on River Flood Hydraulics,* Wallingford, UK, pp. 197–205.

347. Willetts, B. B. and Hardwick, R. I. (1993). 'Stage dependency for overbank flow in meandering channels.' *Proceedings of the Institution of Civil Engineers. Journal for Water, Maritime and Energy,* **101**, 45–54.

348. Willetts, B. B. and Rameshwaran, P. (1996). 'Meandering overbank flow structures.' In Ashworth, P., Bennett, S., Best, J. and McLelland, S. (eds.), *Coherent Flow Structures in Open Channels,* J Wiley, Chapter **29**, 609–629.

349. Wilson, N. R. and Shaw, R. H. (1977). 'A higher order closure model for canopy flow.' *Journal of Applied Meteorology,* **16**(11), 1197–1205.

350. Wolman, M. G. and Miller, J. P. (1960). 'Magnitude and frequency of forces in geomor-phic processes.' *Journal of Geology,* **68**, 54–74.

351. Wolman, M. G. and Brush, L. M. (1961). 'Factors controlling the size and shape of stream channels in course non-cohesive sands.' Paper 282-UG, *US Geological Survey,* Washington, DC.

352. Wormleaton, P. R. (1988). 'Determination of discharge in compound channels using the dynamic equation for lateral velocity distribution.' *Proceedings of the International Conference on Fluvial Hydraulics,* pp. 98–103.

353. Wormleaton, P. R. (1996). 'Floodplain secondary circulation as a mechanism for flow and shear stress redistribution in straight compound channels.' In Ashworth, P.,

354. Wormleaton, P. R., Allen, J. and Hadjipanos, P. (1982). 'Discharge assessment in compound channel flow.' *Journal of the Hydraulics Division, Proceedings of the ASCE,* **108**(HY9) 975–993.

355. Wormleaton, P. R., Allen, J. and Hadjipanos, P. (1985). 'Flow distribution in compound channels.' *Journal of the Hydraulics Division, Proceedings of the ASCE,* **111**(HY2), 357–361.

356. Wormleaton, P. R. and Merrett, D. J. (1990). 'An improved method of calculation for steady uniform flow in prismatic main channel/floodplain sections.' *Journal of Hydraulic Research,* IAHR, **28**(2), 157–174.

357. Yalin, M. S. (1997) "Mechanics of sediment transport", Oxford, UK. Pergamon Press.

358. Yalin, M. S. (1992). *River mechanics.* Oxford, UK, Pergamon Press.

359. Yang, C. T. (1973). 'Incipient motion and sediment transport.' *Journal of Hydraulic Division,* ASCE, **99**(10), 1679–1704.

360. Yang, C. T. (1984). 'Unit stream power equation for gravel.' *Journal of Hydraulic Division,* ASCE, **110**(12), 1783–1797.

361. Yen, B. C. (1973). 'Open-Channel Flow equations revisited.' *Journal of Engineering Mechanical Division ASCE,* **99**(EM5), 979–1009.

362. Yen, B. C. (1995). *Channel flow resistance: centennial of Manning's formula.* Water Resources Publications, Colorado.

363. Yen, B. C., Camacho, R., Kohane, R. and Westrich, B., (1985), Significance of floodplains in backwater computation, *Proc 21st Congress IAHR,* Melbourne, Vol. 3, pp. 439–445.

364. Yoshizawa, A. (1984). 'Statistical analysis of the deviation of the Reynolds stress from its eddy-viscosity representation.' *Turbulence and Chaotic Phenomena in Fluids, Proceedings of the International Symposium,* pp. 433–438.

365. Yoshizawa, A. (1984). 'Statistical analysis of the serviation of the Reynolds stress from its eddy viscosity representation.' *Physical Fluids* **27**(6), 1377–1387.

366. Younis, B. A. (1996). 'Progress in turbulence modelling of open channel flows.' In Anderson, Walling and Bates (eds.), *Floodplain Processes,* Wiley, Chapter 9, pp. 299–332.

367. Younis, B. A. and Abdellatif, O. E. (1989). 'Modelling sediment transport in rectangular ducts with a two- equation model of turbulence.' In, Wang, S. S. Y. (ed.), *Sediment Transport Modelling, Proceedings of the International Symposium,* New Orleans, August, 1989, ASCE, pp. 197–202.

368. Yu, G. and Knight, D. W. (1998). 'Geometry of self-formed straight threshold channels in uniform material.' *Proceedings of the Institution of Civil Engineers: Water, Maritime and Energy Division,* London, **130**, Paper 11398, 31–41.

369. Yuen, K. W. H. (1989). *A study of boundary shear stress, flow resistance and momentum transfer in open channels with simple and compound trapezoidal cross section.* PhD Thesis, The University of Birmingham, UK.

370. Yuen, K. W. H. and Knight, D. W. (1990). 'Critical flow in a two stage channel.' In White, W. R. (ed.), *Proceedings of the International Conference on River Flood Hydraulics,* Wallingford, Wiley, Paper G4, pp. 267–276.

Index